P9-BIW-152

Tourism and the Consumption of Wildlife

Consumptive forms of wildlife tourism (hunting, shooting and fishing) have become a topic of interest – both to the tourism industry, in terms of destinations seeking to establish or grow this sector, and to other stakeholders such as environmental organisations, animal-rights groups, and the general public. Hunting tourism, in particular, has come under fire with accusations that it is contributing to the demise of some species. Practices such as 'canned hunting' (within fenced safari parks) or the use of hounds are described as unethical, and fishing tourism too has attracted recent negative publicity as it is said to be cruel. At the same time, however, many peripheral and indigenous communities around the world are strategising how to capitalise on consumptive forms of wildlife tourism.

This book addresses a range of contentious issues facing the consumptive wildlife tourism sector across a number of destinations in Europe, North America, Africa, India, Arabia and Oceania. Practices such as baited bear hunting, trophy hunting of threatened species, and hunting for conservation are debated, along with the impact of this type of tourism on indigenous communities and on wider societies. Research on all aspects of 'consumptive wildlife tourism' is included, which for the purposes of the book is defined to include all tourism that involves the intended killing of wildlife for sport purposes, and may include the harvest of wildlife products. This includes, among others, the practices of recreational hunting, big-game hunting and safari operations, traditional/indigenous hunting, game-bird shooting, hunting with hounds, freshwater angling and saltwater game fishing.

This is the first book to specifically address the tourism aspects of consumption of wildlife. It will appeal to tourism and recreation academics and students, tourism industry operators, community tourism planners and wildlife managers.

Dr Brent Lovelock is a Senior Lecturer in the Department of Tourism at the University of Otago, Dunedin, New Zealand.

Contemporary geographies of leisure, tourism and mobilty
Series editor: C. Michael Hall

Professor, Department of Management, College of Business & Economics, University of Canterbury, Private Bag 4800, Christchurch, New Zealand.

The aim of this series is to explore and communicate the intersections and relationships between leisure, tourism and human mobility within the social sciences.

It will incorporate both traditional and new perspectives on leisure and tourism from contemporary geography, e.g. notions of identity, representation and culture, while also providing for perspectives from cognate areas such as anthropology, cultural studies, gastronomy and food studies, marketing, policy studies and political economy, regional and urban planning, and sociology, within the development of an integrated field of leisure and tourism studies.

Also, increasingly, tourism and leisure are regarded as steps in a continuum of human mobility. Inclusion of mobility in the series offers the prospect to examine the relationship between tourism and migration, the sojourner, educational travel, and second home and retirement travel phenomena.

The series comprises two strands:

Contemporary Geographies of Leisure and Mobility aims to address the needs of students and academics, and the titles will be published in hardback and paperback. Titles include:

The Moralisation of Tourism
Sun, sand … and saving the world?
Jim Butcher

The Ethics of Tourism Development
Mick Smith and Rosaleen Duffy

Tourism in the Caribbean
Trends, developments, prospects
Edited by David Timothy Duval

Qualitative Research in Tourism
Ontologies, epistemologies and methodologies
Edited by Jenny Phillimore and Lisa Goodson

The Media and the Tourist Imagination
Converging cultures
Edited by David Crouch, Rhona Jackson and Felix Thompson

Tourism and Global Environmental Change
Ecological, social, economic and political interrelationships
Edited by Stefan Gössling and C. Michael Hall

Routledge studies in contemporary geographies of leisure, tourism, and mobility is a forum for innovative new research intended for research students and academics, and the titles will be available in hardback only. Titles include:

1. **Living with Tourism**
   Negotiating identities in a Turkish village
   Hazel Tucker

2. **Tourism, Diasporas and Space**
   Tim Coles and Dallen J. Timothy

3. **Tourism and Postcolonialism**
   Contested discourses, identities and representations
   C. Michael Hall and Hazel Tucker

4. **Tourism, Religion and Spiritual Journeys**
   Dallen J. Timothy and Daniel H. Olsen

5. **China's Outbound Tourism**
   Wolfgang Georg Arlt

6. **Tourism, Power and Space**
   Andrew Church and Tim Coles

7. **Tourism, Ethnic Diversity and the City**
   Jan Rath

8. **Ecotourism, NGOs and Development**
   A critical analysis
   Jim Butcher

9. **Tourism and the Consumption of Wildlife**
   Hunting, shooting and sport fishing
   Brent Lovelock

Tourism and the Consumption of Wildlife

Hunting, shooting and sport fishing

Edited by Brent Lovelock

LONDON AND NEW YORK

First published 2008
by Routledge
2 Park Square, Milton Park, Abingdon, Oxon OX14 4RN

Simultaneously published in the USA and Canada
by Routledge
270 Madison Ave, New York, NY 10016

Routledge is an imprint of the Taylor & Francis Group, an informa business

Reprinted 2008

Typeset in Times by
HWA Text and Data Management, Tunbridge Wells
Printed and bound in Great Britain by
MPG Books Ltd, Bodmin

British Library Cataloguing in Publication Data
A catalogue record for this book is available from the British Library

Library of Congress Cataloging-in-Publication Data
A catalog record for this book has been requested

ISBN10: 0–415–40381–2 (hbk)
ISBN10: 0–203–93432–6 (ebk)

ISBN13: 978–0–415–40381–8 (hbk)
ISBN13: 978–0–203–93432–6 (ebk)

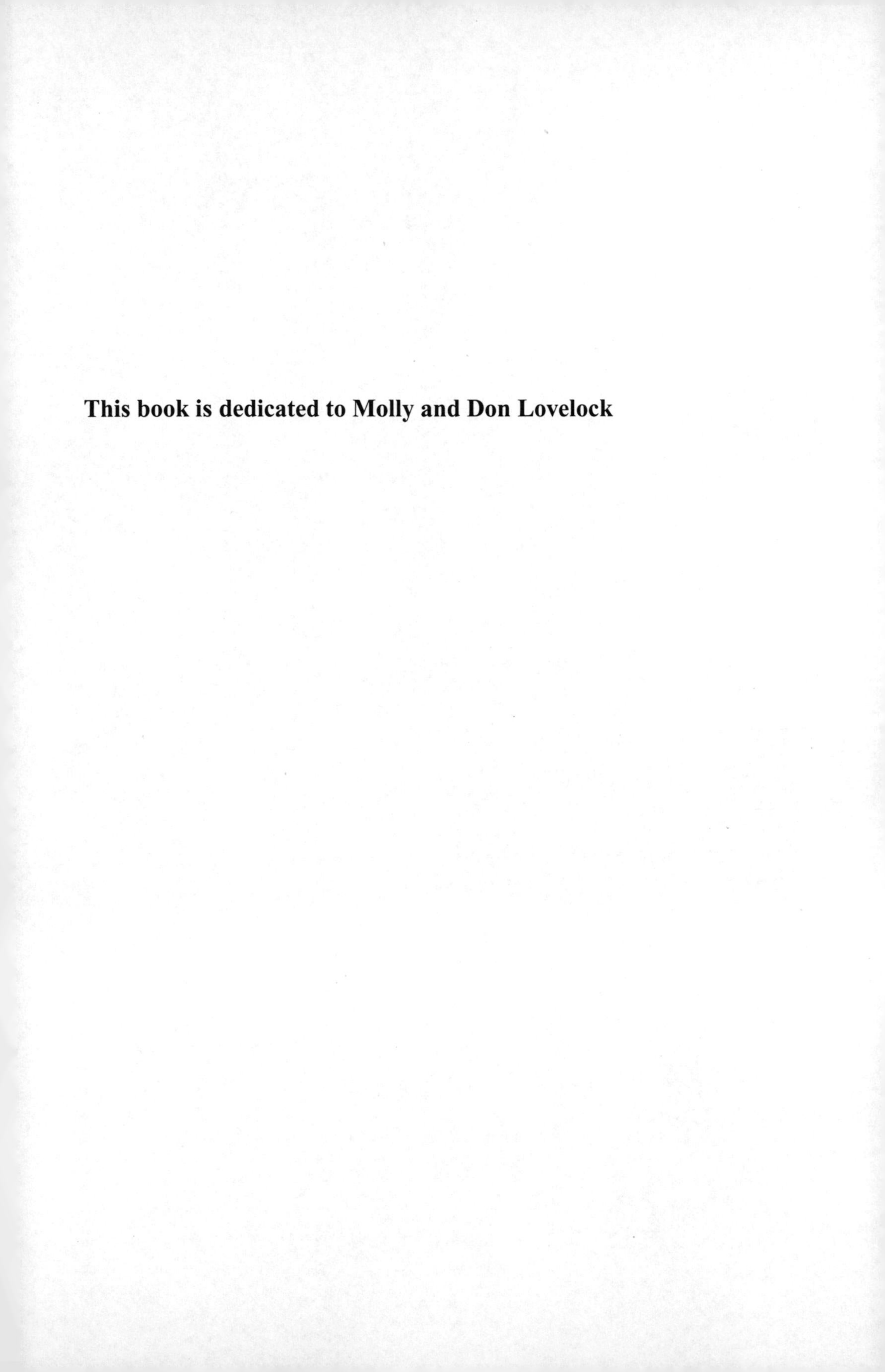

This book is dedicated to Molly and Don Lovelock

Contents

Figures

Tables

Contributors

John S. Akama, Department of Tourism Management, Moi University, PO Box 1125, Eldoret, Kenya. Email: jsakama@yahoo.com.

Jonathan I. Barnes, Economics Unit – Directorate of Environmental Affairs Ministry of Environment and Tourism, PO Box 25942, Windhoek, Namibia. Email: jibarnes@iafrica.com.na.

Mattias Boman, Southern Swedish Forest Research Centre, Swedish University of Agricultural Sciences, PO Box 49, SE-230 53, Alnarp, Sweden. Email: mattias.boman@ess.slu.se.

Runar Brännlund, Department of Forest Economics, Swedish University of Agricultural Sciences, SE-901 83, Umea, Sweden. Email: runar.brannlund@econ.umu.se.

Michael Campbell, Recreation Studies, 112 Frank Kennedy Bldg, University of Manitoba, Canada R3T-3V7. Email: campblm@cc.umanitoba.ca.

Kenneth Cohen, Recreation and Tourism Department, Central Washington University, Ellensburg, Washington 98926-7565, United States of America. Email: cohenk@cwu.edu.

Stephen J. Craig-Smith, School of Tourism and Leisure Management, University of Queensland, Brisbane, QLD 4072, Australia. Email: s.craigsmith@uq.edu.au.

Jackie Dawson, Department of Geography, University of Waterloo, 200 University Avenue West, Waterloo, Ontario, Canada N2L 3G1. Email: jpdawson@fes.uwaterloo.ca.

Gordon McL. Dryden, School of Animal Studies, Faculty of Natural Resources, Agriculture and Veterinary Science, University of Queensland, Gatton QLD 4343, Australia. Email: g.dryden@uq.edu.au.

Göran Ericsson, Department of Animal Ecology, Swedish University of Agricultural Sciences, SE-901 83, Umea, Sweden. Email: goran.ericsson@szooek.slu.se.

Guil Figgins, Department of Geography, University of Otago, PO Box 56, Dunedin, New Zealand. Email: guil.figgins@geography.otago.ac.nz.

Lee Foote, Department of Renewable Resources, University of Alberta, Edmonton, AB T6G 2H1, Canada. Email: Lee.Foote@afhe.ualberta.ca.

Adrian Franklin, School of Sociology, University of Tasmania, Private Bag 17, Hobart, Tas 7001, Australia. Email: Adrian.Franklin@utas.edu.au.

Yvonne Gunnarsdotter, Department of Urban and Rural Development, Swedish University of Agricultural Sciences, Box 7012 S-750 07 Uppsala, Sweden. Email: Yvonne.Gunnarsdotter@sol.slu.se.

Kevin Hannam, School of Arts, Design, Media and Culture, University of Sunderland, Sunderland SR1 3PZ, United Kingdom. Email: kevin.hannam@sunderland.ac.uk.

Bengt Kriström, Department of Forest Economics, Swedish University of Agricultural Sciences, SE-901 83, Umea, Sweden. Email: bengt.kristrom@sekon.slu.se.

Thomas Laitila, Department of Business, Economics, Statistics and Informatics, Orebro University, SE-701 82 Orebro, Sweden. Email: Thomas.Laitila@stat.umu.se.

Frederic Launay, Environment Agency (EAD), PO Box 45553, Abu Dhabi, United Arab Emirates. Email: flaunay@ead.ae.

Brent Lovelock, Department of Tourism, University of Otago, PO Box 56, Dunedin, New Zealand. Email: blovelock@business.otago.ac.nz.

Leif Mattsson, Southern Swedish Forest Research Centre, Swedish University of Agricultural Sciences, PO Box 49, SE-230 53, Alnarp, Sweden. Email: leif.mattsson@ess.slu.se.

Joseph E. Mbaiwa, Okavango Research Centre, University of Botswana. Email: jembaiwa@neo.tamu.edu.

Øystein Normann, Department of Travel and Tourism, Harstad University College, N-9480, Harstad, Norway. Email: Oystein.Normann@hih.no.

Marina Novelli, Centre for Tourism Policy Studies, School of Service Management, University of Brighton, Darley Road, Eastbourne BN20 7UR, UK. Email: mn19@bton.ac.uk.

Anton Paulrud, Swedish Board of Fisheries, Resource Management Department, Economics Unit, P. Box 423, SE-401 26 Göteborg, Sweden. Email: Anton.paulrud@Fiskeriverket.se.

Robert Preston-Whyte, School of Environmental Sciences, University of KwaZulu-Natal, Durban 4041, South Africa. Email: preston@ukzn.ac.za.

Nick Sanyal, College of Natural Resources, University of Idaho, PO Box 441139, Moscow Idaho 83844, United States of America. Email:nsanyal@uidaho.edu.

Philip J. Seddon, Department of Zoology, University of Otago, PO Box 56, Dunedin, New Zealand. Email: philip.seddon@stonebow.otago.ac.nz.

Pia Sillanpää, Mid Sweden University, S-831 25 Ostersund, Sweden. Email: pia. sillanpaa@miun.se.

Carl Walrond, Te Ara – The Encyclopedia of New Zealand, Ministry for Culture and Heritage, PO Box 5364, Wellington. Email: carl.walrond@mch.govt.nz

George Wenzel, Department of Geography, McGill University, Montreal, QC, Canada H3A 2K6. Email: wenzel@felix.geog.mcgill.ca.

Acknowledgements

The editor would like to thank first of all, the chapter authors of this book, for contributing such interesting chapters, and for providing them in such a professional and timely fashion. It has been a pleasure working with you. A number of people at Taylor & Francis assisted in the production of this collection: Andrew Mould, Zoe Kruze, and Jennifer Page must be thanked in particular, as must John Hodgson of HWA for text management.

The editor is indebted to all of the participants in various aspects of his research into consumptive wildlife tourism in New Zealand, Scotland, Sweden and Poland: the hunting and fishing guides, the regional and national authorities involved in game and fish management and the tourism organisations. In New Zealand, special thanks go to the New Zealand Professional Hunting Guides Association, and to Jeff Milham and Kevin Robinson, previous co-researchers.

Michael Hall, editor of the series in which this book appears, former colleague and mentor, has been supportive of this book, and is an inspiration to all tourism researchers. My colleagues in the Department of Tourism, University of Otago all contributed through discussion and encouragement at various stages of production: in particular, thanks to Hazel Tucker, Anna Carr, David Duval, Donna Keen, Eric Shelton and James Higham. Thanks to Diana Evans for moving office around me and to Monica Graham for help with the tricky bits. The editor acknowledges the Department of Tourism, and the School of Business, University of Otago for supporting my work in this field through research grants.

And finally, on a personal note, thanks to my wife and partner, Kirsten, for outstanding advice, assistance and encouragement at all stages – you are a great writer and researcher, and I am certain that you would also make a great hunter. A special thanks to my children – Millie for her vehement protection of Bambi, and to Oscar for sharing his hunting game with me.

Part I

Introduction and conceptual issues

1 An introduction to consumptive wildlife tourism

Brent Lovelock

Introduction

This book explores the field of touristic hunting, shooting and sport fishing. It investigates contemporary trends in the industry, and suggests some possible futures for the sector. Consumptive wildlife tourism, while arguably neglected in current tourism research, has become an increasingly contested domain. Animal rights activists and environmentalists argue that it contributes to the demise of some species, and that practices such as 'canned hunting', 'virtual hunting' (but with *real* game) and the use of hounds are unethical. Concurrently, however, many remote, indigenous or developing communities around the world are strategising on how to capitalise on potentially lucrative consumptive forms of wildlife tourism. This book, through a series of case studies from around the world, considers the argument for growing consumptive wildlife tourism, looking at the relationships between hunting, fishing and local communities, impacts, economies and ecologies.

Consumptive wildlife tourism (CWT), as a niche product, has received relatively little attention from researchers. This may be attributed to a number of reasons, including the relative lack of visibility of this sector not only in terms of its economic scale but also in terms of any large physical infrastructural presence. It is also possible that tourism researchers have tended to treat hunting and fishing as non-touristic activities, leaving the sector to leisure and recreation specialists. A further reason for lack of research may relate to the fact that hunting and shooting are not generally popular pastimes of the educated middle class, and furthermore, that as a field of research the topic falls between the uncomfortable (guns, firearms) and the unforgiveable (killing Bambi). As Dizard observes: 'Nice people don't hunt' (2003: 58). Nice people prefer to drink wine, go on gastronomy tours or visit heritage buildings in Tuscany. No one wants to research people performing unpleasant acts.

Consumptive wildlife tourism in the tourism world

As a niche tourism product, namely a small specialised sector of tourism which appeals to a well-defined market segment, CWT fits into the broader nature-related macro-niche of wildlife tourism (Novelli and Humavindu 2005). Wildlife tourism includes activities classified as 'non-consumptive'; that is, wildlife viewing, photography, feeding and interacting in various ways, as well as 'consumptive' activities. The latter may include killing or capturing wildlife, i.e. hunting, shooting or fishing. The most popular forms of CWT are illustrated in Figure 1.1. Bauer and Herr (2004) use the hunting/fishing dichotomy, and sub-divisions based upon game and/or habitat. Their representation is useful for showing the diversity of forms of CWT. For the purposes of this book, CWT is defined as a form of leisure travel undertaken for the purpose of hunting or shooting game animals, or fishing for sports fish, either in natural sites or in areas created for these purposes.

However, CWT is more than just about killing animals, and participants demonstrate a range of motives with respect to the experience they seek. A typology of hunters (and the same could probably be said for fishers), has been constructed and includes nature hunters, meat hunters and sport hunters (Kellert 1996). Thus, we see a range of purposes and immediately that CWT has some commonalities with eco-tourism and sport tourism, participants thereof who have a range of motivations. Indeed, some definitions allow us to view CWT as a form of sport tourism (e.g. Gibson *et al.* 1997 in Delpy-Neirotti 2003), and the sporting aspect of CWT is strongly apparent in the way that participants score their performances. There are a number of scoring systems employed for hunted and fished species – for example the 'Boone and Crockett' system for big game and the 'Douglas' scoring system for ungulates, while sporting prowess in fishing is expressed in terms of the weight of a fish caught.

But CWT is also a form of cultural tourism, when defined as the '… search for and participation in new and deep cultural experiences, whether aesthetic, intellectual, emotional, or psychological' (Stebbins 1996: 948). There is often a strong sense of cultural exchange between hunters and fishers and their hosts (see Foote and Wenzel, this volume). This may be particularly obvious when CWT is organised by, or engages the services of indigenous peoples, and especially

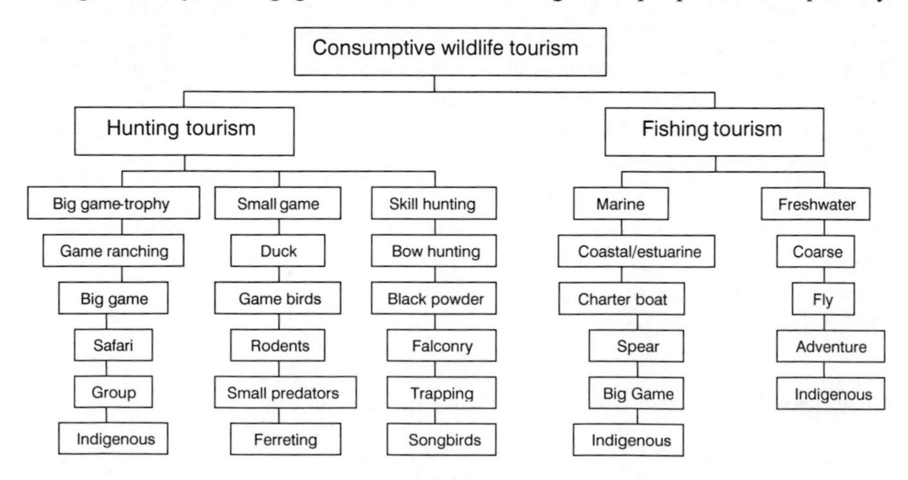

Figure 1.1 Consumptive wildlife tourism activities. *Source*: Bauer and Herr (2004).

so when traditional hunting or fishing practices are used. And in a heritage tourism sense, arguably consumptive wildlife tourists, especially hunters, may seek not just the experience of the hunt, but also to recreate a sense that they are amongst the 'first', are 'pioneers', and imagine in doing this that they are like the Victorian gentleman-hunter. This is especially seen when hunters adopt primitive technologies such as black-powder rifles. For many hunters from the new-world, an attraction of CWT in the old-world, where hunting remains of great cultural significance (Bauer and Herr 2004), may be the rich heritage of hunting evoked through dress, protocol and arcane practices such as 'blooding' the hunter.

It is clear that CWT is a multi-dimensional practice, rather than a simple act of killing. CWT is a sport, and as such is culturally embedded and can be a heritage experience, an adventure and an ecotourism experience. Radder's (2005: 1143) study of trophy hunters suggests that the CWT experience is not driven by a single motive, but by a 'multidimensional set of inter-related, interdependent and overlapping motives' falling within the realms of spiritual, emotional, intellectual, self-directed, biological and social motives. The importance of a number of motives is illustrated in Radder's study, most clearly by the finding that participants valued the concept of experiencing new places, people and culture higher than collecting hunting trophies. And while CWT could also be conceived as a form of adventure tourism, Radder's research shows only weak support for risk as a major motive of hunting and fishing tourists. What we can be assured of, however, is that serious consumptive wildlife tourists are highly motivated – demonstrated by a survey of British sport fishermen which revealed that more than half would rather catch a record-breaking trout or salmon than spend a night with a supermodel (*Otago Daily Times* 2006).

Scale and scope of CWT

Hunting, shooting and sport-fishing are immensely popular recreational activities. Fishing, for example, is one of the most popular forms of outdoor recreation in many countries. Estimates of participation rates in the United States, for example, indicate that up to 16 per cent of the adult population fish (USDOI *et al.* 2002). In Australia the fishing participation rate is estimated at 19.5 per cent (FRDC 2001), while Japanese participation in fishing is slightly higher at 23 per cent (SB&SRTI 2006). A national angling survey in the United Kingdom in 1994 estimated that there were 3.3 million fresh and sea water anglers, while in the wider European Union there are an estimated 25 million recreational fishermen.

Participation in hunting is generally lower however. In New Zealand, the participation rate is put at about 2 per cent of the adult population (Groome *et al.* 1983), and in Australia a mere 0.35 per cent (Australian Sports Commission 2006). In the United States, 6 per cent of the adult population hunt (USDOI *et al.* 2002).

But not all of these recreationists become consumptive wildlife tourists in the traditionally accepted use of the term tourist. And to complicate matters, there is some debate about what constitutes a tourist. The US Travel Data Center considers tourists to be those that take trips with a one-way mileage of 100 miles or more, or all trips involving an overnight stay away from home, regardless of distance travelled. Other definitions rely upon an individual crossing a border to become a tourist. The World Tourism Organisation (WTO 2006) defines tourism as comprising the activities of persons travelling to and staying in places outside their usual environment for not more than one consecutive year for leisure, business and other purposes. They distinguish between inbound tourism (from another country) and domestic tourism, the latter involving residents of a given country travelling within that country. It is generally considered that to be a tourist, an individual must spend at least one night away from their home, however, the WTO also notes the importance of same day visitors to the 'tourism' industry. Clearly, the definition of what constitutes a tourist is somewhat loose and problematic. However, for the purposes of this book, consumptive wildlife tourists are taken to be those that travel to fish, shoot or hunt in a region other than their own.

Unfortunately, accurate figures are not kept by many national tourism organisations on the numbers of inbound consumptive wildlife tourists – a fate of many forms of special interest tourism (McKercher and Chan 2005). So the 'conversion' rate of domestic or recreational hunters and fishers into consumptive wildlife tourists is largely unknown. Furthermore, McKercher and Chan argue that existing data relating to inbound special interest tourism is unreliable in terms of identifying that special interest as a primary activity or motivator. So, data from international visitor surveys such as those undertaken in New Zealand which identifies that fishing was undertaken by 2.6 per cent of inbound visitors (Ministry of Tourism 2006), and in Australia where 4 per cent of international visitors engaged in fishing whilst in the country (FRDC 2006) are interesting but not definitive in terms of identifying if CWT is a primary motive for visiting a destination.

Estimates at this stage, of the total market size, therefore, are fraught with lack of precision. Work within the United States, however, comes closest to estimating market size. Hunters combined with fishers, total a substantial 47 million people who engage in either activity (USDOI *et al.* 2002). Fishing tourism in particular appears to contribute substantially to overall visitor-days, with a very high number (estimated 70 million) of out-of-state fishing days (Ditton *et al.* 2002). Collectively US$20 billion was spent on trip-related expenses for both hunting and fishing (USDOI *et al.* 2002).

Naturally, in this process, some states end up as net gainers and some as net losers in terms of fishing tourism days. On an international level, this is what Hofer (2002) refers to as demand and supply countries, where some destinations gain from inbound CWT. Traditionally North America and Western Europe have been important both in terms of supply and demand for international CWT, although both of these regions have their own substantial domestic CWT markets (or in the case of Europe, domestic plus intra-European markets).

However, new demand and supply countries are emerging, and while this may be producing only marginal effect upon the global distribution of income from CWT, substantial local effects are arising (see Foote and Wenzel's chapter in this volume on hunting in Nunavit, in the Canadian Arctic). A typology of CWT destinations around the world is offered below (Table 1.1) as a broad descriptive tool. It should be borne in mind that this is based upon reported broad trends and not upon comprehensive CWT visitor data.

Recently, Central and Eastern Europe and the Balkans have emerged as growing supply regions for inbound hunting tourism, with growth of inbound CWT to countries such as Poland, Hungary, Bulgaria and Romania. Scandinavian nations too, are to some extent also experiencing the effects of growth of outbound hunters and fishers from Western Europe and the United Kingdom, where CWT is increasingly being seen as an expensive and crowded proposition (see e.g. Gunnarsdotter's chapter in this volume on the impact of European inbound hunters on a local Swedish hunting area). Central Asian countries such as Kazakhstan and Mongolia are also becoming more popular CWT destinations. These new supply nations appear to be competing on the basis of price, novelty and the emergence

Table 1.1 A typology of CWT destinations

CWT Destination type	*Example of countries*	*Characteristics*
Traditional old-world	Germany, France, United Kingdom, Italy	Strong domestic markets, and intra-EU but consolidating. Traditional fish and game species, strong domestic markets.
Traditional new-world	South Africa, Zimbabwe, Botswana, USA, Canada	The mainstay of outbound CWT industry. However, some destinations experiencing stagnation problems and human wildlife conflict and conservation issues.
Growing old-world	Poland, Bulgaria, Romania, Bulgaria	Substantial domestic market. Growing inbound due to competitive pricing (non-Euro) and uncrowded nature of experience. Coupled with cultural tourism products.
Emerging new-world	Chile, Argentina, Mexico Australia, New Zealand, Tonga, New Caledonia Congo, Cameroon, Central African Republic, Ethiopia, Mozambique	Increasing profile as affordable, 'fresh' destinations. Security issues still a concern in some emerging (or re-emerging post-conflict) Africa destinations.
New discoveries	Kazakhstan, Uzbekistan, China, Azerbaijan, Mongolia, Russia.	To hunt/fish species that may be protected elsewhere, or are only locally available. Novelty factor of new species, together with attraction of inaccessibility. Some conservation issues.

of hitherto rare opportunities to hunt desirable species. Many of these post-socialist states are only now beginning to recognise the potential for CWT, and more liberal institutional arrangements, coupled with entrepreneurial spirit and increasing assurances of visitor safety and comfort have meant that many are now in a position to attract hunters and fishers in substantial numbers.

Participation trends

There has been no reported increase in the numbers of hunters and fishers since the 1980s. Indeed, participation has generally remained stable or slightly declined from the 1980s to the current day (USDOI *et al.* 2002; Bauer and Herr 2004). The most substantial research in this area is the *U.S. National Survey of Fishing, Hunting and Wildlife-Associated Recreation* which includes domestic and outbound CWT as well as non-touristic fishing and hunting. Ten-year trends from the early 1990s (1991–2001) indicate a drop in the number of fishers (4 per cent) and hunters (7 per cent), however, expenditures have increased for both groups (14 per cent and 29 per cent respectively) (USDOI *et al.* 2002). Significantly, although the total number of hunters declined, the number of big game hunters remained constant. Big game hunters make up the largest component of outbound and domestic CWT. Total fishing numbers in the United States comprise approximately one-quarter saltwater and three-quarters freshwater anglers. Over the 10-year period indicated above, the latter group experienced a slightly higher decline in numbers compared to their saltwater counterparts (USDOI *et al.* 2002).

Interestingly, the number of people engaged in non-consumptive wildlife watching fell by a greater amount (13 per cent) over the same period, and in particular the number of people taking trips to watch wildlife (down 29 per cent). If the United States can be seen as a barometer for global tourism trends, this then puts paid to the popular perception that nature-based tourism and ecotourism are the fastest growing tourism sectors. By extension, and in light of the lack of firm data, the number of consumptive wildlife *tourists* most probably parallels this trend, assuming a direct relationship between participation in the activity of hunting/fishing and engaging in hunting/fishing tourism.

Anecdotally there is some evidence that there is a shift in gendered participation, with more women participating in CWT. However, this is not supported by research in the United States where participation rates in hunting/fishing are much lower than for men, and barely remaining stable (USDOI *et al.* 2002). Despite low participation rates, some evidence would also suggest that women are at least as successful as men in terms of shooting or bagging their catch, and not solely because women are naturally more stealthy or cunning than men, but because unlike men, women aren't embarrassed about listening to advice from professional guides (Nelson 2006). Ethnic minorities, (non-European) in the United States are also showing falling or stable participation rates in hunting and fishing (USDOI *et al.* 2002).

A further relevant trend is the aging population currently being experienced by many Western nations – most of which are major domestic CWT destinations or demand nations for CWT. Participation in hunting declines in the older age groups (>65years (USDOI *et al.* 2002)) as does fishing. Some destinations, concerned about demographic and socially induced declining hunting participation are investigating recruitment and retention strategies. For example, fishing participation is being targeted by authorities in the United Kingdom who are endeavouring to enhance participation of women and minorities (Leapman 2006). Canadian research (McFarlane *et al.* 2003) also identifies women as an increasingly important constituent for hunting agencies, while youth are being targeted (so to speak) in a number of US states.

Research would suggest that declining hunter participation is a complex issue that cannot be simply assigned to demographics or lack of time. Miller and Vaske (2003) identify the role of commitment and investment into hunting, social networks and situational constraints (e.g. regulations, access to hunting land (e.g. Jagnow *et al.* 2006)) that affect participation. These findings indicate the need for destinations that are serious about developing (or maintaining) CWT as a significant part of their product portfolio, to at least identify and address the situational constraints that exist.

Such constraints may include factors that to the uninitiated could appear unimportant but which can have a profound impact upon the ability of destination to capitalise upon CWT. For example, Sunday hunting is prohibited in eight US states; for a tourist to bring a firearm with them to hunt in the United Kingdom is problematic (because of post 9/11 heightened security arrangements); taking post-hunt game meat or other trophy material from the United Kingdom to other EU nations is difficult (similarly so for elephant tusks from most African destinations); hunters in some European destinations are required to undertake time-consuming in-destination hunting/shooting proficiency and safety tests. While it is not suggested that these regulations be discarded – because they do serve valuable purposes – if destinations are aiming to increase CWT participation, they should look at streamlining and standardising requirements, in order to minimise barriers for the growth of CWT.

Of course the big news in global tourism is the rapidly increasing participation in tourism of the growing middle classes of developing nations, in particular China and India. And although there is not a strong history of popular participation in hunting and fishing (although fishing is popular with the Chinese) within these nations, it will be interesting to observe if the dramatic growth in outbound tourism from these sources has any impact upon CWT.

How 'wild' is the wildlife in CWT?

Wildlife is taken to include all non-domesticated animals (both terrestrial and aquatic). And in line with Higginbottom's (2004) definition, is not restricted to animals that are native or endemic, but includes those that may have been introduced to a destination. It may also include animals that have escaped their domestic confines to become feral. In many destinations, exotic species constitute the basis of their hunting or fishing tourism industries (e.g. trout and deer in New Zealand).

There is also 'wild' wildlife and that which is not so wild, for example some hunting is conducted within fenced boundaries, and using animals that are especially bred for the purpose. On a European game estate, pheasants in a shoot are purpose bred and fed for the day of the shoot. In a New Zealand river or lake, trout or salmon fry are released specifically to restock 'wild' populations for fishing. It is thus accepted that CWT will often rely upon a degree of human intervention and that the actual consumption of the animal may take place in an environment that is somewhat modified from its natural state. However, the activity will generally be undertaken in a natural or semi-natural environment (but sometimes within a wider rural or urban setting) and the target species will generally be self-sufficient (see 'canned hunting' below).

While it is obvious that zoos are excluded from our discussions in this book (zoo visitors are generally not permitted to kill zoo animals for recreational purposes), the exclusion of such practices as bullfighting, cockfighting or even bear-baiting or dog fighting requires qualification: while these practices are definitely consumptive in the sense that the animals involved are either killed or harmed, the origins of these animals are generally domestic or they are held in a captive state.

Consumptive and non-consumptive tourism

This raises the question of what is consumptive or non-consumptive wildlife tourism? As outlined above, consumptive activities are fairly clear – in that they involve the killing or capturing of animals. Freese (1998) defines CWT as a practice that involves animals being deliberately killed or removed or having any of their body parts utilised. It has been argued, however, 'there is little evidence that non-consumptive wildlife tourism involves greater empathy, respect or learning benefits' than consumptive wildlife tourism (Tremblay 2001: 85). As suggested by Tremblay, a continuum of human–wildlife interaction based on concepts of 'intention' and 'purposefulness' (as in Duffus and Dearden 1990) may be more useful, particularly if combined with a measure of the intensity or nature of impacts on the wildlife (Tremblay 2001). Such an approach would allow a better understanding of ambiguous areas – for example the practice of 'bloodless', green or non-lethal hunting that may involve hunting the target species with a tranquiliser gun, often as part of a wildlife management project (e.g. to tag animals or to undertake measurements). In either scenario, the target species may suffer some stress or potential injury.

But how different is hunting from the range of wildlife viewing practices of the many millions of wildlife tourists and 'ecotourists' in hundreds of destinations around the world? Although wildlife viewing and photography is typically viewed as non-consumptive, there are scores of empirical studies documenting very real impacts upon a range of species (for useful works on general wildlife tourism impacts and management issues see: Roe *et al.* 1997; Higginbottom 2004; Newsome *et al.* 2005). Briefly, such impacts include disruption of feeding, breeding, migration and social behaviour, introduction of pathogens, habituation and physical harm from vessel and vehicles.

However, it remains the popular perception that hunting and fishing result in greater impacts upon wildlife (e.g. Reynolds and Braithwaite (2001)) even if the impacts are more intense and concentrated (single animals within populations) but on a smaller scale overall. Despite the above shortcomings, the consumptive/non-consumptive dichotomy that is widely adopted in the literature, while admittedly flawed, is utilised in this book for the purposes of simplicity and consistency.

While contemporary hunting tourism has arguably sustainable intentions, the impact of uncontrolled hunting and fishing tourism in the past has been acknowledged and responded to over time. This was probably best demonstrated during the Victorian era of the gentleman hunter, when vast numbers of game were bagged, often with little thought given to the vulnerability of species. Nowadays, in most parts of the world, hunting and fishing are managed with a view to the long-term sustainability of fish and game populations.

The twenty-first century hunter or angler abides by legislated bag limits, often tightly controlled spatially and/or temporally. The hunter-tourist is often guided in their consumption of wildlife by a strict set of ethics. For example, the Safari Club International, the largest non-profit organisation advocating for hunters worldwide, has over 40,000 members throughout 85 countries, all of whom are bound to a *Hunters' Code of Ethics* (SCI 2006a). Most hunting organisations throughout the world have similar codes and a requirement that hunters attend compulsory education and training sessions to ensure not just hunter safety, but ethical hunting practices in the field. Bauer and Herr (2004) describe the German concept of *Waidgerechtigkeit*, a combination of tradition, rules and guidelines with the ultimate aim of ensuring the game resource is managed in a sustainable way. Similar codes apply in other popular hunting destinations with long histories of hunting, such as Poland (Szpetkowski 2004), but also in new world destinations (e.g. United States, Canada, New Zealand). The irony, in a tourism sense, is that compared to this 'consumptive' form of tourism, arguably no other tourist segment gets the same degree of ethical and practical guidance in terms of limiting their impact upon wildlife species and habitats. This has led some commentators to describe hunters as the ultimate ecotourists (e.g. Haripriya 2004). Schoenfeld and Hendee in their classic *Wildlife Management in Wilderness* (1978) say wilderness hunting may be one of the most ecologically pure human experiences.

In the context of the CAMPFIRE projects in Zimbabwe, hunters are considered a desirable segment not only because of this relative lack of negative ecological impact, but also because of the positive economic (and social) benefits they bring to local communities (e.g. providing bush meat, deterring poachers). For the same level of economic impact, 20 conventional ecotourists would be needed, resulting in 20 times the sewage output, water and imported food requirements, and transport needs (Cheney 2006).

In destinations such as New Zealand, where all game species are introduced, causing much modification to natural ecosystems and where they are legally defined as pests, hunting could quite validly be considered a form of restorative ecosystem management (Lovelock 2006). In this situation, hunting tourism and tourists are at odds with other nature-based tourists (ecotourists!?) who actually enjoy encountering an exotic (non-endemic) game animal in the native forest. Thus we see how the dichotomy of consumptive (non-ethical?) versus non-consumptive (ethical?) tourism collapses when broader ecosystem integrity is considered or when placed under closer inspection generally. This issue is also visited by authors in this book (see chapters by Mbaiwa and Akama).

Figure 1.2 Cover of New Zealand Government Tourist Department promotional brochure [ca 1935] with (introduced) red deer. (Artist: Mitchell, Leonard Cornwall 1901–1971). *Source*: Alexander Turnbull Library, National Library of New Zealand – Te Puna Matauranga o Aotearoa.

Further such complexities include catch-and-release fishing being described as ecotourism (Holland *et al.* 2000). However, the extent to which this practice is consumptive or non-consumptive has been debated, as stress upon the target species results when they are removed (consumed) from their natural environment, albeit temporarily by anglers.

The grounds of this argument, that hunting and fishing are ecotourism, will always be contentious. This is due undoubtedly to a number of factors, not least, that death is unambiguous, and that we humans have a tendency to anthropomorphise game killed for consumption. CWT, therefore becomes ultimately vulnerable to the strong voice against the continuation of 'blood sports'.

CWT and destination competitiveness

While almost all countries have something to offer in terms of actual or potential fish or game species, undoubtedly some are more competitive as CWT destinations than others. To date no comprehensive research has been conducted into what makes some CWT destinations more competitive than others, although comprehensive destination competitiveness models such as those offered by Ritchie and Crouch (2003) or Dwyer and Kim (2003) offer clues as to why some CWT destinations may be more successful. Such models point to the importance of a range of interconnecting factors that combine in such a way to produce 'successful' destinations. Dwyer and Kim, for example, highlight the role of *endowed resources*, but also note the importance of *created resources*, *supporting factors*, *destination management*, *situational conditions* and *demand factors*. In the CWT context, preliminary work by Lovelock and Milham (2006) in New Zealand has indicated that competitiveness may not simplistically hinge upon the presence of sought-after game species, but may depend upon a raft of other factors. While New Zealand has a range of valued game species, including Himalayan Thar, chamois, red deer, fallow deer and wapiti, it is more than the simple presence of these animals that makes the country a competitive CWT destination. Other factors include:

- product awareness, generated through presence at international game fairs, internet promotion and word-of-mouth networks;
- the existence of a user-friendly legislation regarding the importation of firearms;
- the legal status of target species (i.e. unprotected, noxious pest);
- seasonality issues (e.g. Southern Hemisphere destinations providing off-season hunting for Northern Hemisphere visitors);
- most hunting is fair chase;
- relative lack of hunting pressure in uncrowded surroundings;
- complimentary target species within close geographical proximity;
- trophy quality animals through careful game management practices;
- experienced guides/outfitters quality assured through professional body;
- scenic and other natural and cultural attractions;

- low value of the currency – exchange rate advantages;
- a well-developed aerial transportation sector and a good network of trails and back-country huts;
- family-friendly activities for accompanying family members, e.g. adventure tourism, wine tourism, cultural tourism and retail opportunities;
- a strong and positive brand image for the destination as a whole;
- a positive image of the destination as being safe and secure;
- relative proximity to other South Pacific hunting destinations.

Anecdotal evidence would suggest, however, that price competitiveness is a primary factor, and this is borne out by the increasing trend, for example, for Western European hunters to travel to Central and Eastern Europe, or even Central Asia for their hunting holidays. Similarly, Norwegian hunters cross the border in droves to hunt moose in Sweden – a phenomenon linked to both game availability and price. However, when a range of destinations are similarly competitive on price, or as in the case with New Zealand, when a destination is quite peripheral in terms of time and cost of accessing it, other factors increasingly play a role in the destination decision-making of potential consumptive wildlife tourists. For example, many trout anglers choose the long-haul flight to New Zealand because of the uncrowded nature of the country's back-country waters, the trophy size trout and the relative low cost of the fishing.

Price, however, is not a constraint for a significant portion of the CWT market, many of whom are high tourism spenders. Safari Club International members, for example, are typically business owners, professionals or executives, with an average annual income of more than US$200,000. They spend on average, 37 days a year hunting, 21 of which are outside of the United States, spending on average US$61,000 on travel-related costs. Members spend nearly $44,000 per year on hunting, including US$10,000 on hunting and shooting related equipment (SCI 2006c). Similarly, in the European context, Bauer and Herr (2004: 64) refer to a 'powerful, highly organised and economically viable group of hunters', and note that some hunting trips may cost over US$100,000.

For this group, long-haul hunting destination choice is not based upon price but upon the other factors noted above, the availability of trophy-quality animals likely being paramount. At the other end of the spectrum of consumptive wildlife tourists, are those that are more likely to be domestic tourists or at least intra-regional (e.g. within EU), for whom price is important. This is by far the largest part of the market, and overall would have the greatest economic impact.

Economic impact of CWT

Economic impacts and the benefits to local tourism providers, communities and regional economies are increasingly being used by some hunting groups (and fishers) to legitimate their activities, in the face of interest group, or in some cases, wildlife management agency opposition. The scale of such impacts is hard to ignore, with actual worldwide revenues from safari-hunting alone estimated to surpass US$half a billion annually (Van der Walt 2002). If the economic analysis is extended to include all trip-related expenses, in the United States alone, this totals US$20 billion per annum for hunting and fishing together (USDOI *et al.* 2002). However, in what is becoming an increasingly antagonistic world for hunters and fishers, the actual net benefits of CWT have been questioned. A TRAFFIC Europe report on the overall economic relevance of European hunting tourism, where 20–30 per cent of hunters travel abroad for hunting, purports to show that the impact of trophy hunting to a country's income is lower than assumed. The average direct cost for a hunt is about €2,000, with a total spend of €120–180 million, but only a third of the income remains in the supply countries, generating very little towards the GNP of the region: 'Even in Hungary, which supplies the tourist hunting demand with a big chunk of 10,000–20,000 hunts per year, the economic impact of this is limited to 0.0005 per cent of the GNP' (Hofer 2002).

In this respect, CWT differs little from many other forms of tourism, suffering from the economic leakages associated with importing the goods and services needed to support the activity. One of these leakages is on spending on hunting and fishing equipment, which is mainly undertaken in the tourists' home locations. Similarly, spending on food and other supplies is often sourced from home locations. In response to changes in production, with losses of primary production earning capacities, many peripheral areas have been turning to CWT as a potential source of income. The Pacific Northwest is a good example, where changes to environmental legislation, coupled with competing off-shore production, have led to a decline in the timber industry. Although some communities have considered promoting CWT as an alternative regional economic generator, empirical evidence suggests that the returns for such areas from this form of tourism are less than anticipated, largely because of provisioning occurring in the generating region (e.g. Meyer *et al.* 1998; Guaderrama *et al.* 2000). Similarly, Lovelock and Robinson (2005) in their study of hunting whitetail deer on New Zealand's remote Stewart Island point to the irony of the peripheral location of the island being the attraction for hunters, but because of the lack of services and retail sector, the community misses out on economic opportunities – which again accrue to visitors' home locations or gateway communities well outside the hunting region. In peripheral *nations*, the same may apply, with consumers preferring 'First World' goods and services, that, if available, are often provided by non-local or foreign operators. In Southern Africa, for example, a region with a US$75 million dollar hunting safari industry the vast majority of hunts are guided by expatriates as opposed to indigenous African guides (Lewis and Jackson 2002).

However, local communities *can* benefit from CWT, but the extent to which destinations may capture income from CWT in part depends upon the relationship between the hunter, the target species and the destination. In the examples cited above, the consumptive tourist is often a domestic tourist, hunting a familiar species in a fairly familiar environment. In the case of a European hunter in Botswana or an American fisher in Chile, the relationship is less familiar – the visitor does not know the target species, the geography or language, and is much more dependent upon the services of outfitters and local guides, transport operators and accommodation services. Thus there are many more opportunities for local tourism providers to capture CWT spending. This is supported in research (e.g. Child 1995) noting the benefits of CWT to communities in Africa and elsewhere, and is also the subject of discussion in later chapters in this book. What is clear is that the economic impact of CWT is undoubtedly context-dependent and complex.

Demand countries also have the potential to generate income from CWT through the sales of hunting and fishing-related equipment (e.g. US sales = US$14 billion (USDOI *et al.* 2002)), as well as through hunting and fishing expos and conventions. Annual conventions such as the Safari Club International's annual hunting convention in Reno, Nevada and the Western Hunting and Conservation Expo in Utah, are expected to bring between 20,000 and 30,000 attendees each, with an economic impact of over US$40 million for the SCI event (Speckman 2006; Randazzo 2006).

CWT and conservation

Despite the best efforts of the CWT industry to promote the industry as being environmentally friendly, unfortunately, unregulated hunting has had substantial impacts upon some threatened species e.g., Siberian musk deer, lynx and argali sheep (IUCN 2006; The London Zoological Society 2006). There is a saying, attributed to former wildlife biologist of international acclaim (and US Forest Service chief) Jack Ward Thomas, that 'If you want to do a species a favour, get it on the hunted list' (Petersen 2000: 47). And while not all trophy hunting or fishing is sustainable, if certain conditions are satisfied (e.g. scientifically determined wildlife populations, enforceable quotas, honest and competent industry management) then conservation benefits are likely (Baker 1997b).

While the jury is still out on the exact extent to which hunting benefits conservation, its ability to generate funds that can go into conservation programmes is not disputed. The payment of game fees – trophy fees and fishing licences – to public and private bodies has a demonstrated ability to contribute to conservation programmes. However, the extent to which local communities and conservation programmes benefit from CWT depends largely upon the model of revenue collection and disbursement systems adopted within the destination. From her comparative study of six sub-Saharan destinations, Baker (1997a) develops an optimal model for the collection and disbursement of hunting revenue. Such a system hinges upon the establishment of a direct connection between each animal and its benefit to the community. Concession fees would be paid for wildlife programme administration and trophy fees would be paid directly to local communities (rather than through a distant centralised revenue system).

When considering the costs of protected area management in central Africa, Wilkie and Carpenter (1999) note that, typically, government and donor investments meet less than 30 per cent of such costs, with few additional sources of funding available. They conclude that safari hunting could offer a 'significant and sustainable source of financing', noting, for example, that Cameroon would have to attract only 4 per cent of Safari Club International members travelling to Africa to maintain a revenue stream of US$750,000 per year in trophy fees – enough to meet nearly 40 per cent of the management costs of all protected areas in the country.

The linking together of hunting and conservation is a trend increasingly being seen in hunting and fishing organisations. Prime examples are the Safari Club International Foundation, which runs a number of conservation programmes. The organisation, for example, has developed partnerships with wildlife management and hunting companies in Central Asia and Mongolia. The International Council for Game and Wildlife Conservation (CIC) is a European-based NGO that runs hunter-conservation programmes in many CWT destinations. Other examples of 'Sportsmen NGOs' include Ducks Unlimited, the Foundation for North American Wild Sheep and the Wild Turkey Federation, all of which could be described as thinly veiled hunting and shooting clubs. While participation in an outdoor activity (e.g. fishing) is connected with a concern for the resource (e.g. rivers) upon which that activity depends (Jackson 1986), the adoption of species and habitat restoration projects can serve a wider purpose. Such an approach might be considered a deliberate strategy to trigger an associational relationship between hunting and conservation that historically may not have existed or have been strong. This association will be beneficial in terms of assuring continued availability of game species for hunting tourists (in some cases bringing species back from the brink of extinction to the point of now having populations that can sustain a degree of hunting), but also, and most importantly, in enhancing public, government and non-governmental acceptance of the organisations and the CWT practices of their members.

Animal rights movement

But while significant NGOs such as the WWF, the National Audubon Society, and the National Wildlife Federation, accept hunting tourism (Petersen 2000; CIC 2004)) not all environmental or animal rights NGOs condone the activity.

The last legal 'traditional' foxhunt in England and Wales was held on 20 February 2005. This capped a successful campaign by the League Against Cruel Sports and other animal rights interests dating back to 1949. Hunting foxes with dogs was made illegal under the Hunting Act, but hounds can be used to follow a scent and to flush out a fox. The fox can then be killed by a bird of prey or shot – if only two dogs are involved. Amazingly, while this was predicted by advocates of the hunt to be the death knell of a long tradition and way of life, and along with it thousands of hunt-related jobs in supporting services and hospitality, the sky did not fall. A remarkable ability to adapt has been demonstrated by those involved in the hunt, which has continued, albeit in a modified form.

By extension, the anti-hunting debate is an anti-CWT debate, and is of high relevance to the future of CWT. Highly organised and very well resourced animal rights groups from the US Humane Society (with assets of US$96 million (USsportsmen.org)) to the Doris Day Animal League oppose hunting to some degree or other, if not in totality. Petersen (2000) distinguishes between the Animal Rights (where use of animals for any human benefit is wrong) and Animal Welfare groups (supports humane treatment and freedom from unnecessary pain and suffering). Extreme affection for individual charismatic megafauna may result in what Kellert (1978) refers to as 'undue and dysfunctional anthropomorphism' i.e. the so-called Bambi syndrome.

Contemporary concern has moved beyond Bambi, however, to even include animals such as the 'wicked wolf': American NGO Friends of Animals has run high profile campaigns against wolf hunting in Alaska, part of which is promoting a travel boycott against that destination. In 2004, two of the United States' largest anti-hunting organisations (The Humane Society and the Fund for Animals) merged, with a promise to seek an end to bow-hunting as their first priority (SCI 2004). The latter is the most vociferous anti-hunting group in the United States, and raised nearly $7 million in 2003. In response, hunting organisations are organising, and increasingly using the rhetoric of 'conservation-hunting' as noted above, or of heritage hunting as a part of their defence.

However, some forms of CWT look increasingly indefensible, such as the seal-hunting and whaling tourism available in Norway. In 2001, the Norwegian Fishery Minister proposed that 'Seal hunting in the wild Norwegian coastal nature should be sold as an exclusive product to tourists' (Planet Ark 2001). The idea was 'a hit' as the Minister predicted, and Norwegian safari companies now take certified, licensed domestic and foreign hunters, accompanied by guides to shoot leopard seals.

In comparison, fishing-tourism receives relatively little attention from the animal rights movement. This may be because of the long-perpetuated urban myth that fish feel no pain, or because of the assumption that fish are a lower form of life not imbued with the ability to feel pain and emotion as do mammals (such as Bambi and his friends). The fact that fishing is a ubiquitous industry may also serve to desensitise people to this order. As Bauer and Herr (2004) note, there is generally little controversy surrounding fishing and therefore fishing tourism, which is treated with a social indifference that the hunting-tourism industry would welcome. However, a recent innovation in fishing has been 'pointless' fishing – where the curve and point is cut off the hook, a step beyond catch-and-release fishing!

While these battles are fought in the popular media and legislative assemblies throughout CWT destinations around the world, some anti-hunting groups are adopting a more pragmatic approach to ending CWT. An environmental group in Canada recently bought (CAN$1.35 million) the trophy hunting rights to one of the most valuable hunting territories in British Columbia, ending the shooting of bears and wolves by foreign hunters (BC residents will still be able to hunt). The Raincoast Conservation Society, along with First Nations groups intend to develop a new wildlife photography industry to offset the loss of income from foreign hunters (CBC News 2005).

Arguably, the battle for animal rights will also increasingly take place in the woods – on our trails and back-country huts and lodges where consumptive and non-consumptive users will increasingly interact. There is already evidence that hunters and non-hunters hold different social and environmental values, and that this can be a source of conflict (Daigle *et al.* 2002). With increased visitation to the back-country, some protected area management agencies are devoting more resources to researching such conflict – for example in New Zealand, there is potential for this conflict of values to impact upon visitor patterns, behaviour and satisfaction for both consumptive and non-consumptive groups (Lovelock 2003).

Current and future constraints

Undoubtedly the anti-hunting movement is the biggest threat to the CWT sector. As Baldus claims, 'Radical anti-hunting ideologies pose presently the largest danger for hunting as a form of sustainable land use' (1991: 366). For hunting tourism to continue it will increasingly rely upon its proponents clearly defining the benefits of the activity – whether they be in terms of ecosystem integrity, preservation of threatened species, economic return to marginal communities or even simply safety on the roads. It will also depend upon the CWT industry and hunting and fishing groups maintaining their political strength – this may involve greater political organisation, coupled with programmes of hunter/fisher recruitment and retention, particularly in view of changing demographics. It will also involve the CWT sector working alongside and developing collaborative relationships with the broader tourism industry, and notably, with non-consumptive wildlife interests – something that all too few hunting and fishing organisations have been successful at doing (e.g. Lovelock and Milham 2006).

However, there are other less apparent threats to the development of the CWT sector. These fall within the broad parameters of global environmental change. The threats of global warming upon subsistence hunting of peri-Arctic communities, through changes to vegetation and pack ice patterns, have already been highlighted (BESIS 1996; IPCC 2001). While this is currently only a small market within the CWT realm, there are already anecdotal reports of the effects of climate change in other ecozones – for example, upon the water regimes of wetlands, which will potentially impact the much larger sub-sectors, of waterfowl shooting, and is already in evidence (Jackson 2006). Similarly predicted changes to savannah habitat in sub-Saharan Africa may have potentially devastating effects on game populations, distribution and associated incomes (Von Maltitz and Mbizvo 2005).

The fishing-tourism industry will face these threats, along with its particular demons – which include competition with the commercial fishing industry, pollution of rivers, lakes and coastal waters, and increasing competition with agricultural and industrial users for freshwater resources.

Biosecurity breaches also have the potential to impact upon game species and their habitats, through the introduction of pests, parasites and predators. A good example of this is the recent accidental introduction to New Zealand of the aquatic algae Didymo. This has spread through many of the country's prime trout rivers, and has the possibility of lowering productivity, trophy potential, impacting upon the availability of fishing waters, angler movements and consequently on the competitiveness of the country as a trout angling destination. The potential of game-borne diseases such as Avian infuenza (bird-flu) upon game bird populations and associated tourism has yet to be fully analysed.

A further threat to the development of CWT is simply the maintenance of sustainable populations of suitable (e.g. trophy quality) game animals. While this is partly the role of wildlife management, animal husbandry and genetic manipulation, habitat protection is a critical aspect, and along with this, protection from poaching – which threatens both consumptive and non-consumptive forms of wildlife tourism alike.

Ironically, the CWT sector is a threat to itself, not only through uncontrolled over-hunting or fishing, but through the increasing efficiencies of the industry itself. This is particularly evident in the development of cheater practices in both fishing and hunting. The use of 'fish-finders' and global positioning systems, night-vision glasses, heat sensing, hearing enhancers and two-way radios all stack the odds in favour of the hunter or fisher and are frowned upon by purists (Peterson 2000) who promulgate the 'fair chase' approach.

Fair chase and canned hunting

'Fair chase' is a part of the original code of conduct first used by Boone and Crockett Club members in the early 1890s. It referred to hunts being undertaken in such conditions that game animals would have an unrestricted capacity to evade the hunter, thus granting no hunter an advantage over another in terms of potential to bag a trophy. Coincidentally (and perhaps unintentionally), the code also granted some early form of animal 'rights' to the game animals. Currently the fair chase code includes proscriptions against shooting animals from airplanes, boats, land vehicles, against herding animals towards shooters, and the use of cheater technology. Fair chase also proscribes the hunting of fenced-in animals. The latter so-called 'canned hunting' may be on large estates of thousands of acres (e.g. in South Africa or New Zealand's game estates) but is also conducted on numerous small ranches where a range of captive exotic species may be hunted. There are reportedly over 1,000 such establishments in the United States, and in Texas, which has the highest concentration, the average size is only 75 acres. The practice has been described as 'execution-style killing' (Petersen 2000: 45) but attracts hunter tourists because of convenience, guaranteed chance of success and ability to collect an exotic trophy. Ironically, canned hunting may be a practice that will ultimately make a significant contribution to the sustainability of CWT through captive breeding programmes of game species and its role in reducing the pressure of natural populations and habitats. The viability of this practice, at least in the United States, has been opened to debate, however, with the introduction of a bill (the Sportsmanship in Hunting Act 2005) aiming to place a minimum size on game ranches (1,000 acres) and to limit the transportation of animals for the purpose of hunting (SCI 2005).

In an extreme version of canned hunting, coupled with cheater technology, a Texas entrepreneur recently promoted the concept of hunting by remote control. The business involves a Remington .30-06 rifle, linked to video camera, mounted in a game ranch. From anywhere, someone with an internet connection can fire the rifle, in a 'real-time, on-line hunting and shooting experience' (Moreno 2005). While the ethics of this operation have been questioned, it does offer those with mobility problems an opportunity to participate in a form of CWT. But from a tourist-industry point of view, a move towards virtual CWT may be less desirable than the real thing because of the uncertainty of any social, cultural or economic benefits accruing in this operation – at least to the extent that they may in a real CWT scenario. It is noteworthy that the sporting community came out strongly against this business and has advocated for laws banning this activity.

One problem facing destination managers, for example national or regional tourism organisations who wish to market CWT, is how to do this in a sensitive manner that will not alienate large and lucrative segments of their tourist market. While some destinations overtly promoted CWT opportunities some years ago, this may no longer be an acceptable practice given contemporary realities of the animal-rights/welfare movement. Strong voices from this interest sector have argued that state agencies should not advocate hunting (e.g. Rutberg (2001), from The Humane Society of the United States).

Entire nations, however, have established their tourism industries upon CWT. New Zealand, for example, arguably built its early tourism industry on the back of red-deer, wapiti, trout and big-game fishing (the latter courtesy of Zane Grey who spent some time there courtesy of New Zealand's state tourism agency). Interestingly, an historic analysis of the images of animals used in the nation's tourism promotional literature reveals consistent use of images of big-game fishing, angling and hunting only until the 1960s/1970s, a point from which only the angling images prevailed and which also saw the rise of the animal-rights/welfare movement (Mabon 2006). Few National Tourism Organisations today, even those with substantial CWT markets, employ images of hunting and fishing within their websites and promotional literature. Yet despite this censor, the hunting and fishing fraternity appears to be fairly well networked, promotion being undertaken effectively by a less threatened private sector, and with word-of-mouth seemingly the most effective means of promotion (Lovelock and Milham 2006).

These issues and the range of potential threats would point to risk of destinations developing a strong dependence upon the niche tourism CWT market. Such a warning may not necessarily apply to destinations that depend upon non-consumptive forms of wildlife tourism – although biological risks may be similar for both markets. Populations of fish and game vary, and the prosperity of their associated sectors will likewise rise and fall – as aptly demonstrated by the huge drop in income from quail hunting associated with quail population decline in south-eastern United States in the early 1990s (Burger *et al.* 1999). Recently, with the decline of lion populations in Botswana, the heavily hunting-safari dependent industry there has taken steps to try and attract non-hunters (Hale 2004).

Figure 1.3 Woman fishing. Cover of New Zealand tourism promotional brochure [ca 1935] (Artist Mitchell, Leonard Cornwall 1901-1971). *Source*: Alexander Turnbull Library, National Library of New Zealand – Te Puna Matauranga o Aotearoa.

Ultimately the future of CWT will depend in part upon the sector being developed as a complement to, and also being complemented by, other tourism products (e.g. cultural tourism, heritage tourism, gastronomy and wine tourism). Mexico is one example of a destination investigating the feasibility of CWT – not because it will bring in droves of visitors, but because alongside its other tourism offerings, it fits nicely, especially in terms of seasonality issues and rural development potential. Furthermore, it will complement nature-based tourism in general through acting as a catalyst to protect and improve habitats (REDES 2003).

Destinations may also need to resort to non-consumptive practices temporally or spatially when consumptive practices are deemed unsustainable. Such a complementarity of consumptive and non-consumptive uses is seen by some as a critical aspect of competitiveness for wildlife tourism (e.g. Tremblay 2001).

This book

This book is the first collection to specifically address hunting, shooting and sport fishing as touristic activities. While there is a body of research that has considered these pursuits in terms of leisure or recreation, little consideration has been given to them in a tourism context. Consequently, this book, as with all 'new' fields is necessarily eclectic, drawing together current research from a number of disciplinary perspectives.

The intent of the book is to highlight some key issues facing CWT in the contemporary world. The book endeavours to present issues from a broad geographic perspective. As with most forms of tourism, the issues arising are very context-dependent, yet as we will see, a number of central issues and themes emerge from case-studies sourced from North America, Europe, Africa, Scandinavia, India, Arabia and Oceania. Perhaps an equally important goal is for the book to raise awareness of the significance of this sector – not only for researchers, but also for destination managers interested in pursuing the CWT pathway for destination development.

The book is divided into four Parts. Following this introduction, Part I 'Introduction and Conceptual Issues', continues with two conceptual chapters. The first of these by Franklin addresses the 'Animal Question' – a broad debate on the ethics, practice, humanity and environmental implications of our relations with animals. In this chapter, Franklin raises and discusses a number of issues surrounding consumptive wildlife practices (especially sensual and embodied practices), humanity issues (what our proper connections are with wildlife) and environmental issues (how best to produce a sustainable connection with wildlife). Preston-Whyte in the following chapter also addresses the culture–nature divide. He employs an actor-network approach in which agency is attributed to both human and non-human actors, to consider the struggle for dominance between tourist-fisher and fish. Actor-network theory provides a useful framework for considering the very forces that compel and motivate the consumption of this form of wildlife tourism.

Part II, 'Historic Precedents', comprises four chapters and addresses the historic factors that have influenced the development of CWT in a range of settings. From Scandinavia, Sillanpää discusses aspects of the 170-year history of CWT. Since the Victorian period, many British sportsmen in particular travelled to the Scandinavian Peninsula for hunting, shooting and fishing – 'The Scandinavian Sporting Tour'. The chapter focuses on how these early tourists influenced the host communities economically, socially and culturally, and how this early practice can be seen as a precursor to modern tourism in the region. In an African context, Akama provides an historical evaluation of controversies concerning wildlife. Kenya is selected for consideration, an interesting case-study in view of its anti-hunting policy. It is argued that Kenya's pioneer conservation policies were based on conservationists' conceptions that indigenous resource use methods (including hunting) were incompatible with the dominant Western principles of wildlife management and protection. The resultant non-consumptive wildlife policies and tourism programmes that prevail have resulted in severe people–wildlife conflicts.

The chapter from Figgins provides an interesting cross-national comparative study of the historical role of touristic and recreational deer hunting. Figgins adopts a social constructionist approach, and utilises the neo-Marxist concept – the subsumption of nature – to consider how the hunting of red deer has helped to shape the respective social and physical landscapes in Scotland and New Zealand. He also observes how the increased commercialisation and economic value placed on recreational hunting through tourism has led to some tensions within the hunting community. Hannam completes this Part with a chapter on the historic role of tiger hunting in India. Hannam focuses on the importance of tiger hunting in India for the reproduction and maintenance of the British colonial State. Hunting tigers became emblematic of the exercise of colonial state power and reinforced both the claim to rule and the aura of British invincibility. The chapter also discusses how the different methods of tiger hunting fed into socially constructed ideals of masculinity, health and Englishness.

Part III of the book 'Impacts of Consumptive Wildlife Tourism' contains seven chapters addressing various aspects of the ecological, socio-cultural and economic effects of CWT. Part III opens with a study by Foote and Wenzel of conservation hunting of polar bear in Nunavut, Canada. The chapter discusses the concept of conservation hunting, noting how in this case it can be considered a form of ecotourism because of its relatively light environmental footprint, minimal infrastructure needs, high selectivity of harvest and high degree of exchange between hunters and local community members. The chapter also identifies how economic, social and biological impacts differ greatly between non-consumptive polar bear watching and bear hunting parties. That consumptive and non-consumptive wildlife tourists are indeed different, is the topic of the next chapter from Dawson and Lovelock, who consider marine tourists in New Zealand. Their study reveals that non-consumptive sea kayaking and consumptive sea fishing tourists are two distinct user groups in terms of their socio-demographic characteristics and environmental values, supporting previous environmental values research in terrestrial settings.

The nature of consumptive wildlife tourism in Africa, and the challenges it provokes, are addressed in the next chapter, by Mbaiwa. The chapter focuses on Eastern and Southern African countries where the economic benefits of safari hunting tourism are discussed. Mbaiwa also addresses the connection between CWT in these destinations and the promotion of a sustainable harvesting policies. Finally, attention is drawn to ongoing problems in these destinations, such as the decline of wildlife species, poaching and conflict regarding the trade of game products. Barnesand Novelli also address CWT in Africa, discussing the two main forms of consumptive wildlife tourism in Namibia, trophy hunting and recreational shore angling. The economic value, impacts, contribution to development and social and environmental characteristics of these two activities are compared. Trophy hunting is more economically efficient than coastal angling, and is also more socially and environmentally acceptable. The authors argue that this is primarily related to property rights and institutional factors in this context.

The following chapter by Mattsson and colleagues posits that it is necessary to manage wildlife and fish resources efficiently, so that hunting and fishing can maintain or improve their functions from a welfare economics perspective. The authors discuss what role welfare economics can play in solving the problems associated with natural resource management including fish and game. The chapter considers future research requirements relevant to hunting and fishing, noting the potential for research-supported management of CWT resources for increased welfare. The social and cultural impacts of CWT in small, rural communities are considered in the chapter by Gunnarsdotter, who examines moose hunting in the Swedish countryside. The chapter identifies how both the local moose hunting teams and hunting tourists contribute in different ways to viable rural communities. The local hunting teams help to maintain the sense of community and place that has developed over time. Hunting tourism supports the local economy by providing alternative income streams. Gunnarsdotter also discusses the tensions that sometimes appear between these different groups.

Seddon and Launay's chapter concludes this Part, with an exploration of how the economic growth from petro-dollars has brought dramatic changes to the traditional practice of falconry in the Middle East. The chapter reviews the practice of falconry with a focus on the Kingdom of Saudi Arabia. There, the significant amounts of money spent on falconry and associated expeditions, has changed the scale and spread of the practice, increasing its impacts (domestically and internationally) on both falcons and their quarry species. The chapter describes efforts to put in place species conservation measures.

Part IV, 'Current Issues and Destination Development' comprises six chapters that collectively examine characteristics and behaviours of consumptive wildlife tourists and raise issues relating to destination development for CWT. Campbell's chapter starts the section with an examination of bear hunting, a form of CWT that has been strongly influenced by negative public attitudes to the extent that a number of such hunts across North America and Canada have been cancelled. The spring bear hunt in Manitoba is described as an important tourism product as well as an important component of the province's wildlife management strategy. This chapter explores the role of the bear hunt in tourism and wildlife management in the province, and the development of the bear hunt conflict. The Province's approach to resolving the issue is described. While some communities have adopted CWT as an avenue of economic development, others have faced obstacles in doing so. Cohen and Sanyal present a study of three small towns in rural Northern Idaho, faced with the closure of local timber mills, each town is considering a future based to some extent upon CWT. However, as their study reveals, transitioning from a timber extraction economy to a CWT economy takes more than just a vision. The chapter addresses obstacles and opportunities for developing CWT in communities that are diverse, have strong Native American interests in wildlife and are protective of their own hunting and fishing opportunities. Normann, in his chapter likewise considers the challenges involved in developing CWT in remote communities, typically involved in more exploitative resource-based industries. Marine fishing tourism appears to be an attractive option for destination development, however, as Normann's case-study set in the Lofoten Islands of Northern Norway illustrates, marine fishing tourism lies within a complex policy environment. The chapter draws attention to the benefits of the commercial fishing and tourist fishing industries working together in the development of CWT. Commercial fishermen can take their share of the income generated by demand for CWT, while their experience and knowledge will help improve the product sought by tourists.

Still on the topic of fishing, but this time in a freshwater setting, Walrond draws upon research conducted upon backcountry trout angler-tourists in New Zealand, identifying that this segment of tourists have particularly high demands in terms of product quality. Comparing results with earlier research in New Zealand and North America it is clear that this group is among the least tolerant of encounters with other users. The reasons for this are detailed, and implications discussed in relation to destinations maintaining high levels of CWT visitor satisfaction while facing increasing numbers of angler-tourists. The section concludes with a chapter by Craig-Smith and Dryden, who consider the development of a tourism industry based around the hunting of exotic animals in Australia. The authors express the opinion that hunting tourism has the potential to develop into a small but profitable niche market for Australian tourism. However, there are issues concerning the resource base of exotic 'pest' animals. The role of tourists in providing help in exotic animal population control whilst contributing to economic development in regional Australia is discussed.

The final chapter brings together some of the themes and issues identified throughout the book, and draws some conclusions regarding the future of CWT from the perspective of destination managers, developers and other stakeholders with an interest in this sector. Recommendations for areas of future research related to hunting, shooting and fishing tourism are presented.

References

Australian Sports Commission (2006) *Shooting Sports*. <http://www.ausport.gov.au> (accessed 20 December 2006).

Baker, J.E. (1997a) 'Development of a model system for touristic hunting revenue collection and allocation', *Tourism Management*, 18(5): 273–86.

Baker, J.E. (1997b) 'Trophy hunting as a sustainable use of wildlife resources in southern and eastern Africa', *Journal of Sustainable Tourism*, 5(4): 306–21.

Baldus, R. (1991) 'The economics of safari hunting', *Tourismus*, 361–6.

Bauer, J. and Herr, A. (2004) 'Hunting and fishing tourism', in K. Higginbottom (ed.) *Wildlife Tourism: Impacts, Management and Planning*, Altona, Victoria, Australia: Common Ground Publishing and Co-operative Research Centre for Sustainable Tourism.

Burger, L.W. Jr., Miller, D.A. and Southwick, R.I. (1999) 'Economic impact of northern bobwhite hunting in the southeastern United States', *Wildlife Society Bulletin*, 27: 1010–18.

Bering Sea Impact Study (1996) *The Impacts of Global Climate Change in the Bering Sea Region*, Fairbanks, AK: BESIS Projects Office, University of Alaska.

Cat Specialist Group (2002) '*Lynx lynx*', in IUCN 2006. *2006 IUCN Red List of Threatened Species*. <http://www.iucnredlist.org> (accessed 19 December 2006).

CBC News (2005) *Conservationists Pay More than $1 Million to End Bear Hunt*. <http://www.cbc.ca> (accessed 14 December 2005).

Cheney, C. (2006) 'Hunting in Protected Areas', *African Indaba e-Newsletter*, 4(3): 3.

Child, G. (1995) *Wildlife and People: The Zimbabwean Success*, Harare and New York: Wisdom Foundation.

CIC (2004) 'Tourism and Development: The Win–Win Performance', *African Indaba e-Newsletter*, 2(2): 2.

Daigle, J.J., Hrubes, D. and Ajzen, I. (2002) 'A comparative study of beliefs, attitudes and values among hunters, wildlife viewers and other outdoor recreationists', *Human Dimensions of Wildlife*, 7: 1–19.

Delpy-Neirotti, L. (2003) 'An introduction to sport and adventure tourism', in S. Hudson (ed.) *Sport and Adventure Tourism*, Binghampton, NY: Haworth Press, pp. 1–26.

Ditton, R.B., Holland, S.M. and Anderson, D.K. (2002) 'Recreational fishing as tourism', *Fisheries*, 27(3): 17–24.

Dizard, J.E. (2003) *Mortal Stakes: Hunters and Hunting in Contemporary America*, Amherst: University of Massachusetts Press.

Duffus, D.A. and Dearden, P. (1990) 'Non-consumptive wildlife oriented recreation: a conceptual framework', *Biological Conservation*, 53: 213–31.

Dwyer, L. and Kim, C. (2003) 'Competitive destinations: determinants and indicators', *Current Issues in Tourism*, 6(5): 369–414.

Fisheries Research and Development Corporation (FRDC) (2001) *Implementation of the National Recreational and Indigenous Fishing Survey*, Canberra: Australian Government

Freese, C. (1998) *Wild Species as Commodities: Managing Markets and Ecosystems for Sustainability*, Washington, DC: Island Press.

Groome, K.H., Simons, D.G. and Clark, L.D. (1983) *The Recreational Hunter: Central North Island Study*, Lincoln, Canterbury, New Zealand: Department of Horticulture, Landscape and Parks, Lincoln College, University College of Agriculture.

Hale, B. (2004) *Botswana Battles to Lure Non-hunters*. <http://news.bbc.co.uk> (accessed 12 January 2004).

Haripriya, G.S. (2004) 'A note on ecotoursm', *Envisage*, Vol. 5, April 2004. <http://envis.mse.ac.in/newsletter/envisage5.pdf> (accessed 22 January 2007).

Higginbottom, K. (2004) *Wildlife Tourism: Impacts, Management and Planning*, Altona, Victoria, Australia: Common Ground Publishing and Co-operative Research Centre for Sustainable Tourism.

Hofer, D. (2002) *The Lion's Share of the Hunt – Trophy Hunting and Conservation: A Review of the Legal Eurasian Tourist Hunting Market and Trophy Trade under CITES*, Brussels: TRAFFIC Europe Regional Report.

Holland, S.M., Ditton, R.B. and Graefe, A.R. (2000) 'A response to "ecotourism on trial: the case of billfish angling as ecotourism"', *Journal of Sustainable Tourism*, 8: 346–51.

Guaderrama, M.C., Meyer, N., Harp, A. and Taylor, R.G. (2000) *Replacement of Timber Harvest and Manufacturing with Recreational Visitors as Economic Base: Case of Valley County, Idaho*. Moscow, Idaho: Department of Agricultural Economics and Rural Sociology, University of Idaho.

Intergovernmental Panel on Climate Change (IPCC) (2001) *Climate Change 2001: Impacts, Adaptation and Vulnerability*, Geneva: IPCC Secretariat. Jackson, D.Z. (2006) *Duck Hunters Feel Effects of Climate Change*. <http://seattlepi.nwsource.com> (accessed 18 December 2006).

IUCN (2006) *2006 IUCN Red List of Threatened Species*. <http://www.iucnredlist.org> (accessed 19 December 2006).

Jackson, E.L. (1986) 'Outdoor recreation participation and attitudes to the environment', *Leisure Studies*, 5: 1–23.

Jagnow, C.P., Stedman, R.C., Luloff, A.E., San Julian, G.J., Finley, J.C. and Steele, J. (2006) 'Why landowners in Pennsylvania post their property against hunting', *Human Dimensions of Wildlife*, 11: 15–26.

Kellert, S.R. (1978) 'Attitudes and characteristics of hunters and anti-hunters', *Transactions Forty-Third North American Wildlife Conference*, 43: 412–23.

Kellert, S.R. (1996) *The Value of Life: Biological Diversity and Human Society*, Washington, DC: Island Books.

Leapman, B. (2006) *Angling? It's too White and too Middle-aged, Say Ministers as they go Fishing for Women and Ethnic Minorities*. <http://www.telegraph.co.uk> (accessed 16 April 2006).

Lewis, D. and Jackson, J. (2002) *Safari Hunting and Conservation on Communal Land*. <http://www.itswild.org/pdf/safari_hunting_and_conservation_on_communal_land.pdf> (accessed 13 February 2007).

London Zoological Society, The (2006) *Unprecedented Loss of Mongolian Mammals*. <http://www.zsl.org/london-zoo/news> Tuesday 12 December 2006 (accessed 20 December 2006).

Lovelock, B.A. (2003) 'International and domestic visitors' attitudes as constraints to hunting tourism in New Zealand', *Journal of Sport Tourism*, 8(4): 197–203.

Lovelock, B.A. (2006) '"If that's a moose, I'd hate to see a rat!": visitors' perspectives on naturalness and the consequences for ecological integrity in peripheral natural areas of New Zealand', in D.K. Muller and B. Jannson (eds) *Tourism in Peripheries: Perspectives from the North and South*, Wallingford: CABI, pp. 124–40.

Lovelock, B.A. and Milham, J. (2006) *Summary of Results: Hunting Tourism Industry Survey 2005*, Unpublished report to the New Zealand Professional Hunting Guides Association.

Lovelock, B.A. and Robinson, K. (2005) 'Maximising economic returns from consumptive wildlife tourism in peripheral areas: white-tailed deer hunting on Stewart Island/Rakiura, New Zealand', in C.M. Hall, and S. Boyd (eds) *Nature-based Tourism in Peripheral Areas: Development or Disaster?*, Clevedon: Channelview, pp. 151–72.

Mabon, A. (2006) *How Images of Animals are Used in Promoting New Zealand 1901–2006*, Unpublished Report, Department of Tourism, University of Otago, Dunedin, New Zealand.

McFarlane, B.L., Watson, D.O. and Boxall, P.C. (2003) 'Women hunters in Alberta, Canada: girl power or guys in disguise?', *Human Dimensions of Wildlife*, 8: 165–80.

McKercher, B. and Chan, A. (2005) 'How special is special interest tourism?', *Journal of Travel Research*, 44: 21–31.

Meyer, N., Harp, A. and McGuire, K. (1998) *Economic Impacts and Fiscal Costs of Public Land Recreation in Clark County, Idaho*, Moscow, ID: Department of Agricultural Economics and Rural Sociology, University of Idaho.

Miller, C.A. and Vaske, J.J. (2003) 'Individual and situational influences on declining hunter effort in Illinois', *Human Dimensions of Wildlife*, 8: 263–76.

Ministry of Tourism (2006) *International Visitor Survey data set*. <http://www.tourismresearch.govt.nz/ > (accessed 23 November 2006).

Moreno, S. (2005) 'Oh deer, it's … mouse hunting', *Otago Daily Times – World Focus Supplement*, 16 May 2005, p. 10.

Nelson, M. (2006) *Women Fishing Together With Style*. <http://travel.msn.com/Guide/> (accessed 16 October 2006).

Newsome, D., Dowling, R.K. and Moore, S. (2005) *Wildlife Tourism*, Clevedon, Buffalo: Channel View Publications.

Novelli, M. and Humavindu, M.N. (2005) 'Wildlife tourism: wildlife use vs. local gain: trophy hunting in Namibia', in M. Novelli (ed.) *Niche Tourism: Contemporary Issues, Trends, Cases*, Oxford: Elsevier, pp. 171–82.

Otago Daily Times (2006) 'Supermodels not so alluring', *Otago Daily Tiimes*, 7 July 2006, p. 3.

Petersen, D. (2000). *Heartsblood: Hunting, Spirituality and Wildness in America*, Washington, DC: Island Press.

Planet Ark (2001) *Norway tourist seal hunt could be a hit – Minister*. <http://www.planetark.org> (accessed 18 December 2006).

Radder, L. (2005) 'Motives of international trophy hunters', *Annals of Tourism Research*, 32(4): 1141–4.

Randazzo, R. (2006) 'The care and feeding of the Safari Club', *The Business Report of Northern Nevada*, 8 January 2006.

REDES (2003) *Strategic feasibility study of the hunting tourism segment in Mexico*, N.p.: Secretaria de Tursimo.

Reynolds, P.C. and Braithwaite, D. (2001) 'Towards a conceptual framework for wildlife tourism', *Tourism Management*, 22: 31–42.

Ritchie, J.R.B. and Crouch, G.I. (2003) *The Competitive Destination: A Sustainable Tourism Perspective*, Cambridge, MA: CABI.

Roe, D., Leader-Williams, N. and Dalal-Clayton, D.B. (1997) *Take Only Photographs, Leave Only Footprints: The Environmental Impacts of Wildlife Tourism*, London: Environmental Planning Group, International Institute for Environment and Development.

Rutberg, A.T. (2001) 'Why state agencies should not advocate hunting or trapping', *Human Dimensions of Wildlife*, 6: 33–7.

Safari Club International (SCI) (2004) *Anti-hunting Groups Merge*. <http://www.safariclub.org> (accessed 16 April 2006).

Safari Club International (SCI) (2005) *The Sportsmanship in Hunting Act 2005*. <http://www.safariclub.org> (accessed 14 December 2006).

Safari Club International (SCI) (2006a) *Our Hunters' Code of Ethics*. <http://www.safariclub.org> (accessed 16 October 2006).

Safari Club International (SCI) (2006b) *Conservation Efforts: Asian Programme*. <http://www.safariclub.org> (accessed 16 October 2006).

Safari Club International (SCI) (2006c). *The Publications of Safari Club International: Safari Magazine & Safari Times*. <http://www.safariclub.org> (accessed 16 October 2006).

Schoenfeld, C.A. and Hendee, J.C. (1978) *Wildlife Management in Wilderness*, Pacific Grove, CA: Boxwood Press.

Speckman, S. (2006) 'Hunting convention to draw at least 30,000 sportsmen to Utah', *Desert News*, 26 May 2006.

Statistics Bureau and Statistical Research and Training Institute (SB & SRTI) (2006) *Statistical Handbook of Japan*. <http:// www.stat.go.jp> (accessed 20 December 2006).

Stebbins R.A. (1996) 'Cultural tourism as serious leisure', *Annals of Tourism Research*, 23(4): 948–50.

Szpetkowski, K.J. (2004) *Etyka Lowiecka (Hunting Ethics)*, Warsaw: Lowiec Polski.

Tremblay, P. (2001) 'Wildlife tourism consumption: consumptive or non-consumptive?', *International Journal of Tourism Research*, 3: 81–6.

US Department of the Interior (USDOI), Fish and Wildlife Service and US Department of Commerce, US Census Bureau (2002) *2001 National Survey of Fishing, Hunting and Wildlife-Associated Recreation.*

US Sportsmen.org (2006) *The Top Anti-hunting Organisations in US*. <http://www.ussportsmen.org> (accessed 18 December 2006).

Van der Walt, J. (2002) 'Proliferation of game ranches', *Game & Hunt*, 8(10): 7.

Von Maltitz, G. and Mbizvo, C. (2005) *Livelihood Impacts of Climate Change on Biodiversity Based Livelihoods in Africa*, CSIR Natural Resources and the Environment. <http://www.environment.gov.za> (accessed 11 November 2006).

Wilkie, D.S. and Carpenter, J.F. (1999) 'The potential role of safari hunting as a source of revenue for protected areas in the Congo Basin', *Oryx*, 33(4): 339–45.

World Tourism Organisation (WTO) (2006) *Concepts and Definitions*. <http://www.world-tourism.org> (accessed 16 October 2006).

2 The ‘Animal Question’ and the ‘consumption’ of wildlife

Adrian Franklin

Introduction

The so-called ‘Animal Question’ remains one of the hottest and emotional aspects of contemporary biopolitical debate. Within this ‘the intended killing of wildlife for sporting purposes’ is especially contentious. While either side attempts to make their position unassailably clear and compelling it is almost always complex, muddied and confusing when specific issues are considered. This is made all the worse by those whose view it is that there are universal principles or ethics that can resolve this ‘contested nature’ (Macnaghten and Urry 1998). We are in a better position to make decisions when we know where views, values and new demands come from and, in the same way, we can assess better those whose practices we condemn when we know their motives, their impacts and the consequences of their actions.

I will begin with a brief look at the sociological content of the Animal Question. It becomes apparent that our views on the proper treatment of animals in the past two hundred years has been highly fluid, inconsistent and highly charged. Ironically, as Tester (1992) demonstrated, the changing views on the proper treatment of animals reflect the changing nature of *humanism*. I will argue later that the humanist spin in most variations of the Animal Question hold further implications that we should perhaps consider more. Precisely because the entire issue is about how *we behave* and *how we should behave*, the subjectivity and agency of the animal (always in question) is entirely lost or subsumed. The animal in these debates is frequently treated like a minor, someone whose interests remain the proper business of fully responsible (human) adults. Such a view is unfortunate because the hunting and fishing relation contains two parties, hunter and hunted. Both have agency: the actions of one may only be comprehensible through the actions of the other. Together (and certainly never apart) they constitute the hunting relation and this has a bearing on all aspects of the Animal Question.

The humanism of hunting and fishing is important to understand in its own right, precisely because it has been so influential and has responded so transparently to changing social and cultural circumstances in the twentieth century and beyond. I will argue that one unintended consequence of the humanist ontology in its romantic and misanthropic phase, is the proper separation or distancing of humanity from nature which, ironically, accompanies its championing of the non-human, or 'nature' generally. Anti-modernism now drives 'a politics of the present' in which both the historic right (that called for the preservation of *past* tradition) and the historic left (an orientation to the *future* that evoked the enlightenment and the culture of progressive modernism) are largely absent. Instead, politics is now driven by fear rather than values, and by avoiding human intervention in the world rather than making politics the way in which such intervention is organised. The implications of this as far as it concerns hunting and fishing will be addressed in the final section on environmental issues.

I will then consider the practice of hunting and angling to see if they confirm the worst fears of the humanists. Crucially, what discourses of behaviour and propriety guide the practice of hunting and angling? I will consider the neo-Darwinian discourse that lies at the heart of the anti-hunting lobby and the less known Waltonian discourse that more accurately describes the actual motives of the hunters and anglers themselves. I will argue that in recent years the 'embodied turn' which describes the more intensive attention given over to the experience and control of the body and experiencing the world through the senses has disturbed the automatic repugnance extended to hunting and fishing and offers an entirely new relation with the natural world.

Finally, I will consider whether the touristic hunter and angler is necessarily an anti-environmental agent or whether they (or properly speaking a consumptive wildlife/nature orientation) constitute the possibility of an alternative and possibly more effective environmentalism?

The Animal Question

Tester (1992) reminds us that the politics of animals derived from two (related) sources in the nineteenth century. The first, coalescing around the issue of animal anti-cruelty legislation, set up what Tester calls the 'Demand for Difference'. This essentially humanist idea holds that we must demonstrate our difference from animals by our ability to act ethically, to control our violent urges and to extend the range and compass of civility. This early nineteenth-century social movement was very successful, precisely, Tester argues, because it was actively political and discursive. However, their successful campaigns channelled action into new organisations to enforce and enact new legislations designed to prevent animal cruelty and part of the political compromise was not extending this legislation against popular field sports. Thus by the twentieth century the 'Demand for Difference' had become routinised around its (limited) victories against cruelty to horses, farm and domestic animals, rather than protecting wildlife.

The second, focused on vegetarianism, explicitly asked questions about the proper physical as well as moral relation between humanity and animals. It established what Tester calls 'the Demand for Similitude', or the recognition of similarity against humanist assertions of difference and, especially ascendance/superiority as it was perceived in its modern manifestation. This was influential but mainly among intellectuals, many of whom abandoned the cut and thrust of modern political life for a life of quiet contemplation and reclusiveness – pursuing a politics of personal choice over popular persuasion. They found hunting abhorrent but had few means to extend this to a popular political front. Thus it merely simmered in their publications and writings. It was anti-modernist preaching, mainly to the converted.

Part of the problem for the nineteenth-century case against hunting, widely made by those associated with both 'Demands', was that while the blame for many of the horrors committed against animals could be pinned on the advance of modernity and the churning of culture, society, landscape and nature, hunting and fishing were associated, confusingly, with *tradition* and the old ways. In other words these nature practices had a reassuring air of continuity in an age concerned by change. While at this time many anti-enlightenment intellectuals felt strongly opposed to the march of modern progress and change, asserting the casualties of traditional social norms and ways of life, the fact was it was clear to *most* people that the modernity project was essentially well meaning and practical. Indeed, its emphasis on human progress, democratisation, social justice and scientific advances against disease and famine could only be welcomed. Something new had to change before animals and hunting could be thought of in a different, negative way for a popular majority. For that to happen, the fear of modern change had to be reinforced by a new fear: that modernity had lost its essential goodness.

In the first decades of the nineteenth century, English authors and artists of the day (and others who shape opinion and sensibility), could hardly produce enough in praise of foxhunting. It was *the* national sport, before football and other sporting codes emerged in their modern form. But by the first decades of the twentieth century the literary world was seriously divided over hunting. There was still considerable interest in hunting as attested to by the various hunting and animal 'sports' novels by Ernest Hemingway and others, and by a strong showing in the decorative, fine and cinematographic arts.

What effect did these works have? It is almost certain that they took the edge off the pleasurability of hunting (and fishing). First, they shifted from being the prize among the bountiful resources of nature, gifted by God himself, to a limited, cherishable (and endangered) heritage – in need of help. Animal conservation movements had been prompted not so much by hunting itself but by disturbing modern incursions into wilderness and other natural areas. Second, while not emphasising their sentient being, they did change the emphasis from the Darwinian image of animals as bloody in tooth and claw to one of animals as beautiful, healthy and essentially balanced, or in very simple terms *good* (Cartmill 1993). This essential goodness could then be used to suggest, albeit vaguely, the misanthropic trope of human badness and the path back to righteousness. But again, while the First World War shocked many with its callous indifference to the loss of so much *life*, that an entire generation of manhood could be so easily *squandered*, it was notoriously difficult to find an easy target to blame and even its cause remained a question of some debate.

Far from deepening the potential for misanthropy emanating from the First World War, the aftermath of the Second World War is best known for its celebration of the potency of modernity: an even better world could be rebuilt. This seductive message was followed up by delivery in full through a variety of social democratic political formations. The post-War years exceeded expectations, giving rise not only to a baby boom, but a long economic boom, unprecedented advances in life expectancy and well-being, the arrival of mass further education, and the emergence of a new political interest: consumerism. There were environmental concerns and there was a considerable interest in the impact of economic development in the lives of animals. However, as I argued in *Animals and Modern Cultures* (1999) these were seen largely as unfortunate side effects of a greater good, or at least a competing good: the progress of humanity that had known considerable suffering. It was the solid moral impetus underpinning the post-war economic and social development that pushed other considerations, including the moral standing of animals caught up in it, into relative obscurity.

The noble intentions and record achievements could not be sustained and from the 1970s onwards the social democratic welfare state apparatus of the modernity project began to unravel. The state as the moral guardian and legitimate manager of modern ordering lost ground to the demands for privatised provisioning and organisation. The state could no longer afford, or be trusted to deliver efficiently, new standards set by consumerist society.

The old spectre of misanthropy reappeared alongside claims for a risk society and as it did so the balance between humanism and environmentalism shifted profoundly, and is most noticeable in the new sensibilities extended to animals:

> The depiction of human activity as itself a threat to the world tends to endow this species with an overwhelmingly negative status. Instead of positive transformation and progress, civilisation is portrayed as a history of environmental vandalism … [T]he denigration of humanity that is associated with the downsizing of subjectivity enjoys a powerful resonance in contemporary culture. This development is evident in the elevation of the natural world and of animals on to a par with – if not into a relation of superiority to – human beings.
>
> (Furedi 2005a: 95–6)

Prior to the phase of 'liquid modernity' (Bauman 2000), the more solid modernity created blueprints for the way society should properly be. The social sciences joined the natural sciences in determining how such a socio-technical dream could come true. What all the variations of this dream had in common was a regulated humanity where individual freedoms were restricted in favour of collective goals. When this was liquefied in the 1980s and 1990s through neo-liberal reforms, the new highly informed consumer/risk-taker could see only one view of humanity. It was an unregulated, selfish society of individuals acting 'living without a guide book' (Smith 1999). Critically, the pace and direction of economic and social development now had no (fixed) moral rudder and no moral defence: it *was* deliberately disordered. And as Mary Douglas (1966) warned in *Purity and Danger*, where human societies sense a state of disorder, they typically sense also a state of danger and seek a means of purification and reordering.

I have argued before that new sensibilities towards animals that emerged at this time can be usefully related to changing perceptions of humanity. Of course there had been concerns for animals throughout the twentieth century but the issue is why were they muted before the 1970s and so forceful and loud afterwards? The answer is that they were mute all the time modernity was seen as virtuous and individual ambition curbed in favour of the greater good. If animals suffered as a result of this at least they were not suffering for any reason, even if it was regrettable. However, it becomes a different thing again to consider the oppression of animals in the light of the generalised selfish individualism of humanity. When individualism became deregulated and pursued unbounded ambitions for consumption and advancement, the former compromise with nature fell apart. The more this happened the more a misanthropic gloom descended and the more animals (and nature generally) seemed, by comparison with humanity, good, regulated and sane. Animals now demand not only action to be saved from a disordered humanity but they also offer the metaphor of ecological order as a means for its recovery. In addition to a generalised sense of misanthropy, we derive the ethic of sustainability from ecology. Because all this takes place in a heightened sense of risk and danger, we derive the precautionary principle. Taken altogether misanthropy, sustainability and the precautionary principle provide the antidote to liquid modernity and underwrite a vociferous pro-animal politics. Nature generally and wildlife in particular are now a major political issues. Their elevation is all the more important because it coincided with what is perceived to be a collapse of politics.

In the next section I will move from the animal question in general to modern relations with wildlife specifically. Our relations with wildlife are largely structured by emerging competing patterns of *leisure*. Although non-consumptive forms of wildlife leisure would seem most likely to succeed under these general conditions, I would argue that it is by no means clear cut. While non-violent and protective of animal environments, non-consumptive forms derive from relations that create distance rather than proximity, separation rather than interaction and spectacle rather than sensual, embodied relations. It is moot as to whether non-consumptive forms will always produce a sustainable politics of care that is more robust than one based on consumption. While tourism is still a dominant mode of interacting with our world, within it one can discern a frustration with purely visual forms of engagement and growing interest in closer, more embodied interaction.

Wildlife practices

If the Animal Question was about what it is to be properly human, in other words a discourse on how humans should properly comport themselves *with each other*, it was not exclusively so. The 'Demand for Difference' and the 'Demand for Similitude' were influenced by Romantic thinking which was orientated to material as well as metaphorical relations *between* humanity and nature. This manifested itself most in *embodied* discourses on the appropriate way to experience nature physically and sensuously, largely through forms of leisure. However, there was substantial disagreement on what these nature practices should be and particularly on the most appropriate sensual media with which to engage the natural world. The irony of Romanticism is that it coincided with an increasingly urbanised modern world. Urban society was not of course divorced from the natural world (though this has been exaggerated – see Franklin 2002) but its associations were not predicated on *everyday life*. Those Romantics who recommended nature to modern urban society did so mainly as *tourists*, regular visitors or figures of leisure in the natural landscape. Either way, the overwhelming medium for the consumption of nature was *visual* (Macnaghten and Urry 1998).

As a practice gaining ever more appeal to urbanites, the visual consumption of nature introduced a distance or *perspective* between humanity and nature that was historically novel. Although *understanding* might be accomplished by painstaking observation or mental/poetic conjuring, such understandings were not relationships with the nature objects in question. Of all the senses the visual is therefore the most disembodied, particularly when teamed closely with the romantic sensibility which emphasises imagination.

As with its parent practice, science, the objects under scrutiny by tourists were meant to be minimally influenced as a result of their observation. What was meant to be preserved by nature tourism was entirely aesthetic: it was the visual beauty of the view or sighting. And for whom was it supposed to be preserved if not those with honed romantic sensibilities, those able to appreciate it *properly*. Much wildlife conservation can thus be linked to the fact that it constituted a source of visual leisure for an exclusive touristic class and their wish to preserve it against development (Franklin 2006a). Early wildlife campaigns were specific to known walked tracts of country rather than campaigns to preserve wildlife *generally*. It is obviously also the case that native wildlife had been caught up in nation formation processes in the nineteenth and early twentieth centuries and had thus achieved totemic status in most nation-states around the world. In new colonial nations wildlife served both as a way of consolidating social solidarity in new nations and as a signifier of separation from the colonising powers (see Franklin 2006a). Conserving native animals was at the same time an act of national solidarity and nation building. It is here that the hunting of native species could be seen in an unfavourable light whereas the hunting of introduced species that competed with native species could be lauded, encouraged and approved (see Franklin 1996 for the case of Australia where this process occurred). Such scenarios demonstrate the highly contingent nature of animal politics and the fragility of essential *ethical* positions.

Cartmill (1993) characterised this strictly hands-off wildlife culture somewhat vaguely as tender-hearted Romanticism and compared it with another romantic variant inspired by the neo-Darwinian view of what it was to be properly human. According to this view, humanity had been de-natured by modern urban living which was too far removed from our proper, healthy life as hunter-gatherers or 'killer apes', an evolutionary construction that was favoured at the end of the nineteenth century when a modern hunting culture took root. Hunting and fishing not only drew on evolutionary themes, it too visited the potent source of legitimacy through nationalism: hunting and fishing and a good knowledge of woodland craft were essential for a country prepared to defend its territory. Hunting and fishing were also cast as heritage practices inherited from the much admired pioneering founders of a nation. Indeed, it was their creation of a nation out of a wilderness that rendered wilderness such a potent cultic object: a sacred cathedral of nation that demanded pilgrimage and ritual enactments.

It is possible to identify the influential writings of Isaak Walton (1962) as an alternative to the killer ape hypothesis. If properly addressed through the sensual pleasures of angling and the angling landscape, nature for Walton provided a necessary counter balance to the one-dimensionality of the capitalist social order. Nature was an appropriate altar for Christian worship, and angling was a fitting opportunity to reintegrate the humbled body into the purifying materiality and rhythms of creation. In *The Compleat Angler* (1962 – originally 1653), the restful, still and quiet body required for angling are made to contrast favourably with the noise and strife of the commercial city; the direct, beneficial body contact with the undisputed materiality of nature is made to contrast with the abstractions and deceits of business and progress.

Walton preached the therapeutic virtues of acquiring natural history and an appreciation of the aesthetics of wild flora and fauna; the benefits of quiet contemplation and meditation that the intensely engaging patience of angling provides; and the healthiness of an outdoor pursuit requiring an early rise, a healthy jaunt through the countryside and plenty of fresh air. In Walton it is the physicalities of angling that are emphasised. The benefits of contagion with soil, banks, flowers, dew. The rewards of handling natural materials in the construction of rods, baits, lines and tackle. The healthiness of fresh air, fresh fish, pure water and outdoor companionship. Angling and hunting books have extolled these embodied virtues ever since and, we must conclude, they have held their attraction with most discernible groups of city dwellers from businessmen and professionals through factory and office workers, commuters and the new service class.

In a recent paper (Franklin 2001) I reviewed hunting from the perspective of the writer-hunter. What emerged as a distinctive emphasis was not the thrill of killing but the thrill (and excitement) of being sensually tuned into specific, highly sensitive and difficult tension balances with quarry species – fish or game (see Elias 1993). It was concluded that of the two neo-Darwinian discourses considered, the Waltonian comes closest to describing what the modern hunting aesthetic and what hunters actually experience. As Cartmill (1995) shows, this has a long history:

> But the most literate hunters, the ones who are apt to write books and columns about the joys of hunting generally agree that the chiefest of these joys is the pleasure of a temporary union with the natural order. 'I must know,' writes one sporting columnist, 'that I am part of and have common bond with, the wilderness' (Simpson, 1984). Another calls the hunt 'a Promised Land' that

> keeps the huntsman from being 'isolated from the natural world' (Holt, 1990). Valerius Geist describes hunting as an 'intercourse with nature' (Geist, 1975, p. 153). 'The human being,' wrote the hunting philosopher Ortega y Gasset, 'tries to rest from the enormous discomfort and all embracing disquiet of history by "returning" transitorily, artificially to nature in the sport of hunting' Hunting said Ortega 'permits us the greatest luxury of all, the ability to enjoy a vacation from the human condition'(Ortega y Gasset 1972, p.139).
>
> (Cartmill, 1995: 784)

This is why we must be a little wary of accounts that reduce relations with animals and nature to one of representations of the human conditions, purely symbolic, humanist accounts. Sure, there is no escaping the complex way nature is called upon to represent social relations and conditions, and animals are key representational devices. However, simply because this is true does not mean that it excludes other dimensions to the human–animal relation, in every circumstance. In a way, through their written reflections, hunters and anglers seem to suggest that our relating with animals offers a way out of the human-centred world, an opportunity to live less by our intellect and more by our senses, and an opportunity to relate intimately to other species as opposed to our own. The possibility of doing this and its significance in producing a radically decentred sensibility that is less fixed around the notion of a proper humanity and more open to human *potential* or *becoming* has been explored by the philosophers Deleuze and Guattari, notably through their concept *becoming animal* (Deleuze and Guattari 1999) and is being taken very seriously in recent studies of the human–animal relation and with nature generally (see also Baker 2002; Haraway 2003; Thrift 2001). Equally, recent theoretical work by these post-humanist scholars has explored the significance of the non-human, particularly the agency of non-human animals. It seems to me that the very nature of hunting and angling is predicated not only on the co-agency of humans and animals but the delicate tension balance that their agency produces. The humanist accounts of hunting and angling that centre only on human meanings, representation and agency ignore totally the agency of the animals themselves, preferring to render them completely passive before an 'over-determining' humanity. This is not supported in accounts of what hunters actually do and what they actually feel, since the literature is as much about their failure (as hunters) and about the sensory superiority of their quarry. Accounts of hunting failure demonstrate that the object of their 'sport' is not so much killing and domination, as *participation* in a working and healthy ecology.

The embodied turn in leisure and tourism

In the same way that tourism in the nineteenth and twentieth centuries was dominated by the visual gaze, perspective, distance and the fleeting visit, in recent years we have seen this being broken down both in terms of practice and in our theoretical understanding of tourism. I would argue that this will have, and perhaps already has had, an influence on attitudes and practices of wildlife leisures: while watching and other visual practices of wildlife leisure will remain important, it is possible that consumptive forms (possibly new ones) may become more attractive because they offer a more embodied and intimate *relationship with* the natural world. Bryan Turner (1991) has described contemporary societies as *somatic*. The relative shift in emphasis away from thought, representation and mentality to body and sensuality has many origins, not least the shift from health to fitness, the therapeutic revolution and the growth of new technologies/techniques of the body (see Thrift 2001), kinaesthetic leisures (climbing, surfing, orienteering, mountain biking), and the shift from state to individual regimes of risk management (especially as it affects health, fitness and well-being).

This has been one of the key changes in tourism over recent years. The so-called 'embodied turn' has seen tourists getting 'up close' to objects that were hitherto held at arm's length. There is also more emphasis on senses other than vision and activities that involve kinaesthetic techniques and practices. As opposed to watching the emphasis is now on doing. This can be seen in the various manifestations of 'Shamu' at *Sea World* (see Franklin 1999), the engagement with wild dolphins (Bulbek 2005) and the growth of fly fishing in North America. It was also obvious to me (although I was taken aback at the time), during recent research in Kakadu National Park and the Great Barrier Reef in Australia, that our attitudes to animals can vary with our practices.

Kakadu and Hamilton Island

Recent fieldwork conducted by the author in the wildlife tourist industry in the Kakadu area and Hamilton Island on the Great Barrier Reef revealed an interesting paradox. Most of the activities on offer were either wildlife focused or contained a high proportion of wildlife content. Mostly, the values that informed the manner in which wildlife was discussed by guides and other sources of information could be described as environmentalist: Kakadu and its animals were a fragile and endangered entity and required conservation and constant care. Tourists took this in with the seriousness and reverence reserved for sacred species, and I anticipated this, but it was quite surprising to find that once they were back at their hotels and sitting down for the evening meals at local eateries and restaurants, the star-billed foods also happened to be largely local wildlife: kangaroo, crocodile, barramundi and buffalo, for example, featured widely. Maybe, like the Aborigines themselves a totemic relationship with animals does not necessarily preclude tourists from eating them. This was certainly true of the Aboriginal wildlife tour I participated in at Kakadu. This was a 4WD trip into Aboriginal -owned land and staffed by the owner a young white guy, and his partner in the business, a local Aboriginal woman. It was a fabulous day spent finding bush foods and seeing a staggering amount of wildlife. We were given the chance to find and eat witchetty grubs, green leaf ants, local water lily tubers and other bush vegetables. The high point of the day was cooking food on a fire at sundown, on the shores of one of the larger billabongs. The centrepiece was a shot magpie goose brought along by the Aboriginal woman, and the party were invited to help pluck, butcher and cook the bird, alongside the cooked vegetables we gathered, damper and billy tea. These were a group of people who had spent a lot of their other time watching magpie geese and other local birds at one of more than 20 vantage points in the park and so it seemed strange that none of them passed any remarks on the irony of the occasion. Precisely the same pattern was noted on Hamilton Island where the key activity was diving on the Great Barrier Reef. While during the day tourists were in awe of the tropical fish and frequently reminded of the precarious state of the reef, during the evening it was interesting to note how much of the local menus contained the self-same animals and fish. Coral trout, for example, were a much-admired living part of the reef and much in demand at table. The same dissonance was noted on a pleasure fishing trip out of Hamilton Island. The party I joined were hoping to catch some of the larger species such as Spanish mackerel, tuna and kingfish but after spending most of the day trolling for these species the skipper finally anchored to allow some bait fishing. We were told that the fish we were really hoping to get was coral trout but in the event we didn't. What we did catch was a lot of very small brightly coloured tropical fish that only the previous day, whilst on a dive trip, I had been encouraged to think of as a natural biodiversity under some threat. Some of these became *bait* themselves and had large hooks unceremoniously pulled through them and cast out as live bait for a bigger fish. This underlines the *contingent* nature of our views on animals. We pass between a variety of viewpoints and discourses on them and they become different objects as we do so. A coral trout can embody environmental anxiety one minute, a craved-for dinner item the next and the object of a disappointing fishing trip the following day.

The most significant aspect of this fieldwork concerned the degree of excitement and engagement across the range of wildlife activities. In relation to fish, insects and birds, tourists were far more engaged, active and excited, and seemed to have more pleasure in the more consumptive forms of activity. The non-consumptive forms (watching, photographing and diving to see) were also clearly pleasurable but there was a good deal less energy, less engagement and less intensity in their interactions with non-humans. Often they were more passive relative to an expert or guide and it seemed to me that they were more easily distracted and bored. I think that despite being one of the most outstanding and unusual wildlife destinations in the world, merely *being an observer of* a landscape for days on end is a lot less engaging than being active *doing something in* a landscape (see Franklin 2006a). Clearly, such observations could in future be augmented by more measurable metrics of engagement.

Hunting, fishing and environmental relations: a conclusion

The figure of the tourist has never been heroic, never entirely loved. Bauman (1998; see also Bauman in Franklin 2003b) has also argued that the tourist has become a metaphor for problematic contemporary social relations: like the tourist's relation to places and people (including natures?) visited, contemporary social bonds are loose, non-committal and 'until further notice'. Extending this idea we can argue that one problem with our touristic relationships with animals through non-consumptive means is the absence of ongoing relationships or commitments with them or with their environment. The amount of knowledge about their condition and well-being that they take in on tour is likely to be limited and framed in such a way that the tourists' *pleasure* (which is paramount) is unimpeded and often the allusion of authenticity is maintained.

Although managers of wildlife tourism destinations have the main responsibility for care, these days consumer power is very important, politically. The typical consumer of national parks and wildlife destinations is a tourist whose visit is typically fleeting, shallow and a one-off affair. They may be attracted to such areas in greater numbers than in the past but their relationship with each one is extremely loose and ephemeral and 'until further notice'. There is nothing but their own pleasure and interest binding them to the place and once that begins to wane, as it typically does after a relatively short period, the tourists take their leave. Over time any one tourist may visit many such destinations and hear a common discourse on wildlife fragility and vulnerability. Yet there is nothing about these experiences that galvanises a longer-term relation of care and, indeed, the proliferation of destinations clammering for attention may only serve to create a blasé indifference to any one location, even if, overall, environmental concerns are raised.

National parks and wildlife reserves are vulnerable politically and non-consumptive tourists are not likely, given this relationship with wildlife, to remain solid and effective supporters. In comparison, consumptive practices require more depth of knowledge about the precise state of wildlife populations, and not just quarry species specifically. Further, they are more likely to form longer-term relationships with given areas and thus become more embedded in their political as well as natural affairs. This is surely why the single largest political rally in the UK, in August 1997, was mounted by hunters and anglers in response to proposals to ban foxhunting and perhaps other field sports. Although organised to address a national issue, this rally was constituted by largely local organisations, each with highly specific and localised hunting environments.

In comparison with other tourists, the consumptive practices of hunters and anglers tend to form around known places, even if they are not locals themselves. While there are certainly those hunter and anglers who do travel widely, it also true that both traditions place great emphasis on knowing their country, nature and landscape. In their writings, hunters and anglers tend to emphasise their knowledgeability and love of place in terms of its *particularities*. These tend to be embodied experiences and are expressed in visual terms, colours, landscapes, light and shade, but also in terms of smells and tactile experiences. It is also about knowing where things are (local bush foods, birds' nests, water, wallows, snake infested areas, etc.) and how they change over the season. Hunters cultivate an association with particular areas because in hunting the knowledge of these particularities becomes greater than a sum of the parts and are a tangible factor in the successful hunt.

Hunting and angling are wilderness activities in some areas but mostly they are practised in areas between the wilderness and the city. These are not merely farming landscapes; they are the strips of river and the patches of scrub or heath; the disused canals, the undisturbed coastal flats, the gravel pits and reservoirs created as towns expanded. Such areas are typified by specific populations of flora and fauna and game and fish populations can flourish. Rabbit populations in the UK have grown dramatically in response to a declining rural working class; dramatic increases in pigeon populations have resulted from extensive agriculture. Many wetlands close to industry teem with wildfowl and fish.

In introducing Leopold's *A Sand County Almanac*, Finch underlines this point:

> By and large these shack essays recount experiences that are Adamic and archetypal: of Leopold, the weekend explorer, discovering, recognizing, and celebrating the richness and drama of a new world. The fact that these discoveries take place, not in some unspoiled Adenic wilderness but on a run-down sand farm that has been abused and misused through human greed and ignorance for over a century, makes his achievement even more impressive. Leopold conveys a sense of natural abundance where most would only see a diminished environment.
>
> (Finch 1987: xviii)

Leopold himself speaks for many hunters and anglers in underlining the treasures that are locally theirs by virtue of their familiarity and knowledge:

> Few hunters know that grouse exist in Adams County for when they drive through it, they see only a waste of jack-pines and scrub oaks … Here, come October I sit in the solitude of my tamaracks and hear the hunters' car roaring up the highway, hell-bent from the crowded counties to the north … At the noise of their passing, a cock grouse drums his defiance. My dog grins as we note his direction.
>
> (Leopold 1949: 56)

Similarly, Gierach a self-styled trout 'bum' fishes more or less permanently in the very best rivers and creeks in Colorado and Montana. However, his first love is the ordinary little stream, the St Vrain, which he has got to know in great depth.

> Most of the fishermen I know … have a creek like this somewhere in their lives. It's not big, it's not great, it's not famous, certainly it's not fashionable, and therein lies its charm. It's an ordinary, run-of-the-mill trout stream where fly-fishing can be a casual affair rather than having to be a balls-to-the-wall adventure all the time. It's the place where, for once, you are not a tourist.
>
> (Gierach 1986: 59)

This embeddedness in nature with its long tradition of conservation-mindedness seems a long way from the assumptions made by those who feel hunting and angling sports oppose the ethic of environmentalism. However, a perfectly sensible argument could be made for consumptive wildlife leisures as consistent with the ethic of environmentalism, even if it would have to position itself as an alternative in many places. Or *alternatives*, because there are many examples of this. In Norway, for example, as the writings of Arne Naess (see Naess and Rothenberg 1990) demonstrate, the much encouraged relationship with nature is touristic: it emphasises the building, occupying and regular revisiting of particular tracts of country through their passion for wilderness huts away from the everyday (see also Hylland Eriksen 2001). It also idealises berry and mushroom picking and fishing and hunting. Indeed, variants of this elsewhere are far more common vacation styles than is often appreciated (Franklin 2002). In Norway and Scandinavia generally, the landscape is not divided into private inaccessible lands on the one hand and strictly managed national reserves on the other as it is in New Zealand and Australia (see Shoard 1987). Instead of creating a sense of 'true nature' in the reserved wilderness areas, and thus attracting a problematically high flow of touristic visitation, the Scandinavians endorse an ancient 'freedom to roam law' (*Allemansret* or 'everyman's rights') that allows access across most categories of ownership and encourages a more even and consumptive presence of people in the landscape (see Franklin 2006b).

Australian Aboriginal ways of living on the land, often referred to as *country*, also places a great value on a *nourishing* landscape (Head 2000) and could, through growing tourism enterprises, promote this further particularly in a country where introduced animals are perceived as a threat to the environment and where population explosions of some native animals are common. In many places consumptive wildlife tourism, especially shooting, could be tied more to wildlife management as it is, for example, in the Netherlands (see Franklin 1999: 115).

In *Contested Natures*, Macnaghten and Urry (1998) found that the most effective biopolitical understanding and commitment emerge from direct experience and locality. They warn of the assumption that 'environmental concerns exist *a priori*, waiting to be revealed through sample surveys. By contrast, our research points out how people make sense of environmental issues within particular localised and embedded identities ... Moreover, people were able to understand environmental issues in terms of how they impinged on their identities, for example, as mothers, as rural dwellers, as global or as British citizens' (Macnaghten and Urry 1998: 245).

If this is true it might be better to encourage identification with nature through more specific and territorialised practices such as consumptive wildlife leisures than non-consumptive wildlife tourism where the relationship is so loose that identification and embedding rarely occur.

References

Baker, S. (2002) *The Postmodern Animal*, London: Reaktion Books.
Bauman, Z. (1998) *Globalisation*, Cambridge: Polity.

Bauman, Z. (2000) *Liquid Modernity*, Cambridge: Polity.
Bulbek, C. (2005) *Facing the Wild*, London: Earthscan.
Cartmill, M. (1993) *View to a Death in the Morning*, Cambridge, MA: Harvard University Press.
Cartmill, M. (1995) 'Hunting and humanity in Western thought', *Social Research*, 62(3): 773–86.
Deleuze, G. and Guattari, F. (1999) *A Thousand Plateaus*, London: Athlone.
Douglas, M. (1966) *Purity and Danger*, London: Routledge and Kegan Paul.
Elias, N. (1993) 'An essay on sport and violence', in N. Elias and E. Dunning (eds) *Quest for Excitement*, Oxford: Blackwell.
Finch, R. (1987) 'Introduction' to A. Leopold (1949) *A Sand County Almanac*, New York: Oxford University Press.
Franklin, A.S. (1996) 'Australian hunting and angling sports and the changing nature of human–animal relations in Australia', *Australian and New Zealand Journal of Sociology*, 32(3): 39–56.
Franklin, A.S. (1999) *Animals and Modern Cultures*, London: Sage.
Franklin, A.S. (2001) 'Neo-Darwinian leisures: the body and nature in modernity', *Body and Society*, 7(2): 57–76.
Franklin, A.S. (2002) *Nature and Social Theory*, London: Sage.
Franklin, A.S. (2003a) *Tourism*, London: Sage.
Franklin, A.S. (2003b) 'The tourism syndrome: an interview with Zygmunt Bauman', *Tourist Studies*, 3(2): 205–18.
Franklin, A.S. (2006a) *Animal Nation: The True Story of Animals and Australia*, Sydney: UNSW Press.
Franklin, A.S. (2006b) 'The humanity of the wilderness photo', *Australian Humanities Review*, 38, March 2006: 1–16.
Furedi, F. (2005a) *The Politics of Fear*, London: Continuum.
Furedi, F. (2005b) *The Culture of Fear*, London: Continuum.
Gierach, J. (1986) *Trout Bum – Fly-Fishing as a Way of Life*, New York: Simon and Schuster.
Haraway, D. (2003) *The Companion Species Manifesto*, Chicago: Prickly Paradigm Press.
Head, L. (2000) *Second Nature*, Syracuse, NY: Syracuse University Press.
Hylland Eriksen, T. (2001) *Tyranny of the Moment*, London: Pluto.
Leopold, A. (1949) *A Sand County Almanac*, New York: Oxford University Press.
Macnaghten, P. and Urry, J. (1998) *Contested Natures*, London: Sage.
Naess, A. and Rothenberg, D. (1990) *Ecology, Community and Lifestyle: Outline of an Ecosophy*, Cambridge: Cambridge University Press.
Shoard, M. (1987) *This Land is Our Land*. London: Paladin.
Simpson, J.Y. (1984) *Man and the Attainment of Imortality*, New York: Doran.
Smith, D. (1999) *Zygmunt Bauman*, Oxford: Blackwell.
Tester, K. (1992) *Animals and Society: The Humanity of Animal Rights*, London: Routledge.
Thrift, N. (2001) 'Still life in nearly present time: the object of nature', in P. Macnaghten and J. Urry (eds) *Bodies of Nature*, London: Sage.
Turner, B.S. (1991) 'Recent developments in the theory of the body', in M. Featherstone, M. Hepworth and B.S. Turner (eds) *The Body: Social Process and Cultural Theory*, London: Sage.
Walton, I. (1962) *The Compleat Angler*, London: J.M. Dent and Sons.

3 The lure of fly-fishing

Robert Preston-Whyte

Introduction

Fly-fishing incorporates at least three features that it shares with activities like hunting and shooting. First, it has a long history, probably extending over thousands of years (Herd 2005), during which fly-fishing transformed from a survival strategy into a consumptive recreational pursuit. Today, fresh and salt-water fly-fishing plays an important role in the tourism consumption of wildlife with fishers travelling to destinations where environmental conditions favour the occurrence of trout or other fresh or saltwater gamefish in South and North America, Australia, New Zealand, Africa, Asia and Europe (World Angler 2006; On Fly Fishing Directory 2006). Second, this process of change was shaped by emerging technologies and innovative strategies in the creation of rods, flies, lines and reels and how to use them, environmental knowledge about fish and their ecosystems and ethical concerns that emerged from notions of humane behaviour towards animals. Finally, fly-fishing has retained, and possibly even expanded, its following of enthusiastic supporters and their economic contributions help to sustain a network of fly-fishing services such as specialist shopping outlets, fly-fishing books and magazines, fly-fishing tours and accommodation providers and the owners of fishable waters.

Fly-fishing is a form of niche tourism within the umbrella of wildlife tourism consumption. The nature of this niche, and how it differs from other niches within wildlife tourism, lies in the nature of the dynamic network of embedded environmental, social, cultural and technological interactions that epitomize the sport. Even so, fly-fishing shares many common features with hunters, shooters and other fishers. These include the need to travel to sometimes distant favourite locations, costly equipment, the willing expenditure of time, the acquisition of specialist knowledge and skills, search for solitude, the escape from stressful urban environments and the excitement and anticipation of the hunt.

The transformation of fly-fishing from a subsistence activity to a form of niche tourism within wildlife consumption appears as another triumph of human ingenuity over nature. From the conception of the rod, the plaiting of appropriate line, the invention of the reel, the tying of suitable flies, developing casting skill and employing appropriate environmental knowledge, the history of fly-fishing appears as an increasingly successful manipulation of the material environment by fishers in terms that describe the domination of culture over nature (Herd 2005). However, historical observations tell only part of the story. While it is true that technology and knowledge have increased the chances of experienced fishers succeeding with a fly, the challenges, needs, desires and enticements that sustain the fly-fisher's passion are equally important. Challenges arise out of coping with uncontrollable events and adverse environmental conditions. Rod tips break, water temperature changes and fish stop feeding. Changes such as these fall outside the manipulating influence of the cultural domain and permit natural events and environmental factors to affect outcomes. It is overcoming obstacles such as these that many fishers recognize as contributing to the enduring appeal of the sport.

If nature plays a significant role in determining why fly-fishers spend time and money and often endure discomfort, to sustain their passion, then a philosophical approach that locates fishers and fish on either side of a culture–nature divide seems inappropriate. A more constructive approach may be one in which our widely held assumptions of the domination of culture over the material environment is suspended in favour of a neutral stance towards human and non-human actors. Actor-network theory provides this view, given that it 'maps relations that are simultaneously material (between things) and semiotic (between concepts)' (Wikipedia 2006: 1). The work of actor-network theorists such as Callon (1986, 1991), Latour (1986, 1987, 1988, 1993, 1994, 1998) and Law (1986, 1991, 1992) provide examples of how material-semiotic networks form coherent wholes in which actors/actants continually strive to bring other actors into alignment with their own interests. Tourism researchers are also beginning to engage with these ideas. Franklin (2004: 277) uses the notion of relational tourism and Foucault's ideas on governance to show how the heterogeneous assemblage of things 'to be seen, felt, interpellated and travelled' make up the tourist world. For Jóhannesson (2005), the utility of actor-network theory as a methodological approach in tourism studies is attractive because of the way it deals with relational materiality while providing a way of grasping multiple relational orderings. In relation to wildlife issues, Whatmore (1998, 2000) employs actor-network ideas in her work on the geographies of wildlife and spatial formation of wildlife exchange.

By assuming a material-semiotic network in which agency is attributed to both human and non-human actors, actor-network theory allows one to think about how fishers come to acquire power in a network of linkages that connect the wide array of material objects, devices and settings associated with fly-fishing. It becomes possible to bridge the culture–nature divide by, on one the hand, viewing the way fishers attempt to manipulate and order the entities that comprise the actor-network by employing tools, skills and appropriate knowledge and, on the other hand, appreciating how they may be frustrated in their endeavours by resistances imposed by material objects in the network. In this way, actor-network theory provides a framework that locates the heterogeneous assemblage of actors that constitute the fly-fishing network, traces the struggle for dominance between fisher and fish and provides insights into the forces that sustain the enthusiast's passion for the consumption of this form of wildlife tourism.

An actor-network perspective

The use of the network metaphor in different ways in the social sciences makes it important to distinguish between them. Its use to explain the organizational practices that link people, objects and events, have been the topic of extensive discussion particularly in relation to networking within and between firms (Knoke and Kuklinski 1982; Cooke and Morgan 1993; Amin and Thrift 1995; Castells 1996). In pursuit of this objective, it provides a way for assessing structures and strategic interactions between nodes in the network where action is attributed solely to human agency. It also describes prescribed, compulsory or directional circulations between nodes in two- or three-dimensional surfaces that employ Cartesian notions of layers, levels, structures and systems.

In contrast, actor-network theory views material-semiotic networks and their interactions between nodes in an entirely different manner. First, networks are constructed from a heterogeneous assemblage of objects and settings that include humans, sentient non-humans and inanimate objects. It is the insistence that all entities have the potential to become actors (i.e. something that acts) and should initially be treated as equal in the network, that sets actor-network theory apart from other forms of social enquiry and provides a different perspective to questions that interrogate the emergence of power in the network. Using this approach, fly-fishers begin as actors surrounded by a heterogeneous array of other potential actors. The *a priori* distinctions between natural and social entities, between fish and fisher, are abandoned and the interactions within the network then describe how certain actors emerge with the power to order and control. To understand what holds the network together, and how it becomes stable, effective and predictable requires what Latour (1998: 2) calls 'background/foreground reversal: instead of starting from universal laws … it starts from irreducible, incommensurable, unconnected localities, which then, at a great price, sometimes end into provisionally commensurable connections'. This argument rejects any form of reductionism that portrays the ultimate source of agency privileging either fishers or material entities on the assumption that human and non-human entities share the same framework of interactions in which either could prevail.

Second, actor-network theory is concerned with the mechanics of power. Networks are visualized as fluid and constantly changing as actors strive to establish their dominance in an often-contested environment. Because of this, no network remains unaffected for long without entities attempting to resist the ordering influence of actors aspiring to power. In such settings fly-fishers with power are:

> those able to enrol, convince and enlist others into networks in terms which allow the initial actors to 'represent' the others. Powerful actors speak for all the enrolled entities and actors, and control the means of representation.
>
> (Murdock 1995: 748)

Successful fly-fishers exercise this control by understanding the nature and uses of their tools, demonstrating virtuosity in casting and rod handling and acquiring and applying knowledge about the material environment that sustains fish. In this way, they impose order and stability on the network. This level of network stability, however temporary and provisional, signals the emergence of the fisher as an actor able to organize, control and manage other actors in the network that then become compliant intermediaries.

Third, the emergence of influential actors in the network is described ontologically by flows and connections between entities. By tracing and relating the behaviour of a heterogeneous array of actors in which 'nodes have as many dimensions as they have connections' (Latour, 1998: 2), the emphasis on human agency ceases to dominate and it becomes possible to view the contest between fish and fisher free of the constraints required by assumptions of culture and nature. By doing away with this and other duality assumptions, the working of the network becomes clearer and so do the reasons why some fly-fishers fail while others succeed.

Fly-fishers gain authority and influence by enrolling others into the actor-network. This power struggle takes place through a process called 'translation'. Through this process, focal actors define identities and interests of other actors that conform to their own goals and objectives. It is 'about attempting to gain rights of representation, to speak for others and to impose particular definitions and roles on them' (Burgess *et al.* 2000: 123). Actors, entities and places 'are persuaded to behave in accordance with network requirements' (Murdock 1998: 362), by reinterpreting and transforming notions of identity, interests and courses of action available to them. In the case of fly-fishing, much of this persuasion is sustained by innovative developments in equipment and the transfer of knowledge about how to control fishing networks.

Given that fly-fishing takes place in a spatial context, the employment of the network metaphor must not conflict with dualities informed by distance. In this case, Murdock (1998) comes to our assistance with two spatial typologies. The first describes what he calls 'spaces of prescription', by which he means a network where the heterogeneous elements of the entities comprising the network have been stabilized through successful translations to enable 'the enrolling actor (the "centre") to "speak" for all' (Murdock 1998: 362). This leads to networks that are stable, well integrated, relatively irreversible, and with predictable forms of behaviour. The second network type describes what Murdock (1998: 363) calls 'spaces of negotiation'. Such networks are poorly integrated, fluid, reversible and unstable as ongoing negotiation and re-negotiation between contending actors leads to changing network shape. By incorporating these notions of space, and more specifically the stabilizing or destabilizing effect of various forms of translation, a useful framework is provided to consider the development status of fly-fishing. Transfers of knowledge, innovation and learning are regarded as particularly important attributes in the toolbag of fly-fishers aspiring to create spaces of prescription. So too is the integration of specialist fly-fishing outlets, tours, accommodation and transport providers contained in the network.

The mechanics of power

The history of fly-fishing is one of fisher participation in a network that includes techniques, settings, devices, fish, innovation and environmental knowledge construction. However, between the desire to land a fish and the reality lies a challenge, as most fly-fishers will confirm. To meet the challenge, fishers must submit to trials and tests before they can expect anything other than serendipity in their contest with fish. Initially, they enter into a network in which they are on an equal footing with other actors. There they find that they are not the source of agency but part of a heterogeneous network made up of fisher + rod + reel + fly + trout + knowledge + skill + water temperature … in a network of collective action. To become powerful in this network demands the acquisition of appropriate knowledge and skills in which all the component entities are translated into a 'space of prescription' in which the fisher may emerge triumphant.

According to Herd (2005), Aelian, writing in about AD 200 in his *Natural History*, provided the first reference to fly-fishing. He tells how fishers in Macedonia used an artificial fly, made to resemble a local insect, to catch what Herd (2005) assumes are trout in a river called the Astraeus. However, Herd (2005) suspects that fly-fishing had been continuing unrecorded in many parts of the world for many years prior to AD 200 promoted by the effectiveness of the technique, dispersed by travelling caravans and nomadic shepherds and later by the globalizing efficiency of the Roman Empire.

Little is known about fly-fishing from these early times to the seventeenth century, although a glimpse into fly-fishing in medieval times is recorded in *The Treatise of Fishing with an Angle* (Herd 2005). The author of this early text discounts hunting, hawking and fowling as burdensome, exhausting and unpleasant activities, not to be compared with the pleasure of fly-fishing,

> For he can lose at most only a line or a hook, of which he can have plenty of his own making, as this simple treatise will teach him … And if the angler catches fish, surely then there is no happier man … Also whoever wishes to practice the sport of angling, he must rise early, which thing is profitable to a man in his way … As the old English proverb says: 'Whoever will rise early shall be holy, healthy, and happy'.
>
> (Herd 2005)

The ability to make and use suitable lines and hooks shows the fisher gaining influence by enrolling others into the network. Knowledge of fish feeding behaviour is implied and the importance of rising early to catch trout is included under the banner of spiritual duty, a healthy body and personal satisfaction.

Our understanding of how fly-fishing tools, strategies and knowledge progressed improves in the seventeenth century as anglers began to write about the sport. From then until the present, fly-fishing developed relatively rapidly as fishers experimented with different materials, adopted new strategies and applied innovative skills as they improved their control over fishing networks. Fishing line developed from twisted, tapered horsehair and silk lines to coated nylon. Reels evolved from spike foot, clamp-foot and plate-foot reels in the 1800s, through many innovations that brought in multiplying reels, interchangeable spools and disc drag mechanisms. Rods changed from being homemade in the seventeenth century, to constructions from greenheart and split cane in the nineteenth century, glass fibre in the 1950s, and lightweight carbon fibre in the 1970s. Techniques included the adoption of false casting, along with debates about the benefits of fishing upstream or downstream. Increasing ease of transport, such as the nineteenth-century development of railways in Britain and elsewhere, made it possible for people to travel to fishable waters to indulge in their sport.

Innovation and knowledge construction is an interactive process that links fly-fishing with a developing commodity-network. Some commentators regard knowledge as central to the economy and society and speak of a 'knowledge society' (Bell 1973), an 'information society' (Lyotard 1984) and a 'risk society' (Beck 1992), while Hughes (2000: 179) draws attention to Appadurai's (1986) contention that, 'there are three sorts of knowledge at work in the social life of a commodity. First, there is knowledge that goes into its production. Second, there is knowledge required to consume it. And, mediating between these knowledges, is that which fuels commodity circulation and exchange.' This suggests that knowledge circulates and, using the network metaphor, promotes the notion of flows of ideas and information between producers and consumers that, according to Appadurai (1986), can at once be technical, social and aesthetic.

The success of fly-fishing as a niche activity within wildlife tourism consumption, owes much to the innovative production of new rod, reel and line materials and its enthusiastic consumption by fly-fishers. Specialist fishing shops, tour guides and facilities that provide transport and accommodation are necessary parts of an expanded network that draws on innovative strategies to capture the fly-fisher's attention. Texts, in the form of books and magazines, play a vital role in the transfer of innovative ideas as Law (1992: 385) explains:

> Thoughts are cheap but they don't last long, and speech lasts very little longer. But when we start to perform relations – and in particular when we embody them in inanimate materials such as texts and buildings – they may last longer.

By acquiring knowledge and skills through experience and the use of texts, by using transport systems to reach desirable fishing locations, by having at hand appropriate information and acceptable accommodation, fishers are able to translate what initially are collections of unrelated entities into an ordered network of interactions in which they have a chance to succeed. How to make a fly to resemble an insect and when to use it, how to cast, what line to use and where to obtain it, depict the actions of an actor striving to gain control of the network. Their interactions within the network do not end there. Fishers soon discover the need to use the best equipment. They obtain information within the network about the complexity of fish ecosystems, the importance of time of day, water temperature, turbidity and the knowledge disseminated by fly-fishing magazines and other forms of information transmission. This, in turn, provides the setting in which they are able to manipulate and be manipulated by forces that make it possible to put their passion for fishing into practice. In this way, they may temporarily acquire power.

Nurturing passion

Actor-network theory offers a way of thinking about fly-fishers as part of an integrated material-semiotic network in which they emerge as powerful actors having enrolled other network entities through their ability to innovate, manipulate and control. The remainder of this chapter attempts to deduce what drives their quest for power.

A fantasy replay provides an insight into the moment every fly-fisher seeks. As the morning sun edges over the horizon casting a golden glow over a body of still water, a male fly-fisher, clad in a warm coat, hat and waders, stands in shallow water near the shore. He is keenly aware of the frost-encrusted grass behind, the steaming mist rising off the mirror-like water surface, the rose-tinged sky, the silence. A trout breaks the surface taking an insect on the water surface but betraying its position. The fly-fisher focuses his eyes on the spot from which small decaying waves spread concentrically outwards. With firm clean movements of his left hand, he draws in the tapered line through the eyes of his carbon-fibre rod, allowing it to fall into a basket at his waist. The fly at the end of the trace comes into sight. He wonders if he should change it from a Walkers Killer to a Connamara. He decides not. With skill and grace, he draws back the rod, allowing the line to unravel from the basket and fly over his shoulder. It runs out through the eyes in the rod briefly forming a graceful curve behind his back. He flicks the line forward, but does not allow it to touch the water, before sweeping it back behind his shoulder with an easy fluid movement. Once, twice, he repeats the movement, each time allowing more and more line into the air. The third time he allows the fly to settle gently on the surface, almost exactly on the spot where the trout had risen. He draws in the line as before. Without warning, the line twitches and the tip of the rod bends as a trout investigates the fly. The fisher ignores the electrifying signal but his heart rate quickens. Again. This time the fisher strikes. The rod tip bends sharply. The trout is hooked. A thrill of excitement surges through him. He holds the line taut keeping the tip of the rod well up. The trout moves back and forth, struggling desperately to eject the barb from its mouth. Slowly, it tires. Skilfully, he works the trout towards the bank. At the last moment, he unhooks a net hanging from his belt and shakes it open. The trout is now almost within reach. He works it closer, and then with his left hand holding the net, scoops the fish into the net. He wades back to the shore and lays the rod on the grass bank. He lifts the struggling fish from the net and carefully removes the hook. Gently, almost reverently, he places the trout back in the water. For a moment, it lies still, exhausted, traumatized. Then abruptly, it darts away. The fisher smiles and stands erect. He takes a deep breath, savouring the smell of clean air and conscious of the beauty of the awakening world. A swirl on the water surface betrays the position of another trout. He picks up his rod and strides purposefully to a place where he can reach it with a well-placed cast.

This fantasy replay, backed up by interviews with fishers, suggests that the successful use of tactics, techniques and devices lead them to a feeling of euphoria and suspension of time when a fish is hooked and landed. Like many fantasies of real, imagined or desired events in a fly-fisher's life, the moment of hooking and landing the fish is both the focal point and the basis of their passionate engagement with the sport. To this end, fly-fishers are prepared to devote time, spend money and acquire knowledge in order to gain power in the network. However, there is a limit to the amount the network can be manipulated. Fly-fishers are unable, for example, to influence fish to feed and to predict the moment of the strike. At such times frustration and satisfaction, manipulation and being manipulated become two sides of the same coin. These being the case, both are necessary to prepare fishers for their moment of triumph. Thus the profane activities of striving to learn fishing lore, acquiring appropriate skills and developing the patience necessary to curb frustration is punctuated by sacred moments when a fish is caught. In such moments, the passionate involvement with fly-fishing is nurtured and enhanced.

Sometimes the explanation of 'what happens' can be explained by action theory (deliberative or situated) in situations where actions are circumscribed by specific networks that are stabilized to 'work', as in the case of scientific experiments. In the case of fly-fishing, however, we are interested in that moment when, without warning, a fish decides to investigate the fly. It is the moment the fisher has been waiting for and 'just occurs' without promise or prediction. It is the culmination of a long period of preparation in which fishers, as part of a network collective, enter into a new network state that is transformed by the 'event'. It is the moment towards which they have been working through physical preparation in ordering the network and the intellectual effort involved in learning *how* to order it. The hooking of a trout represents the climax of these efforts and fishers are prepared to endure discomfort, expend resources and consume time towards achieving it.

Once a trout takes the hook, its world and that of the fisher is changed. This event has an authority of its own. It becomes an intervention in the network that shapes and composes what will happen (the trout may die, be eaten by the fisher or released and survive) and cannot be reduced to a simple interaction between the fisher and trout. However, it does serve the purpose of nourishing the fisher's passion through 'the abandonment of forces to objects and the suspension of self' (Gomart and Hennion 1999: 227).

Abandonment to passion is not an immediate state of being. Instead, it is a process of arriving through a series of states, in which fishers must establish appropriate conditions, consider certain strategies, employ a range of devices and be patient and receptive to environmental information while being sensitive to unseen investigation of the fly by the trout. The equipment must be in working order, the type of fly selected must be appropriate to the environmental conditions such as wind, water temperature and time of day. In other words, passion is neither the driving force of events nor an outcome. It is the product of careful preparation and long hours of failure punctuated by moments of triumph.

At the moment of the strike, and while the fish is being fought, the fisher passes from a state of patient expectation to excited activity. Deeply buried primeval urges surface that glory in the lust for the hunt. Time comes to a standstill. The fisher inhabits a liminal space between the moments before the fish takes the hook until after its successful capture. The collective network is electrified by the event before settling back into the previous state of being.

The passionate enthusiasm of individual fishers defies explanation using only the language of social construction. It is also inappropriate to direct attention solely to the satisfaction provided by gaining authority by enrolling others into the fly-fishing network. This alone does not drive the process towards passionate involvement in fly-fishing. This comes instead from interweaving the sense of abandonment to an external force at the moment the fish strikes and during the fight for its life, with the delight and satisfaction afforded by casting virtuosity, hunting skills, and local environmental fish lore. Fly-fishers are at once active and passive in the manner described by Gomart and Hennion (1999), taking delight in manipulating the network and, at the same time, allowing themselves to be manipulated by it.

Conclusion

Actor-network theory is used to analyse the geometries of power that shape the outcome of interactions between the heterogeneous assemblage of material objects, ideas, skills and social and environmental settings that constitute a fly-fishing network. The exploration of why systems work or fail because of changes in the material-semiotic network that influences its integrity, is not restricted to fly-fishing and may usefully be applied to describe similar associations and interactions in hunting, shooting and fishing. Although controversial, the notion that interactions and alliances between human and non-human actors take place in an environment in which all actors are assumed to have the same ability to act, facilitates our understanding of how networks are constructed and stabilized in the 'blind spot' where society and matter meet (Latour 1994: 41).

The utility of actor-network theory as a way of understanding *how* fly-fishing networks are formed and held together, draws on the manner by which power is orchestrated to produce stable, predictable and well-connected linkages and interactions. Successful fly-fishers are the ones able to stabilize the network by identifying the problems that need to be solved and convincing, enrolling and mobilizing the heterogeneous array of actors into a forum over which they hold sway. This is achieved through technical innovation, skills training and environmental knowledge, and the orchestration and assignment of roles to the heterogeneous assemblage of objects, entities, knowledge and settings which then become intermediaries in the network. However, networks are dynamic entities and stability is not guaranteed. Instead, networks require constant attention by those seeking control to prevent rebellious actors from disrupting network coherence.

Although the purchase of appropriate devices, the honing of skills and the acquisition of knowledge improves the chance of fishers, their passionate involvement in fly-fishing is not governed solely by the moment the fish strikes, despite its electrifying importance. It is about an extended process that leads step by step through the careful construction of a fishing network to the 'event' which leads to the sacred moment of passionate connection with a fish fighting for its life. Gomart and Hennion put it well:

> The paradox is that passion is entirely orientated towards an idea which is not the realization of the self, nor the realization of an intention, but the inverse: to let oneself be swept away, seized by something which passes. This active process of conditioning so that something might arrive is a central theme to passion.
>
> (Gomart and Hennion 1999: 244)

References

Amin, A. and Thrift, N. (1995) 'Institutional issues for the European economy', *Economy and Society*, 24: 121–43.

Appadurai, A. (1986) 'Introduction: commodities and the politics of value', in A. Appadurai (ed.) *The Social Life of Things: Commodities in Cultural Perspective*, Cambridge: Cambridge University Press, pp. 3–63.

Beck, U. (1992) *Risk Society: Towards a New Modernity*, London: Sage.

Bell, D. (1973) *The Coming of Post-industrial Society: A Venture in Social Forecasting*, New York: Basic Books.

Burgess, J., Clark, J. and Harrison, C.M. (2000) 'Knowledges in action: an actor-network analysis of a wetland agri-environment scheme', *Ecological Economics*, 35: 119–32.

Callon, M. (1986) 'Some elements in a sociology of translation: domestication of the scallops and fishermen of St Brieuc Bay', in J. Law (ed.) *Power, Action, Belief*, London: Routledge & Kegan Paul, pp. 19–30.

Callon, M. (1991) 'Techno-economic networks and irreversibility', in J. Law (ed.) *A Sociology of Monsters. Essays on Power, Technology and Domination*, London: Routledge, pp. 132–61.

Castells, M. (1996) *The Rise of the Network Society*, Oxford: Blackwell.

Cooke, P. and Morgan, K. (1993) *The Associational Economy*, Oxford: Oxford University Press.

Franklin, A. (2004) 'Tourism as an ordering: towards a new ontology of tourism', *Tourism Studies*, 4(3): 277–301.

Gomart, E. and Hennion, A. (1999) 'A sociology of attachment: music amateurs, drug users', in J. Law and J. Hassard (eds) *Actor-Network Theory and After*, Oxford: Blackwell Publishers, pp. 220–45.

Herd, A.N. (2005) *A History of Fly Fishing*. <http://www.flyfishing.com> (accessed 14 May 2005).

Hughes, A. (2000) 'Retailers, knowledges and changing commodity networks: the case of the cut flower trade', *Geoforum*, 31: 175–90.

Jóhannesson, G.T. (2005) 'Actor-network theory and tourism research', *Tourist Studies*, 5(2): 133–50.

Knoke, D. and Kuklinski, J. (1982) *Network Analysis*, Beverly Hills, CA: Sage.

Latour, B. (1986) 'The powers of association', in J. Law (ed.) *Power, Action, Belief. A New Sociology of Knowledge*, London: Routledge and Kegan Paul, pp. 264–80.

Latour, B. (1987) *Science in Action: How to Follow Scientists and Engineers Through Society*, Milton Keynes: Open University Press.

Latour, B. (1988) *The Pasteurization of France*, London: Harvard University Press.

Latour, B. (1993) *We Have Never Been Modern*, Hemel Hempstead: Harvester Wheatsheaf.

Latour, B. (1994) 'Pragmatogonies', *American Behavioural Scientist*, 37: 791–808.

Latour, B. (1998) *On Actor-Network Theory: A Few Clarifications*. <http://www.amsterdam.nettime.org/Lists-Archives/nettime-1-9801/msg00019.html> (accessed 29 March 2005).

Law, J. (1986) 'On power and its tactics: a view from the sociology of sciences', *Sociological Review*, 34: 1–38.

Law, J. (1991) 'Introduction: monsters, machines and socio-technical rations', in J. Law (ed.) *A Sociology of Monsters. Essays on Power, Technology and Domination*, London: Routledge, pp. 1–24.

Law, J. (1992) 'Notes on the theory of the actor-network: ordering, strategy, and heterogeneity', *Systems Practice*, 5: 379–93.

Lyotard, J.F. (1984) *The Postmodern Condition*, Manchester: Manchester University Press.

Murdock, J. (1995) 'Actor-networks and the evolution of economic forms; combining description and explanation in theories of regulation, flexible specialization and networks', *Environment and Planning A*, 27: 731–57.

Murdock, J. (1998) 'The spaces of actor-network theory', *Geoforum*, 29: 357–74.

On Fly Fishing Directory (2006) <http://www.onflyfishing.com> (accessed 7 July 2006).

Whatmore, S. (1998) 'Wild(er)ness: reconfiguring the geographies of wildlife', *Transactions of the Institute of British Geographers*, NS23: 435–54.

Whatmore, S. (2000) 'Elephants on the move: spatial formations of wildlife exchange', *Environment and Planning D: Society and Space,* 18: 185–203.

Wikipedia, the Free Encyclopedia, (2006) 'Actor-network theory'. <http://en.wikipedia.org/wiki/Actor-Network_Theory> (accessed 7 July 2006).

World Angler (2006) <http://www.worldangler.com/destinat.htm> (accessed 7 July 2006).

Part II

Historic precedents

4 The Scandinavian Sporting Tour 1830–1914

Pia Sillanpää

Introduction

Hunting as a touristic activity today attracts quite a number of foreign visitors to Sweden. This type of tourism constitutes an important source of income in many rural areas in Sweden (Gunnarsdotter 2005), and is generally considered to have great potential for the future use of Swedish forests (Hörnsten-Friberg 2004). Although one might think that wildlife tourism is a relatively recent phenomenon in Swedish society, this is not the case. More than 170 years ago British people in particular came to certain parts of the Scandinavian Peninsula for hunting, shooting and fishing. Some of the visitors either leased, bought or built sporting lodges in the Scandinavian backwoods, thereby becoming regulars in the local peasant communities.

From a historical point of view, the British are well known for their devotion to travel and exploration. Also, the hunting, shooting and fishing tradition has, for a long time, enjoyed a distinctive role as a pastime in certain layers of British society. From the 1830s until the outbreak of the First World War in 1914, certain aspects of this sporting tradition became closely associated with the Scandinavian backwoods. The sporting travels undertaken by Britons developed over time into a phenomenon with its own peculiar characteristics: 'The Scandinavian Sporting Tour' (Sillanpää 2002). The presence of the British influenced the host communities economically, socially, culturally, and in some cases even architecturally, and moreover, can be seen as a precursor to modern tourism. From the 1870s onwards, the Scandinavian sporting grounds increasingly became a politically contested arena comprising several actors, not only on the individual and local levels but also at the national level.

The lure of Scandinavia

British 'sporting gentlemen' initially seem to have found their way to the Scandinavian wilds around the year 1830. The overcrowding of the Scottish sporting grounds together with the rising prices for recreational sport in the home country turned the eyes of many sportsmen towards more secluded spots. If Scandinavia could offer space in plenty, and lakes, rivers and forests full of fish and game, it also enabled the sportsmen to engage in recreational fishing, hunting and shooting quite cheaply. An additional asset, it seems reasonable to conclude, was the fact that Scandinavian flora, fauna and landscape much resembled that of the Scottish Highlands, while at the same time the area was considered exotic.

If the visitor to Scandinavia of the 1870s could travel comfortably by steamship directly from Hull to Norway, the traveller of the 1830s had a lengthy and obviously rather fatiguing journey ahead of him, as the existing means of communication required that the journey be made by way of Hamburg in Germany. However, in accordance with the ethos of the time, a 'true sportsman' was supposed to be adventurous and prepared to 'rough it'. The Scandinavian Sporting Tour was characteristically connected with some hardship, and so skill and performance were important. In other words, Scandinavia offered an arena for performing 'manly' acts.

The first few pioneering sportsmen primarily explored the Norwegian salmon rivers. In their wake, during a period of approximately 90 years, followed an accelerating number of their fellow countrymen who gradually discovered the potential of the Swedish mountain areas. According to Admiral Sir William Kennedy (1902: 227) Scandinavia offered, amongst other things, 'elk [the term commonly used in Scandinavia for the species *Alces alces*, i.e. moose] shooting in the forests, reindeer stalking on the high fjelds, or "still" hunting for red deer – "the higher branches of sport".' Norway's numerous rivers were abundant with salmon and trout, and the Norwegian and Swedish mountain areas offered ptarmigan and other game birds in plenty. Ptarmigan – commonly referred to by the sportsmen as 'ryper' – was the exotic Scandinavian equivalent of the coveted Scottish grouse. The capercailzie, on the other hand, had been extinct in Britain since the eighteenth century (Brusewitz 1967: 172). The wild reindeer of Norway substituted for the royal Scotland stag, as did the red deer (on the island of Hitteren in Norway) and the mythical moose – 'the noblest animal of the deer tribe' (Kennedy 1902: 230). Also, by tradition, bears had always been desired game (Brusewitz 1967: 110, 151). Edward North Buxton (1893: 257) wrote that '[n]early every Englishman who takes a gun to Norway has a latent expectation of shooting a bear'.

Cultural clashes

Gunnarsdotter (2005; also see Chapter 13 in this volume) in her article on contemporary hunting tourism in the community of Locknevi in southern Sweden has studied how local moose hunters and other locals experience the presence of foreign huntsmen, mostly Germans and Danes. Her study shows that some cultural clash is a typical ingredient in the encounter with the hunting tourists as a group, and consequently the visitors are commented upon and criticized.

It is interesting to draw direct parallels from twenty-first century hunting tourism in Locknevi to the Scandinavian Sporting Tour of the nineteenth and early twentieth century. An elementary and at the same time intriguing aspect of the Scandinavian Sporting Tour as a phenomenon and an experience concerns the relationship between the British visitors and their Scandinavian hosts. The presence of the British sporting gentlemen has led to the emergence of myths and legends connected with these visitors, their beings and doings. Typically, stories depicting the extraordinary habits of the British have survived in the local communities to this date.

Many British sportsmen wrote about their hunting, shooting and fishing expeditions in Scandinavia in the form of diaries, articles or books. And typically, *their* memoirs abound with comments on local people and local circumstances, and how these in one way or other affected the successes or failures in the hunting or fishing arenas.

In many cases, the relationships between British sportsmen and their local gillies and stalkers were characterized by mutual respect, even friendship. The British genuinely appeared to have liked and admired the Scandinavian character: ingenuity, perseverance and trustworthiness. Yet, the visitors represented a culture very different from that of the Scandinavian peasant communities. In addition, in all likelihood the features of a British sporting estate in the Scandinavian backwoods were prone to reinforce the notion of social closure. The sportsmen, representing the upper and upper middle layers of society, felt superior in their relations with the Scandinavian peasant class, and the sporting estate with its British traditions and everyday life concretely manifested this social superiority by physically shutting out the locals (Sillanpää 2002).

The British sporting gentlemen looked upon the Scandinavian landscape with totally different eyes and had completely different motives for using the wilds as compared to the local peasants. In other words, the landscape was the same but it represented different functions. At the outset, for the locals, the forests and waters provided food, whereas for the sportsmen, the same landscape offered relaxation, amusement and adventure. Later on, as British leisure hunting, shooting and fishing extended as a phenomenon, the same landscape became interesting as a source of extra income for the local landowners, a development that led to the successive commercialization of the landscape. Thus, Löfgren's argument (1989: 183, 205) that not only different generations but also different groups of people look upon and use the same landscape in different ways seems directly applicable to the Scandinavian Sporting Tour.

Divergent interests

Indeed, the appearance of British sporting gentlemen in the Scandinavian backwoods at first seems to have given cause to both wonder and amusement amongst the locals. An illuminating example of the misunderstandings that could occur between local people and the sportsmen due to their divergent interests is given by Lord Walsingham (1927: 89), who depicts a sporting expedition in Norway in the 1870s. A young schoolmaster volunteered as their guide, but – obviously due to the author's poor knowledge of Norwegian – mistook the intention of the British guests. Instead of showing them to good shooting grounds, he made them climb to a top of a hill to admire the landscape, a misunderstanding that Walsingham found quite amusing. He also depicts another occasion, where during a deer stalk he wondered about the sad countenance of one of their local stalkers (1927: 100). It turned out that the stalker was in grief over the fact that he had to spend his birthday on a remote mountain instead of celebrating with his family. In the latter case, Lord Walsingham shows no understanding for the stalker's feelings. Indeed, many examples in this material imply a lack of sympathy or compassion on the part of the sportsmen for those who did not share their opinion that there is nothing more important than recreational hunting. It did happen, however, that the locals occasionally protested against the apparently somewhat tiresome enthusiasm of the British sportsmen, for example by simply refusing to act as their stalkers or guides (e.g. Seton-Karr 1890: 9).

In the early days of the Scandinavian Sporting Tour, the local people had not yet woken up to realize the potential of leisure hunting and fishing for their own pockets, and thus these activities could be enjoyed very cheaply if not even for free. Commonly, it sufficed if the foreigners handed over the fish that they were not able to consume themselves to the locals – since the British, in accordance with the tradition of their leisure activity, caught much more fish than they could possibly eat (e.g. Permansson 1928: 68). Indeed, the idea of someone being prepared to pay high sums of money for the rights of fishing and hunting seemed strange to the peasants (e.g. Pottinger 1905, I: 39). Williams (1859) notes that the locals were mightily amused by the British and their fascination for leisure fishing, since fishing was:

> one of the vulgar occupations by which men obtain a livelihood. Our laundresses would be similarly amused if Chinese mandarins were to migrate annually to England and pay large sums of money for the privilege of turning their mangles.
>
> (Williams 1859: 66)

The concept of 'salmon lords', commonly used by the local Scandinavian inhabitants when referring to the British sporting gentlemen, seems to indicate that the overall image of these visitors was (and still is) that they were rich and titled – which is why money was no problem if they really wanted something. It also happened that the locals made fun of the British. The mockery, could amongst other things, relate to their curious clothing or equipment as well as to their peculiar habits, such as that of slavishly following the clock when fishing, no matter whether or not the fish were biting (e.g. Williams 1859; Cappelen-Smith 1935; Berg 1938).

All in all it seems apparent that the British visitors were objects of the natives' curiosity. Walsingham, giving an account of his river and lodge, states that the river and its torrent are not very grand, and thus do not attract British tourists. However:

> many of the native inhabitants on their way up and down the coast call at the port and take a *liten tur* [Norw. for *short trip*, my translation] up the valley to see my foss [Norw. for *stream*, my translation] and to look on (in admiration

> we will hope) at the fisherman endeavouring to induce the reluctant salmon to take the fly.
>
> (Walsingham 1926: 100)

Since the Scandinavians traditionally had considered fishing and hunting as a way of acquiring food, it took a while before they started to show any interest in trying out fly-fishing for themselves, and thus began to imitate the foreigners. The eventual emerging interest in fly-fishing, together with the development that as time went by, the locals came to realize the potentials of tourism, and, as a consequence, learnt to 'exploit' the visitors in various respects, resulted in clashes between the hosts and the guests. In other words, occasional disputes were part of the natural course of events as the phenomenon of the Scandinavian Sporting Tour became more widespread. With respect to the local people's ability to imitate the British sportsmen or the locals' knowledge of fishing in general, the British attitudes seem to differ. Some state that the locals learnt fast, whereas some did not think very highly of the locals as sportsmen, and considered them generally ignorant of anything connected to their art (Bilton 1840; Sandeman 1895; Kennedy 1902). Yet this is, in fact, rather contradictory since, in most cases, the British were totally dependent on local gillies and stalkers and their skills and knowledge of the landscape and the fauna for their sporting successes.

British superiority

Morgan and Pritchard (1998: 15) claim that '[t]ourism is a cultural arena in which ... hegemonic ideas of superiority and inferiority are continuously played out'. Such a concept, it is argued here, can be applied very usefully to the analysis of the Scandinavian Sporting Tour: all in all, the paternalistic and elitist attitudes of the British towards the local people, and the notion of social closure are conspicuous. The British sportsmen largely seem to have considered their sporting activities in the Scandinavian backwoods as a British perquisite and would thus have preferred to keep it their own privilege. The local peasants were not supposed to be engaging in leisure hunting and fishing, instead, their role was to serve and obey their employers – the British sportsmen – without protest. Kennedy (1903: 62), for example, mentions a Norwegian boatman who, like so many other boatmen, 'had the audacity to take command and order about the Englishman'. He called this particular boatman a 'sigher', a name he had invented for those boatmen who rowed unwillingly and emitted long sighs.

Particularly from the 1870s onwards, it appears that various types of confrontations between the British visitors and the Scandinavians developed into an unavoidable ingredient in the Scandinavian Sporting Tour. Confrontations between hosts and guests not only occurred on an individual, but also on a state level, including the commercialization of the sporting grounds and the introduction of game laws. Even though some authors admit that British sportsmen may indeed be at fault for the necessary introduction of some restrictions, these developments were mostly regarded as unfair impediments. The idea that Scandinavians somehow interfered in their leisure hunting and fishing activities hugely annoyed the sportsmen, and anything that could destroy, spoil or intrude on their pursuits, or in any other way collide with their interest, was frequently lamented upon. Nor were comments made only in the personal accounts of the sportsmen: the issues were also openly debated on the pages of, for example, *The Field*, a popular sporting journal of the time.

Commonly, the comments of British sportsmen indicate a notion that local people were not supposed to hunt as they pleased since their methods were no good and thus destroyed the fauna and opportunities for hunting and fishing. As to the sporting activities of the British themselves, on the other hand, it was clearly not regarded that they were doing any harm (e.g. Kennedy 1902). Another thing that is often lamented upon is that the Scandinavians do not abide by the existing game laws. Buxton (1893: 348), for example, feels that there are no worse 'poachers and pot hunters in the world' than the Norwegians. Admiral Kennedy (1902: 236), on his part, apparently does not trust that the locals will follow the law regulating moose hunting in Sweden, for he notes that 'elk [moose] ought to be increasing, but I am not sure if such is the case, as the law is difficult to enforce in remote districts'. In 1902, an association for game preservation was founded in the county of Jämtland in Sweden, with the specific task to prevent illegal moose hunting (County governor's five-year report 1901–05). Yet it seems that according to the sportsmen themselves, they are not compelled to follow the regulations to the letter. On one occasion, Admiral Kennedy and his brother Edward shot a reindeer stag off-season, and then managed to deceive the local official in charge of game licences, when making an unexpected call, into believing that he was offered mutton instead of reindeer for supper (Kennedy 1902). Though the story is rather amusing, it also points at another side of British leisure hunting: if the temptation got big enough, it was possible that the sportsmen did not automatically choose to abide by the Scandinavian game laws. The interests of sport were perhaps given the highest priority.

The differences between the British and the Scandinavians in their ways of looking at sport also manifested themselves in other ways. Løchen (1991) mentions examples of various restrictions imposed by British lessees of the Stjørdal river – in the vicinity of the city of Trondheim in Norway – on the natives through the lease contracts. For example, fishing with nets was explicitly forbidden during the lease period, and the lessor was compelled 'to place some stones in the river in order to prevent netting of the same' (1991: 108).

An institutionalized travel phenomenon

The increase in prices that came into play as the stream of travellers and tourists to Scandinavia grew was a development for the worse that was frequently remarked upon. This aspect of counter-exploitation is an interesting phenomenon that is highly relevant in discussions on the impact of tourism in general. The question may be posed as to whether the local people – as time went by and they woke up to the realization that the visitors were prepared to pay dearly for certain types of privileges, services or goods – created an image of a British tourist whom they wanted to exploit. The increases in prices not only involved goods such as souvenirs or hotel accommodation but also impinged upon the prices of sporting rights.

Sometimes the prices demanded by the local landowners indeed exceeded the willingness of the sportsmen to pay, as was the case for Lord Walsingham (1927), when in 1879 he tried to negotiate the lease of a sporting ground for red deer hunting on the island of Hitteren. This issue of a tourist industry demanding high prices, annoying and scaring off potential buyers, developing a greedy image, and even pricing itself out of the market remains, of course, an important issue for contemporary tourism. Many of those Britons who complained about the conduct of the locals predicted that, as a result, the British sportsmen would eventually disappear from the scene (e.g. Sandeman 1895).

Lumley House, 34, St. James's Street, London, S.W. 157

Swedish Elk and Ryper Shootings and Trout Fishings

LEGAL SEASONS FOR SHOOTING GAME IN SWEDEN.

*ELK.—1st to 15th Sept.	PARTRIDGE—10th Sept. to 11th Nov.
DEER—1st Sept. to 31st Dec.	BECASSINE—DUCK—10th July to 30th Dec.
ROE—15th Sept. to 16th Nov.	BLACKCOCK—20th Aug. to 15th March.
RYPE—16th Aug. to 14th March.	SNIPE—1st Aug. to 15th March.
HARE—20th Aug. to 15th March.	

**Dates in Sweden vary according to locality and yearly regulations.*

EDOSSAN HOUSE, SWEDEN

No. 2261.

SWEDEN—EDSOSSAN ESTATE, comprising about 60,000 acres of excellent Ryper and Mixed Game Shooting, showing a good and steady Game Bag. The ground is computed to be good for an average of about 300 brace in ordinary seasons. The Angling Rights on both banks of the River afford excellent Trouting (the catch in 1900 was 700 Trout, including fish up to 3½ lbs., and is all Fly-Fishing), also seven good Trouting Lakes.

Residence built by English owner contains 3 Sitting-rooms, 11 Bed-rooms, and Kitchen, etc. *See Illustration.*

Railway Station within 400 yards. Post near and Telephone in House. Plate and Linen included. Photos may be seen at 'Lumley House.'

The Property, including the Freehold Ground and Houses, with the Furniture, etc., therein, and the Sporting Leases on the remainder of the Estate are also for Sale. Price, £2,500.

Figure 4.1 An advertisement for a sporting estate in Sweden (Source: *Norwegian Anglings 1907*, p. 157. By courtesy of Stockholm University Library)

One result of the increased commercialization of hunting, shooting and fishing was the appearance of agents who made their living by leasing and selling sporting rights. These agents were not solely Scandinavian, but also British. The immediate result of this commercialization, it seems reasonable to assume, was that the prices of sporting grounds were forced up at the same time as it grew more difficult to find sporting grounds the leases of which could be negotiated directly with local peasants. Admiral Kennedy (1902: 273) felt that due to commercialization, sport in Scandinavia had deteriorated. He complained about agents who let their rivers to Britons 'at four times their original cost', concluding that those who have managed to secure long leases on their own rivers, and thus remain independent of the agents, can count themselves fortunate. Considering the fact that one of the attractions of Scandinavia, as a sporting ground for the British, was originally the freedom and the cheap prices it could offer, it is obvious that this change for the worse would be criticized. In addition to the agents who made sporting activities increasingly dear, the introduction of game laws and other regulations that in one way or another restricted leisure hunting and fishing appear, indeed, to have been a direct consequence of the institutionalization of the Scandinavian Sporting Tour. For example, in 1877 a new game law requiring a licence for shooting on government grounds – which included a large proportion of the best sporting grounds in the country – was introduced in Norway (Baedeker 1879). In 1878, the import of dogs to Norway was prohibited (Walsingham 1927), thanks to which a new type of business would apparently emerge: the leasing or selling of Scandinavian sporting dogs (Kennedy 1903). Obviously, the fact that one could no longer bring one's own dog to Scandinavia spoilt the former charm of Scandinavian hunting and shooting for many (Kennedy 1903; Walsingham 1927). Lord Walsingham (1927) was convinced that, in essence, it was the jealousy of local sportsmen, not the fear of hydrophobia (i.e. rabies) – the official reason given – that resulted in the prohibition on the importation of sporting dogs.

Further restrictions were progressively imposed in both Sweden and Norway. For example, from 1893, a licence for shooting reindeer or other game on public ground in Norway now cost £11, a restriction that according to Buxton (1893) was above all aimed at British sportsmen. Such developments were indeed quite controversial, with leisure hunting and fishing now entering the political arena, and many issues being debated for years. In an article in *The Field* (7 January 1888, p. 21), dealing with the ongoing 'war' against British sportsmen in Norway, parts of a letter written by the Norwegian Inspector of Fisheries, Mr Landmark have been published. His letter contains an abstract of his latest proposal for the alteration of the laws affecting salmon and trout fisheries. Pointing out that Norwegians are neither willing, nor could afford, to pay the same high prices as the British for leisure fishing, he goes on arguing *for* the presence of the British, as not only does the lease of rivers to sportsmen greatly help to preserve and develop the salmon fisheries: the visitors also bring much money into the country. Landmark estimates that between 1876 and 1879, the rents alone paid by foreigners for salmon waters amounted to 70,000 Norwegian kroner yearly.

It seems evident that, where the presence of British sportsmen was concerned, the local people were divided in, at least, two parties: one group opposing the British invasion of the Scandinavian sporting grounds, another one valuing its positive economic impact, which in their view exceeded the negative effects.

A spill-over from Norway into Sweden

Admiral Kennedy (1902) expressed his anger towards the Norwegian government who restricted leisure hunting by instituting game laws that almost excluded British sportsmen from ryper shooting. In his opinion, they were being unfair towards the British and, moreover, he thought that this was a very short-sighted and narrow-minded policy. In line with Landmark's letter mentioned above, the Norwegian farmers, Admiral Kennedy argued, had benefited from the influx of

Figure 4.2 An advertisement for recreational hunting, shooting and fishing in the Swedish mountain areas (Source: Cook [ed.], 1901, p. 271. By courtesy of the Thomas Cook Archives)

Britons, but because of all the restrictions introduced in Norway, the sportsmen were now increasingly crossing the border into Sweden where the laws were less stringent. This – the so-called push-effect of a particular locale and the pull-effect of another – is an important feature of the Scandinavian Sporting Tour, at the same time as it is an important aspect of tourism development in general. The high prices and overcrowded grounds in Norway resulted in a spill-over of British sportsmen into the Swedish wilderness, a development which is discernible particularly from the 1880s, and which, apparently, is directly linked to the much less stringent regulations in Sweden, in combination with improved infrastructure between the two countries (e.g. Baedeker 1879).

However, even though, in contrast to Norway, the Swedish sporting grounds of the 1880s still seemed temptingly free and relatively cheap, a similar kind of development was more or less inevitable. As early as 1889 (that is, one year after the above mentioned introduction of a new Norwegian game law concerning partridge shooting), there seems to be an internal debate going on in the Swedish Touring Club about the necessity of restricting leisure hunting and fishing. Apparently, the Club had two years earlier issued a prize competition in order to promote these activities in Sweden, an issue that had met with harsh critique partly due to the scarcity of game. The proposal was therefore put forward that the Club should cooperate with the local game management associations in order to make hunting and shooting as good and lasting a source of income as possible for the local riparian farmers and their communities, for example by introducing hunting fees. Norway was considered a bad example of how foreigners had destroyed the fish and game (Svenonius 1889).

Letters from readers in *The Field* form an interesting source of information on developments in Sweden. For example, in *The Field* of 28 April 1888 (p. 606), under the heading 'Purchase of land by aliens in Sweden', the author of the article – 'A.D.' – wanted to warn the readers against answering advertisements offering Swedish sporting properties for sale: according to him, foreigners were obviously no longer welcome to purchase land in Sweden, a fact which he had personal experience of after he had tried to buy a property in Jämtland. Everything was settled with the vendors and the only thing still required was the licence which foreigners needed to acquire for the purchase from the Crown. This licence used to be a formality, but for some reason, after several months' delay, A.D.'s application was refused: a policy on the part of the Crown, he argued, that in the future would be the order of the day. An answer to this letter was published in *The Field* of 19 May 1888 (p. 707), where it was claimed that A.D.'s application must have been somehow incomplete. This particular writer, a resident of Sweden, points out the necessity of presenting good references as to one's character from well-known British citizens, and ends up by stressing the fact that Britons are welcome throughout Sweden. The matter does not, however, quite end here: A.D. in reply to these comments, claims in the issue of the following week (26 May 1888, p. 747) that the necessary formalities were complied with, and this with the help of a Swedish lawyer. Yet, the answer to subsequent inquiries had been that Britons would no longer be allowed to buy land in Sweden. The above example indicates that in Sweden as well, voices were raised against the 'invasion' of land and water by British sportsmen.

In this context, it is interesting to consider the fact that in Britain – or in the Norman and Anglo-Saxon society – royalty in the past introduced game laws to keep the common people out. Equally, in the British colonies, game laws prohibited the local inhabitants from hunting. In Scandinavia, on the other hand, Britons first exported these British traditions into the Scandinavian backwoods by acquiring private sporting grounds to protect their sport from others. As time went by, this, in turn, resulted in Scandinavian game laws, which were introduced to protect the common people and their rights to the natural assets against the wealthy and numerous Britons.

The end of the British era

The Scandinavian Sporting Tour came to a more or less abrupt end with the outbreak of the First World War in 1914. The war naturally had many implications for British society, socially as well as economically. Until the outbreak of the war, despite the problems discussed above, British leisure hunting, shooting and fishing seemed to have continued to be an industry to reckon with in many parts of Norway and Sweden. From the 1880s onwards, different types of summer tourism were developing in the area in parallel with the Scandinavian Sporting Tour. In addition to members of the British sporting community, Norwegian coastal towns such as Trondheim were visited by an increasing number of people who may be labelled as mass tourists. These visitors travelled to Norway by sailing yachts on tours such as those arranged by Thomas Cook, who organized his first Scandinavian tour in 1875 (e.g. *The Excursionist*, 12 June 1875). Cook's tourists wanted to gaze at the beautiful, romantic Norwegian mountain and fjord scenery, as well as to experience the North Cape with its Midnight Sun. In the mountain areas of mid-Sweden an increasing number of people flocked to breathe the invigorating air (Kilander, forthcoming). These tourists were commonly referred to as 'air guests'. Hikers, mountaineers and botanists formed additional groups of travellers. It remains, however, that the British sportsmen were the first international travellers of any significant numbers to enter the Scandinavian scene.

Today, the most tangible proof and legacy of the Scandinavian Sporting Tour (apart from the activity of recreational hunting, shooting and fishing – and especially fly-fishing) are the sporting lodges once owned and inhabited by British sportsmen, which still exist in the midst of the Scandinavian backwoods.

Current issues in Swedish consumptive wildlife tourism

In 2003, the Rural Economy and Agricultural Society in the county of Västerbotten, Sweden mapped out Swedish hunting tourism, concluding that there are about 260 companies operating within this line of business. More than half of these are operating in the three northernmost counties. It is common that companies offering hunting tourism complement these activities with other types of business;

in fact, in most of the cases hunting tourism accounts for a small part of the total business. Roughly speaking, more than half of the companies are oriented towards other tourist categories, most importantly fishing tourists. These entrepreneurs are typically located in the north of Sweden. About one third complement hunting tourism with agriculture and forestry, and these companies are typically located in the southern parts of Sweden. The remaining 20 per cent, mostly situated in the central parts of the country, combine hunting tourism with conferences and education of various types (Turistdelegationen 2004).

The entrepreneurs themselves are optimistic as to the potential of Swedish hunting tourism. Thanks to, amongst other things, the variety of game and a low price level it is believed to have great possibilities for development. Also, the demand for hunting amongst Swedish and foreign huntsmen alike is huge. Some of the inhibiting factors mentioned by the entrepreneurs, on the other hand, include the presence of predators as well as the ignorance of the importance of hunting tourism for rural and sparsely populated areas (Turistdelegationen 2004).

As far as Swedish fishing tourism is concerned, though still poor in efficiency, it is gaining ground. A federation has been established for companies offering fishing tourism, with the aim being to support industry development in the field. One important goal of the federation is, through training, to increase the professionalism of the entrepreneurs. The Swedish government has commissioned the Swedish Environmental Protection Agency together with the Swedish Board of Fisheries to investigate the opportunities and obstacles for an expansion of fishing tourism (Turistdelegationen 2005).

Summary and conclusions

A scrutiny of the hunting, shooting and fishing tours undertaken from the 1830s until 1914 by Britons to Norway and Sweden reveals that in many respects these tours, as a travel phenomenon, may be regarded as a precursor to modern tourism in Scandinavia. The phenomenon of the Scandinavian Sporting Tour influenced the host communities not only economically but also socially, culturally, and even architecturally. Moreover, from a legislative point of view, the Scandinavian Sporting Tour seems to have had far-reaching implications. It is interesting to note that certain issues seem to recur in the discussions of the pros and cons of wildlife tourism in Scandinavia. As was the case more than 100 years ago, it is obvious that also for many present-day rural areas, the presence of hunting and fishing tourists means a much-welcomed source of extra income. At the same time, the visitors still represent a foreign and thus in some respects disturbing element in the communities concerned.

From the point of view of tourism development, the example of this particular travel phenomenon illustrates that how a hunting or fishing destination develops over time is closely connected not only with factors that can be influenced, such as, for example, the price level, or relevant game law, but also more complex factors, most pertinently, the personal and local level responses of the host community.

References

Baedeker, K. (ed.) (1879) *Norway and Sweden: Handbook for Travellers*, Leipzig: Karl Baedeker; London: Dulau and Co.

Berg, E. (1938) 'Jämtländska storviltjägares äventyr. Erik Göransson berättar', in *Östersunds-Posten*, 14 November 1938.

Bilton, W. (1840) *Two Summers in Norway*, 2 vols, London: Saunders and Otley.

Brusewitz, G. (1967) *Jakt. Jägare, villebråd, vapen och jaktmetoder från äldsta tider till våra dagar*, Stockholm: Wahlström & Widstrand.

Buxton, E.N. ([1892] 1893) *Short Stalks: or Hunting Camps North, South, East, and West*, 2nd edn, London: Edward Stanford.

Cappelen-Smith, D. (1935) 'Om engelska sportfiskare i Jämtland', in Jämtlands Läns Sportfiskeklubb *Festskrift till 25-årsjubiléet 1935*, Östersund [no publisher mentioned], pp. 43–64.

Cook, T. (ed.) (1901) *Cook's Handbook to Norway with the Principal Routes to Sweden and Denmark etc. with Maps, Plans and Vocabulary*, 4th edn, revised and enlarged, London: Thomas Cook & Son.

Excursionist, The, 12 June 1875, 'Scandinavia', p. 2.

Excursionist, The, 29 June 1875, 'Cook's Tours to Denmark, Sweden, Norway, Finland, and Lapland', p. 3; 'Itinerary and Dates', p. 3.

Excursionist, The, 10 July 1875, An advertisement, p. 3.; 'Detour No. 1', p. 4.

Field, The Farm, The Garden, The Country Gentleman's Newspaper, The, 7 January 1888, 'Foreign Sport, Colonial Estates, Farms, &c.', p. 1; 'Norwegian Game Laws', p. 14; 'Fishing Laws in Norway', p. 21.

Field, The Farm, The Garden, The Country Gentleman's Newspaper, The, 28 April 1888, 'Purchase of Land by Aliens in Sweden' [signed 'A.D.'], p. 606.

Field, The Farm, The Garden, The Country Gentleman's Newspaper, The, 19 May 1888, 'Purchase of Land by Aliens in Sweden' [Fredk. Stoddard's answer to A.D.], p. 707.

Field, The Farm, The Garden, The Country Gentleman's Newspaper, The, 26 May 1888, 'Purchase of Land by Aliens in Sweden' [A.D.'s answer to Fredk. Stoddard], p. 747.

Gunnarsdotter, Y. (2005) 'Vad händer i bygden när den lokala älgjakten möter jaktturismen?', in S. Åkerberg (ed.) *Viltvård, älgar och jaktturism. Tvärvetenskapliga perspektiv på jakt och vilt i Sverige 1830–2000*. Umeå: Hållbarhetsrådet, pp. 106–38.

Hörnsten-Friberg, L. (2004) *Naturturismen i Värmland – markägande och turistföretagande*, European Tourism Research Institute (ETOUR) U 2004: 21, Östersund: ETOUR.

Kennedy, E.B. (1903) *Thirty Seasons in Scandinavia*, London: Edward Arnold.

Kennedy, W. (1902) *Sport in the Navy and Naval Yarns*, Westminster: Archibald Constable & Co Ltd.

Kilander, S. (forthcoming) *Luftburen hälsa. Om industrialiseringsprocessen i västra Jämtland 1880–1920.*

Løchen, G. ([1970] 1991) *Laksefiske i Meraker*, 2nd revised edn, Elverum: Norsk Skogbruksmuseum.

Löfgren, O. (1989) 'Landscapes and mindscapes', *FOLK – Journal of the Danish Ethnographic Society*, vol. 31: 183–208.

Morgan, N. and Pritchard, A. (1998) *Tourism Promotion and Power: Creating Images, Creating Identities*, Chichester: John Wiley & Sons.

Norwegian Anglings 1907 (1907) ed. by James Dowell, London: J.A. Lumley & Dowell.

Permansson, P. (1928) *Engelska jägare i Frostviksfjällen*, Stockholm: Åhlen & Åkerlunds Förlag.

Pottinger , H. (1905) *Flood, Fell and Forest*, 2 vols, London: Edward Arnold.
Sandeman, F. (1895) *Angling Travels in Norway*, London: Chapman & Hall Ltd.
Seton-Karr, H. W. (1890) *Ten Years Travel & Sport in Foreign Lands or, Travels in the Eighties*, 2nd edn, with additions, London: Chapman and Hall Limited.
Sillanpää, P. (2002) *The Scandinavian Sporting Tour: A Case Study in Geographical Imagology*, Doctoral thesis. European Tourism Research Institute (ETOUR) V 2002: 9, Östersund: ETOUR.
Svenonius, F. (1889) 'Bör Turistföreningen befatta sig med våra jagtförhållanden?', in Svenska Turistföreningen (1889) *Svenska Turistföreningens Årsskrift 1889*. Facsimile edn from 1982. Stockholm: Svenska Turistföreningens förlag, pp. 83–5.
Turistdelegationen (2004) *Jaktturism i Sverige* [a report by the Swedish Tourist Authority].
Turistdelegationen (2005) *Innovationsprogrammet 2002–2005* [a report by the Swedish Tourist Authority].
Urry, J. (1995) *Consuming Places*, London and New York: Routledge.
Walsingham, Lord (John de Grey) (1926) *Fish*, London: Philip Allan & Co. Ltd.
Walsingham, Lord (John de Grey) (1927) *Hit and Miss: a Book of Shooting Memories*, London: Philip Allan & Co. Ltd.
Williams, W. M. (1859) *Through Norway with a Knapsack*, London: Smith, Elder & Co.

Manuscript materials

Landsarkivet [the Regional Archives], Östersund: Femårsberättelser för Jämtlands län [The County governor's five-year reports], 1901–1905.

5 Controversies surrounding the ban on wildlife hunting in Kenya

An historical perspective

John S. Akama

Introduction

In order to put the current controversies concerning the ban of wildlife hunting in Kenya into perspective (since 1977 to the present, no form of wildlife hunting is allowed in and outside the country's wildlife park and reserves), it is important to provide an historical evaluation of the country's wildlife policies and tourism programmes as they relate to wildlife hunting. The arrival of Europeans in the rural African landscape, in the early nineteenth century, and Kenya's incorporation into the global market economy was a turning-point in its nature–society relationship. Many of the contemporary issues and problems concerning wildlife conservation and the development of wildlife-based tourism can be traced back to that period. Since 1977, it has been an offence to kill wildlife whatever the circumstance, punishable with imprisonment.

Within the broader historical context, it is important to note that, over the years, the development of wildlife conservation policies and tourism programmes in Kenya has been greatly influenced by powerful interest groups that are headquartered in Western capitals. These groups have tended to alienate resource user-rights from the rural African communities; and the proprietorship and user-rights of wildlife resources have been transferred to the state, conservation organizations and tourism developers who tend to be inherently opposed to the introduction of any form of game hunting either by local people or sport-hunters. This conception is perhaps based on an unsubstantiated belief that allowing wildlife hunting will lead to large-scale extermination of wildlife, especially the much sought after mega-species such as elephants, lions, leopards, giraffe and cheetah. Within this scenario, local subsistence hunting has come to be termed as 'poaching' (Akama 1998). Indeed, the onset of colonial rule set in motion social and economic processes seeing the gradual removal of indigenous decision-making through state wildlife policies and programmes. Rural people's natural resource user rights were therefore weakened vis-à-vis those of the state, international conservation organizations and tourism developers.

This chapter provides an historical evaluation of controversies concerning wildlife conservation, in general, and wildlife hunting in particular. It is argued that Kenya's pioneer conservation policies were based on conservationists' conception that indigenous resource use methods were incompatible with the principles of Western philosophy concerning wildlife management and protection. This conceptual and philosophical under-pinning has persisted to the present time. It is further argued that wildlife policies and tourism programmes which derive from these conceptions, coupled with increasing human population in semi-arid lands surrounding Kenyan parks and other wildlife areas, have resulted in severe and accelerating people-versus-wildlife conflicts.

Pre-colonial African wildlife resource use

It is important to note that indigenous African communities had evolved various methods of wildlife management and other forms of resource use strategies. These resource use methods were based on the indigenous people's socio-economic and cultural understanding and perception of the territorial and social landscape (Campbell 1986; Akama 1998, 2003). Some of the indigenous natural resources use strategies including pastoralism, shifting cultivation and hunting and gathering of wild fauna and flora. Pre-colonial Kenyan societies acted upon and modified rural landscapes, flora, fauna and social strata through such resource use strategies.

Recent research on the historical ecology of wildlife resources in Kenya indicates that most rural Kenyan communities had governing regulations concerning hunting and use of wildlife products. These were community hunting regulations that subsistence hunters were supposed to follow. For instance, in most Kenyan communities, it was taboo to hunt and kill certain wildlife species that were held in great respect and veneration (Akama 1998). The killing of such animals was perceived as a bad omen believed to bring natural disasters, such as drought, famine and disease to the community.

Furthermore, wildlife formed an integral part of the socio-economic and cultural experience of pre-colonial Kenyan communities. Wildlife featured prominently in various indigenous social and cultural activities and routines. Different Kenyan communities had wildlife animals that were recognised as community totems and were held in great esteem and were therefore protected from wanton destruction. These were animals that symbolised a clan or local community, and thus had ritualistic or religious value to the community. Animals that were totems among Kenyan communities such as Kikuyu, Maasai, Meru and Gusii include elephant, cheetah, lion and leopard. In most rural communities, folklore based on various aspects of wildlife was an important mode of imparting cultural and social values to the youth. Stories of wild animals featured prominently to the extent that the youth accepted them as part of their rural environment. Thus, as children grew up, they were taught how to identify different animals, which animals were dangerous, and the habitats of different wildlife species.

The era of big game hunting in Kenya

The declaration of the East Africa Protectorate (colonial rule) on 15 June 1895, and the arrival of European settlers, amateur and professional hunters and other trophy seekers led to rapid extermination and decline of wildlife populations and destruction of wildlife habitats (Akama 1998; 2003). Furthermore, the creation of colonial institutions of governance engendered conditions of relative socio-political stability and the maintenance of law and order, which encouraged pioneer travellers and adventure seekers to venture into the East Africa hinterland.

A major recreational activity undertaken by Westerners who ventured into the East Africa hinterland was big-game safari hunting. In fact, the period from 1900 to1945 in East Africa is generally referred to in popular literature as the 'Era of Big Game Hunting' (Anderson 1987). It has been noted that during the initial period of European colonialism in Africa and other parts of the Third World, the phenomenon of big-game hunting was perceived as a major symbol of European dominance over nature in general and society in particular. As a consequence, big-game hunting was a major determinant of class and socio-political power (Anderson 1987; Akama 1998). Thus most of the pioneer Westerners who undertook safari hunting expeditions in Africa were mainly affluent travellers, high-ranking government officials, politicians and members of the aristocracy.

Famous pioneer travellers and big-game safari hunters to visit East Africa include prominent personalities such as Theodore Roosevelt and John Muir (Akama 1998). For instance, in his most widely published safari to East Africa, which lasted between April 1909 and March 1910, the then US President Theodore Roosevelt travelled with over 200 trackers, skinners, porters and gun bearers. Roosevelt shot, preserved and shipped to Washington DC more than 3,000 specimens of African game.

Most of the pioneer safari hunters provided detailed accounts of their hunting exploits when they returned to the West. Others wrote adventure books based on their big game hunting exploits (Nash 1982). For instance, a British aristocrat and a professional hunter, Abel Chapman, wrote an adventure classic in 1908 entitled *On Safari*, where he recounts his spectacular hunting escapades in the East Africa savannas. He argues here that the big game traveller-sportsman was the best customer of the East Africa colony and game was the best asset (Nash 1982). In the following year, 1909, an American big-game hunter, William Baullie, wrote another hunting classic entitled *The Master of the Game*, with an introduction by Theodore Roosevelt. Part of the book's introduction read, 'there were still a few remote places (on the face of the earth) where one had to hunt in order to eat and where the settlers had to wage war against the game in the manner of the primitive man' (Nash 1982: 354). These safari hunting classics are still popular, and continue to reinforce Western perceptions and images of East Africa as a wildlife 'Eden'.

The start of anti-hunting campaigns

During this period of accelerated wildlife destruction, pioneer Western conservationists realised that if excessive destruction, particularly of larger wild animals, was not checked, the end result would be extinction. Thus, pioneer conservationists raised concern about excessive exploitation of the savanna wildlife. By the turn of the century, there was growing interest in the West for wilderness conservation in frontier territories worldwide, particularly in the Third World. A social class of naturalists and anti-hunting lobby groups had emerged who advocated for wilderness conservation and appreciation of the aesthetic and ethical value of pristine natural areas (Nash 1982). These were people who were generally affluent and were not living at the economic margin and were thus able to organise safari expeditions to Kenya and other parts of the Third World.

The concern of the pioneer naturalists was fuelled by the realisation that pristine natural areas in most frontier territories were rapidly shrinking due to increased human populations with attendant settlement, industrialisation and uncontrolled hunting practices. The pioneer conservationists started to organise conservation awareness campaigns throughout Europe and North America. The campaigns were aimed at sensitising the public in general and the government in particular, on the social and ecological value of wildlife conservation. The conservationists put pressure on those governments such as Britain, France, Germany and Italy who had colonies in Africa and other parts of the Third World, to initiate policies and programmes of wildlife protection.

For instance, in 1903, British conservationists formed the Society for the Preservation of the Fauna of the Empire whose main aim was to sensitise the general public and to urge the British government to initiate and implement policies and programmes of wildlife conservation and protection in the East Africa Protectorate and other colonies. The society urged the British government to establish adequate nature reserves before the country was completely settled by farmers and ranchers and the opportunity for otherwise doing so be lost forever. The society sent a committee to Kenya to investigate the game situation and make future recommendations (Akama 1998).

In 1913, naturalists from 16 European countries and North America held a conference in Basel to formulate conservation guidelines and to agitate for the protection of nature and wildlife areas worldwide, particularly in colonies where there still existed relatively large undisturbed blocks of land. Eventually, in 1928, pioneer conservationists established an international office in Brussels. Its main functions were to gather systematic information on the status of wildlife conservation and formulate wildlife conservation policies and programmes. in October 1933, representatives of European governments, with colonies in Africa, held a convention in London to review the status of wildlife conservation and protection in Africa. Members of the convention re-affirmed their governments' commitment to the establishment of natural parks and game reserves in Africa.

Specifically, concerning Kenya, a distinctive development during this period was the beginning of organised and institutionalised development and promotion of wildlife safari tourism involving both the public and private sector. The government, for instance, started to formulate and promulgate various legislation aimed at the protection of Kenya's unique wildlife resources, and promotion of organised recreational activities in protected wildlife parks and reserves (Kenya Wildlife Society 1957; Achiron and Wilkinson 1986; Akama 1998; 2003). Thus it was realised that the diverse arrays of African savanna wildlife had great potential for tourism development. In consequence, the government created pioneer national parks in Kenya including Nairobi in 1946, Amboseli in 1947, Tsavo in 1948 and Mt. Kenya in 1949. According to state legislation, the parks were to be protected public lands, 'set aside for the propagation, protection and preservation of objects of aesthetic, geological, prehistoric, historic, archaeological or scientific interest for the benefit and advantage of the general public' (Simon 1962; Lusigi 1978; Akama 1998).

It was around this particular period that various interest groups from the private and public sector coalesced to promote policies that were primarily aimed at the protection of the diverse array of African wildlife against any form of perceived threat. It was felt that in order to effectively protect the wildlife, the hunting, killing, or capturing of fauna must be prohibited (Simon 1962).

Thus the initiation of the pioneer wildlife conservation policies and tourism programmes in Kenya was aimed at protecting wildlife from the perceived destructive forces of humans. Lobby groups including wildlife conservationists, government officials and tourism developers felt that, for wildlife in the East Africa Protectorate to be adequately and effectively protected, wildlife conservation areas had to be established and boundaries demarcated which separate wildlife from development activities. Thus, the pioneer state wildlife policies and programmes to be promulgated in Kenya were aimed at protecting the savanna game from:

> (a) The skin hunters who seek and kill game solely for their skin, leaving carcasses for vultures.
> (b) Natives who cannot be made to understand the advantages of a closed season.
> (c) The wanton sportsmen who shoot females and who kill large numbers of males on the chance of securing a good specimen trophy.
>
> (Simon 1962)

It can be argued that, in part, these forms of wildlife management and tourism policies and programmes were a consequence of conservation and administrative officials' Western experience and environmental values. Due to rapid transformation of nature and disappearance of most wildlife in the West, particularly during the industrial revolution, the general perception among pioneer naturalists was that most human land use practices were incompatible with the principles of nature conservation in general, and wildlife protection in particular.

Moreover, the underlying concept among the pioneer conservationists, tourism developers and government officials was that indigenous resource use methods were destructive to wildlife and other natural resources. Officials were faced with unfamiliar natural resources and land utilisation methods, such as subsistence hunting, pastoralism and shifting cultivation, and they had difficulties in evaluating and understanding these resource use practices.

Consequently, most often the conservationists, tourism developers and government officials classified African modes of natural resource use as at best 'unprogressive' and at worst 'barbaric' and to be eliminated. These perceptions of African methods of natural resource use as retrogressive set in motion top-down government intervention policies and programmes to change African resource use strategies (Lusigi 1978; Akama 1998; 2003). When natural resource use problems such as wildlife destruction, deforestation and soil erosion were noticed by state officials and naturalists, the problems were simply defined as caused by irrational land use practices of rural African communities. However, resource degradation was primarily caused by state land use policies and programmes including alienation of land for European settlement, confinement of Africans in restricted native reserves and sedentation programmes which prevented pastoral communities from utilising diverse grazing ranges in different ecological zones (Anderson and Troop 1985; Blaike 1985, 1989).

It was with these environmental perceptions that the pioneer conservation and tourism development policies and programmes were initiated. Indigenous resource use methods were perceived as incompatible with the principles of wildlife conservation and tourism development. Thus when the state established the first national parks, not only was traditional subsistence hunting banned, but rural communities were prohibited from entering the parks and utilising resources such as pasture and fuel wood collection.

A worst case scenario of the impacts of state wildlife protection policies on Africans can be enunciated by what happened to the Walianguru community. With the establishment of Tsavo National park in 1948, the Walianguru mode of subsistence hunting was perceived by wildlife conservationists and tourism developers as incompatible with the principles of wildlife conservation and tourism development in Tsavo National Park (Akama 2003). While subsistence hunting was made illegal and came to be termed as 'poaching', sports hunting for pleasure, an entirely Western phenomenon of wildlife utilisation was permitted to continue in the parks.

In the 1950s, there was a rapid decline of the elephant population in Tsavo. The immediate response of the government officials and naturalists towards the problem of elephant population decline was to intensify anti-poaching measures against the Walianguru subsistence hunters. The poaching problem on the Tsavo plains that was mainly caused by Kamba, Giriama and European amateur and professional hunters, was defined as a 'Walianguru problem'. With intensification of anti-poaching campaigns by the colonial government, most Walianguru males (every male adult was a hunter) ended up in prison with hard labour. The Walianguru people as a culture nearly became extinct, much the same as what happened to the Ik in northern Uganda for much the same reason (Gomm 1974).

Post-colonial wildlife conservation and tourism policies

The colonial policies and programmes of wildlife conservation and tourism development have outlived the political structures that brought them into being in Kenya. When Kenya gained its independence in 1963, it inherited four national parks and six reserves from the colonial government. There are now over 13 national parks and 24 reserves that cover over 10 per cent of the country (Akama 1998). The national parks are exclusive state protected lands and are managed entirely for the protection of wildlife, whereas national reserves are created on any type of land, and usually, with the consent of the local authority (County Councils). The parks have become important centres of tourism attraction. As the case with most Third World countries, the conservation of wildlife and the development of wildlife-based tourism in Kenya is greatly influenced by Western cultural and environmental values. Most wildlife conservation and tourism projects in Kenya have been initiated with the assistance of conservation and development organisations that are based in the Western world.

Thus, Western conservationists and tourism developers still play a significant role in the conservation of Kenya's wildlife and the development of the country's tourism industry. A number of Western conservation organisations have established offices in Kenya which act as watchdogs and assist the government in wildlife conservation and tourism development. These organisations include the International Union for the Conservation of Nature (IUCN), the World Wildlife Fund, the Max Plank Institute and Frankfurt Zoo. These conservation organisations recognise the remaining high concentration of tropical wildlife in Kenya and other Third World countries as a 'world heritage' which should not be allowed to disappear but should be protected for the benefit of future generations.

The activities of such conservation and tourism lobby groups have been strengthened by the emergence of new forms of tourism that are centred upon the increasing concern of environmental harm that are attributed to traditional forms of tourism and are looking for ways to prevent or mitigate these negative environmental impacts. Similar to the conservation groups, the activities of the lobby groups advocating for new forms of tourism are based on the increasing desire to preserve the environment; areas of the so-called wilderness and virgin territories where nature can be experienced by 'discerning' new tourists (most of them from the First World or the West).

These new tourist lobby groups and wildlife conservation groups are currently spearheading the development and implementation of tourism and wildlife conservation programmes and policies that are centred upon sustainability. As Mowforth and Munt (1998) argue, Third World countries (e.g. Kenya) have become a major focus for environmental conservation and the development of sustainable tourism partly as a result of their spectacular environments, landscapes and the mega wildlife species such as elephants, lions, giraffe and rhinoceros. It is further argued that the concept of wilderness that is imbibed and promoted by the new strands of tourism currently represents ecological purity; areas that are free from human interference and development, and which are generally devoid of any form of consumptive human activity. This forms a powerful marketing and advertising tool that is repeatedly conjured up in various forms of media such as brochures and other forms of print and electronic media.

The wilderness conservation and tourism promotion goals of the lobby groups that promote new forms of tourism are framed and dominated by Western ethical and environmental values, and Western scientific philosophies. The Kenya government, as is the case with most Third World governments, follows international guidelines and philosophies of wildlife conservation and tourism promotion. As the country's wildlife conservation legislation states, the main objective of national parks and reserves is to preserve in a reasonably natural state examples of the main types of habitats which are found in Kenya for aesthetic, scientific and cultural purposes (Akama 1998).

As a consequence, wildlife and tourism policies in Kenya continues to emphasise law enforcement to protect the wildlife resources. The main focus of the state has been on the enactment of tougher conservation legislation, reorganisation of the wildlife conservation and tourism department, retraining of the personnel, the prevention of rural peasants and pastoralists from entering and utilising park resources, and the intensification of anti-poaching campaigns in the national parks. Thus, for instance, in 1976, after a re-examination of the deteriorating situation of wildlife resources, the government decided to amalgamate the functions and responsibilities of the Game Department and the National Park Service under a single government department – the Wildlife Conservation and Management Department (WCMD). In 1977, in an attempt to control the problem of poaching, which was widespread in the country's national parks and reserves, the government banned all forms of hunting. In the following year, through an act of parliament, the selling of all forms of wildlife products was banned (Kenya Government 1978).

However, the promulgation of legislation did not prevent further deterioration of the country's wildlife and tourism resources. In recent years, increased poaching activities, especially in the 1980s, have taken their toll on Kenya's wildlife population. In the 1980s for instance, elephants and rhinoceros were nearly brought to extinction by poaching. It was estimated that in 1973 Tsavo National Park (the largest wildlife park in the country) had an elephant population of over 38,000 animals. This was probably one of the largest concentrations of elephant herds in the world. But by 1989, the elephant population at Tsavo had been depleted to less than 5,000 animals. Country-wide, the number of elephants declined from 130,000 to 20,000 due to four decades of heavy poaching (*Daily Nation*, 15 May 1996). Elephants were particularly targeted by poachers due to existing high demand for ivory in the existing illicit international market. In the 1960s the rhinoceros population in Nairobi, Amboseli and Tsavo National Parks was estimated at 8,000 animals. Similar to elephants, there was great demand for rhino horn on the black market in different parts of the world. At present their number has been reduced to less than 500. However, it should be noted that, currently, the poaching problem in Kenya's national parks has been minimised mainly due to increased anti-poaching campaigns.

In subsequent years, it was argued that the WCMD was professionally, physically and institutionally incapable of executing its legal obligations (Kenya Wildlife Service 2004). There was further deterioration and breakdown of protected areas' infrastructure, and low morale and commitment among staff and lack of accountability. The main government response towards this situation was further re-organisation and restructuring the wildlife and tourism department. In this regard, the Kenya Wildlife Service (KWS) was established in 1990 under the Wildlife Conservation and Management (Amendment) Act of 1989. The principle goal of KWS was to protect the natural environments of Kenya and their fauna and flora for the benefit of present and future generations and as a world heritage.

In an attempt to introduce an element of consumptive utilisation of wildlife resources, in 1991 the KWS, through a new wildlife management policy framework, introduced a pilot wildlife cropping programme in selected wildlife ranches in areas adjacent to the major wildlife parks such as Nairobi, Amboseli, Samburu and Tsavo. The cropping of wild animals was mainly conducted through safari hunting where professional and amateur sport hunters were allowed to hunt the game in private ranches for a fee. In tandem with the new wildlife management strategy, selected private land owners were allocated cropping quotas by the KWS, depending on the density of wildlife population in their respective ranches. Hence, land owners were required to conduct regular census of the main herbivore species on their land and submit reports to the KWS. The reports were used to allocate the number of animals to be harvested, on each ranch, per annum. The game meat that was acquired through cropping was sold to tourist restaurants where it was used to prepare exotic cuisines. Thus, private ranch owners derived direct revenue through wildlife cropping initiatives and, as a consequence, they were expected to develop appreciation of wildlife as an economic resource.

However, this new policy framework was stopped in 2003 due to what was said to be difficulties encountered by the KWS in monitoring adherence to quotas, which led to the abuse of quota allocation. There was inaccurate submission of cropping data returns to the KWS by authorised croppers (Kenya Wildlife Service 2004). This included falsification of reports by private ranchers to show an upward trend of wildlife species that in most instances was not the case. There was also an increase of illegal trade in bush meat especially in areas adjacent to the parks. This led to further decline of wildlife numbers (especially antelopes, buffalo, elephant, zebra and giraffe) both inside national parks and adjacent areas.

Thus, it can be argued that although the government of Kenya is currently committed to the preservation of the country's wildlife resources, parks at present confront many problems including accelerated destruction of wildlife habitat, the continued decrease of wildlife species both inside and outside park boundaries, land use conflicts between the local people and wildlife, and the local people's suspicions about and hostilities towards the state policies and programmes of wildlife conservation and tourism development. The present policies and regulations on wildlife conservation and tourism development do not correspond with the socio-economic, cultural, political and ecological realities of the regions where the parks are situated. Most park managers are narrowly preoccupied with protecting park fauna instead of with conserving whole ecosystems of the parks and the surrounding areas as healthy, self-sustaining ecological units.

For instance, over 90 per cent of the park officials (game ranchers and wardens) interviewed in most Kenyan national parks, in the 1990s, indicated that their main work duties included the collection of gate fees from international tourists, providing security to visitors, patrolling the parks to control problem animals and against poachers (Akama 1998). However, none of the park officials mentioned duties outside the national parks (i.e. taking part in community wildlife conservation programmes and tourism projects, or having dialogue with the local people on matters related to park management and tourism development). Consequently, a social and ecological disequilibrium has developed between the national parks and surrounding environments which are experiencing rapid human population growth. The human populations have exceeded the carrying capacity of the land.

Local people's responses to wildlife conservation and tourism development

Rural communities surrounding protected wildlife areas have little or no influence on decision making or the institutions of wildlife conservation and tourism management. Their cultural and environmental values contrast dramatically with those held by conservation officials, tourism lobby groups and international tourists. Local people preoccupied by meeting their subsistence needs, confront poverty issues that are often compounded by destruction of their property by wildlife (Akama 2003). They therefore cannot afford to grant aesthetic value and the goals of long-term wildlife conservation and tourism development a high priority. As the following statement shows, the perceptions of the local people are at variance with those of conservation managers and tourism developers:

> You cannot interest a Maasai in seeing and photographing a giraffe any more than you can interest a New Yorker in a taxicab. Similarly, the restrictions of grazing and farming in an African park or reserve is as perplexing to natives as a law that prevents a New Yorker from living in and using ten square blocks of Manhattan would be.
>
> (Nash 1982: 344)

Thus, the socio-economic and cultural orientation of most Kenyans is quite different from those of international tourists. The social and environmental conditions which led to increasing public awareness and public support for wildlife conservation and the appreciation of wildlife's aesthetic value in Western countries are non-existent in most of rural Kenya. As is the case with most of the Third World, Kenya has not undergone massive urbanisation and industrialisation – the socio-economic processes that encouraged the creation of national parks as centres of wildlife conservation and the promotion of wildlife-based tourism in the Western world (Lusigi 1978; Nash 1982; Akama 1998). It has been estimated that over 80 per cent of Kenya's population reside in rural areas and earn a livelihood through subsistence agriculture (Kenya Wildlife Service 2004).

The small urban middle class that has evolved in Kenya has different socio-economic characteristics from those of the Western countries that spearheaded the national park movement and the appreciation of the aesthetic and ethical value of wildlife. Most of Kenya's middle class may have spent most of their childhood in rural environments and are likely to be of peasant parentage. Consequently, they have strong social ties to the rural areas which they perceive as their homelands and places of ancestral origin. In this social and environmental condition natural areas may not evoke images of an exotic environment full of possibilities for wildlife observation and adventure.

The concept of setting aside wildlife areas as protected parks may at best be inconceivable and at worst repulsive to rural African cultures. Also, the park concept still conjures images of the harsh colonial legacy of wildlife preservation (Lusigi 1978; Akama 2003). The rural peasants' negative perceptions and attitudes towards state-sponsored wildlife conservation and tourism programmes may also be accentuated by the fact that they receive very few direct conservation and tourism benefits. Unequal distribution of the costs and returns from wildlife management and tourism is perhaps the most important conservation and tourism development issue in Kenya (Akama 2003). While revenues from reserves are shared between local and national government, those from national parks go entirely to the national government and tour operators. And while the tourism industry achieves considerable profit, few financial resources are allocated for local development. Conservation benefits to households or community are uncertain and possibly non-existent. Most of the costs of wildlife conservation, such as property damage and the foregone opportunity of not using protected land for agricultural production, or game species for food, accrue almost exclusively to rural peasants.

The most extreme example of shifting the cost of wildlife conservation and tourism is that cultivators and pastoralists cannot protect themselves or their property from wildlife despite considerable injury and severe damage to farms and livestock (Akama 1998). State law prohibits any form of destruction and killing of wildlife. Consequently, peasants are reduced to guarding crops and livestock by making noise, beating drums and lighting night fires so that someone else may make a profit from tourists willing to view and photograph an animal that local opinion would wish dead. Hence, local people's attitude toward protected wildlife areas varies from that of indifference to intense hostility.

Conclusion

Kenya's wildlife conservation and tourism programmes have received international acclaim. Also an increasing number of tourists, especially from Europe and North America, are attracted to Kenya to participate in wildlife viewing and photographing. However, from the time of colonial rule the underlying socio-economic trend of wildlife conservation and tourism development has been the taking away of wildlife-resource-user rights from rural peasants. With the establishment of state-protected wildlife parks, the use of wildlife resources came to be controlled by the state, conservation organisations and tourism groups. In most cases, subsistence hunting by rural peasants came to be seen as poaching.

Kenya's goals and policies of wildlife conservation and tourism are still framed and dominated by Western environmental values and tourism development models that are opposed to the introduction of any form of consumptive use of wildlife resources. But, as the recent history of wildlife conservation and tourism development has shown, these forms of tourism and wildlife policies and programmes have led to increasing land-use conflicts, and accelerated destruction of wildlife habitats and a decrease in wildlife population. Moreover, most of the views and perceptions of the local people on wildlife conservation and tourism clash directly with those held by the state and tourism groups.

Kenya is unlike other African countries, especially South Africa, Zimbabwe and Namibia where various forms of consumptive uses of wildlife resources such as sport hunting, harvesting of selected species for bush meat, and formal and informal trade in various wildlife products is allowed. In these countries, over the years, formal trade in wildlife products has contributed substantially to foreign exchange earnings, whereas informal trade in bush meat has provided food security to rural communities.

Consequently, consumptive utilisation of wildlife has proved to be a viable conservation tool in many other parts of rural Africa.

It is important to note that wildlife practices, whether conservation or the promotion of safari tourism should be viewed in the context of existing social and economic conditions of the local people in areas adjacent to wildlife parks and reserves. In this regard, unlike in the developed world where conservation of wildlife can be justified in terms of aesthetic beauty and ethical values, in rural Africa where there is rampant poverty and social deprivation, wildlife practices should be justified in terms of their social and economic outcomes (i.e. assisting rural people to improve their livelihoods).

Consequently, policies and institutional mechanisms need to be put in place that encourage local participation in the design, implementation and management of wildlife conservation and tourism programmes. At very least, local communities need to be empowered to decide what forms of tourism programmes and wildlife management strategies they want to be initiated in their respective communities, and how the tourism costs and benefits are shared among different stakeholders. To achieve these changes will require the decentralisation of tourism and wildlife conservation authority and decision-making from the national level to legitimate and democratically elected regional and grass-roots institutions and organisations such as welfare societies, local church organisations, indigenous institutions and women's groups. Also, tourism should foster small-scale, locally controlled tourism projects which are sensitive to indigenous cultures and the local environment including the introduction of consumptive use of wildlife resources by allowing elements of subsistence hunting and sport-hunting activities in the parks and adjacent areas.

Perhaps more importantly, there is urgent need to reinstate the consumptive use of wildlife resources as part of the wildlife conservation strategy. At present, wildlife conservation policies emphasise the non-consumptive use of wildlife resources through the promotion of wildlife viewing and photographing by tourists. However, it appears that the overall economic value of wildlife resources can be tremendously enhanced by combining consumptive and non-consumptive use of these resources. This consumptive use can include sport hunting of wildlife by tourists, traditional subsistence hunting and harvesting of target non-endangered wildlife species, such as antelopes, buffalo and wildebeest to provide meat for local consumption and commercial industry.

In this regard, it is gratifying to note that, in recent years, there is increasing lobbying by various interest groups in Kenya to introduce elements of consumptive use of wildlife resources both inside and outside the wildlife parks and reserves. For instance, in a recent stakeholders' conference organised by the KWS in 2004, there was overwhelming support for the introduction of consumptive use of wildlife resources in wildlife parks and adjacent areas (i.e. dispersal zones and migration corridors). The stakeholders concluded that:

> considering the role of landowners and communities in wildlife conservation, it is necessary that commensurate economic benefits of wildlife resources should accrue to these guardians of wildlife. Sustainable utilization of wildlife resources (including hunting) should be considered an integral part of wildlife conservation and management.
>
> (Kenya Wildlife Service 2004: 7)

However, the resolutions that were arrived at in the conference did not yield much, since a new wildlife Bill aimed at reintroducing hunting in Kenya's wildlife parks and reserves was soon vetoed by the president of the republic in the same year.

References

Achiron, M. and Wilkinson, D. (1986) 'The last safari: will Africa's wilderness turn into a string of glorified game parks?', *Newsweek*, 32: 20–3.

Akama, J. S. (1998) 'The evolution of wildlife policies in Kenya, *Journal of Third World Studies*, 5(2): 103–17.

Akama, J. S. (2003) 'Wildlife conservation in Tsavo: an analysis of problems and policy alternatives', *Journal of East African Natural Resource Management*, 1: 1–15.

Anderson, D. (ed.) (1987) 'The scramble for Eden: past, present and future in Africa', *Conservation in Africa: People, Policies and Practice*, Grove, NY: Cambridge University Press.

Anderson, D. and Troop, D. (1985) 'Africans and agricultural production in colonial Kenya', *Journal of African History*, 26: 327–40.

Blaike, P. M. (1985) *The Political Economy of Soil Erosion in Developing Countries*, London: Longman.

Blaike, P. M. (1989) 'Environment and access to resources in Africa', *African Studies Review*, 59(10): 19–35.

Campbell, D. J. (1986) 'The prospect for desertification in Kajiado district, Kenya', *The Geographical Journal*, 152(1): 44–55.

Gomm, R. (1974) 'The Elephant Man', *Ecologist*, 4: 53–7.

Kenya Wildlife Society (1957) *Government Press*, Nairobi: Kenya.

Kenya Government (1978) *Government Policy on Wildlife Management: Sessional Paper No 3*, Nairobi: Government Printers.

Kenya Wildlife Service (2004) 'Stakeholder conference: wildlife utilization and management', Mombasa, Whitesands Hotel, 19–21 May 2004.

Lusigi, W. J. (1978) *Planning Human Activities on Protected Natural Ecosystem*, Ruggell: A. R. Gantner.

Mowforth, M. and Munt, R. (1998) *Tourism and Sustainability*, New York: Routledge Press.

Nash, R. (1982) *Wilderness and the American Mind*, London: Yale University Press.

Simon, N. (1962) *Between the Sunlight and the Thunder*, London: Collins.

6 Game estates and guided hunts

Two perspectives on the hunting of red deer

Guil Figgins

Introduction

This chapter grew out of an ongoing interest in the way that human geographers conceptualise nature–society relationships and draws on research (in progress) on the hunting of red deer in New Zealand and Scotland. The chapter is largely a descriptive account of how hunting the same animal has helped to shape social and physical processes in two quite different contexts. Following a brief overview of constructionist ideas about nature–society interactions within geography, this chapter will explore how the hunting of red deer has helped to shape the respective social and physical landscapes in Scotland and New Zealand.

The 'construction' of nature

A social constructionist conceptual framework is useful when focusing on recreational and touristic hunting. This framework enables an understanding of the constructed 'nature' of the nature–society relationships and challenges the common sense assumption that 'nature' is somehow outside of the social (Kong and Yeoh 1996; Olwig 1996; Gerber 1997). As Cronon observes in his examination of the term 'wilderness' in the context of the American West:

> It's as important to reflect on what we think about nature and about our complex physical relations with the natural world, as it is to reflect on those physical relations themselves. The nature we carry in our heads is as important as the nature that is all around us, because in fact the nature inside our heads is often the engine which drives our interactions with physical nature, transforming both ourselves and nature in the process.
>
> (Cronon 1996: 8)

Here the 'wilderness' is not outside the social, rather it is a social construct formed in a particular context through a discourse between people. Tracing the formation of this social construct enables us to identify the contemporary understandings of the place, value and meaning of non-human nature. As Cronon illustrates for the Amercian West, historically wilderness was seen and described as something that was threatening and needed to be subjugated. Whereas today, the idea of wilderness reflects the discourses of a modern American society where non-human nature is valued because it is fast disappearing (Cronon 1996). Cronon's example of 'wilderness', viewed within the social constructionist framework reveals physical environments, are mediated by humans and in this process can be transformed into reflections of society that can and do change over time.

Geographical research has also focused on the impact of socio-economic processes on the physical environment and how 'non-human nature' is subsumed economically. Marxist geographers have had a strong input into this field of inquiry, as Castree points out:

> Under capitalism humans relate to nature in a specific way, through commoditization of natural products, and in doing so actively appropriate, transform and creatively destroy it. The 'natural regions' of say, the mid western United States cannot be understood simply as pre-existent natural grasslands, as the traditional notion of 'first nature' would imply. Instead – and this is the point of Marxist ideas of social nature – they must be seen as *constructed natural environments* evolving out of decades of intensive, profit driven conversion into what they presently are.
>
> (Castree 1995: 19)

Thus, societies transform non-human nature physically through the socio-economic processes of the capitalist system. It is argued that the relationship between society and nature consists of society 'remaking' nature within the capitalist system for profit with this altered or created nature, in turn, affecting the way that human society develops. The most apparent examples of the use of non-human nature for commercial or industrial purposes (nature-based industries) are in the 'extractive' sectors, such as mining, fishing, sealing and whaling. However, it is in farming and forestry that the material *production* of nature is most apparent. Boyd *et al.* (2001) have proposed a framework to examine the 'industrialisation' of non-human nature within these two sectors, drawing from Marxist concepts of formal and real subsumption of labour. In the *formal subsumption of nature* capitalist firms use non-human nature as an external resource of material properties and bio-physical processes that can be used as inputs into production. In the *real subsumption of nature* biologically based industries are able to use bio-physical properties and processes as not only sources of profit but to also increase productivity. The primary difference between the two is in the distinction between natural and non-natural biological systems and the capacity to manipulate biological productivity. Under formal subsumption, 'capital is forced to circulate around nature' according to the environment and bio-physical processes, for example, in the commercial fishing industry and the need to deal with both seasonal movement of fish and unpredictable ocean conditions (Boyd *et al.* 2001: 563). On the other hand, with real subsumption, systematic increases in productivity can be achieved through the use and modification of internal factors, such as the manipulation of genetic material through traditional breeding programmes and the application of DNA bio-technology, and external factors, the application of pesticides, herbicides, fertilisers and environmental control.

As we shall see, both formal and real subsumption of nature are evident in the hunting-tourism destinations such as Scotland and New Zealand. This framework is potentially useful when considering the place of hunting and hunting tourism in any given society. In employing these conceptual tools, this chapter explores how hunting is socio-culturally, politically, economically and spatially situated within these destinations.

Red deer

Red deer are one of the most widely distributed deer species in the world. Their natural range spreads from northern Europe through to the eastern Mediterranean and into parts of central and northern Asia. This natural range, however, is fragmented because of local level extinctions. The *Cervidae* family consists of 12 subspecies spread throughout this natural range that can differ in appearance, size and behaviour through local food and habitat differences.

Red deer have also been released extensively into the wild throughout the world and are present in thriving populations in Morocco, the United States, Argentina, Chile, Peru, Australia and New Zealand. This ability to adapt to different environments outside their natural range gives an idea of their outstanding overall success as a species. Red deer are at least 20 million years old and its modern form, the genus *Cervus*, is recognisable from 12 million years ago. The largest land mammal in many parts of their range, including the British Isles, the success and survival of red deer has been closely interlinked with man. Humans removed predators, such as wolves and bears from their environment, but they have also used red deer as an extremely important resource of food (Lever 1985; Inskip, 2000).

'The Monarch of the Glen' – red deer in Scotland

Sir Edwin Henry Landseer's famous 1851 painting 'The Monarch of the Glen', depicting an impressive red deer stag (Figure 6.1), perhaps preparing to face a challenger, is the most well-known portrayal of the red deer in Scotland. It conveys the place that this animal has in Scotland; that of the largest mammal and magnificent wild animal that is an integral part of Scotland's natural ecosystem. Red deer have also, however, played an extremely important role in the socio-economic and political history of Scotland, particularly in the Highlands, shaping the physical environment and socio-economic landscape over time.

The Victorian era saw a rise in the popularity of the Highlands as a tourist destination, with sporting pastimes such as walking, shooting grouse and deer stalking the major attractions. By 1839, 28 'deer forests' (actually cleared open land) had been formed, mostly bought by the aristocracy or the new super-rich that had appeared with the industrial revolution. It became something of an extravagant

Figure 6.1 Sir Edwin Henry Landseer's famous 1851 painting 'The Monarch of the Glen', with red deer stag

fashion statement to own a deer forest where one could invite family, friends and business partners to hunt, emulating the traditional land-owning aristocracy.

The lease and subsequent purchase of Balmoral Estate in 1852 by the royal family led to a rapid increase in the popularity of deer forests and, by 1885, 104 had been formed (Orr 1982). By 1906 over 3.3 million acres of land in the Highlands and islands of Scotland were being used as massive hunting and fishing playgrounds for the elite of British society (Wightman 2004). This transformation in land use led to striking socio-economic changes. Land that had been part of a peasant farming economy was now transformed into a capitalist game estate economy controlled by an outside elite, whose primary objective was not to run a productive agricultural enterprise but recreation based around private access to hunting and the social status this achieved (Orr 1982; Jarvie *et a*l. 1997; Wightman *et al.* 2002). 'Balmorality' is a term first coined in the 1930s to describe the outcome of this process of socio-economic and cultural change. Older cultural icons of the Highlands, such as the wearing of tartans, were 'resurrected and developed quite consciously by landowning and social elites to form a new cultural genre' (Wightman 2004: 7). Today Scottish deer forests or the Highland sporting estate, are still one of the dominant land use forms in the Highlands and Islands of Scotland. There are an estimated 340 sporting and game estates, which cover some 5.2 million acres of land comprising 50 per cent of all privately owned land in the Highlands and 30 per cent of the total for Scotland (Wightman *et al.* 2002; Higgins *et al.* 2002).

Under Scottish law all deer are wild animals and belong to no one. Deer freely roam across property boundaries between game estates, public land and farms. However, property owners have the right to shoot deer on their property for recreation or control, and tenants on private property can do so if the deer are causing a problem through damage to forestry and crops. The duty that comes with this privilege is that private landowners are responsible for the welfare of deer on their property. From the 1960s to the late 1980s the red deer population in Scotland doubled in size from 150,000 to 300,000 and the most recent research into red deer numbers in 2003 showed their numbers still on the rise with an current estimated population of 400,000 (Hunt 2003). This rapid increase over the last 40 years has been attributed to mild weather, an increase in forest cover and, most importantly for the management of game estates, the under-culling of hinds (female deer). Red deer numbers are now high enough that they pose a significant threat to the environment and to themselves. It has become apparent that it is critical that the population is managed and this responsibility falls largely on the shoulders of private landowners, game estates being the most important (DCS 2000).

Under the Deer (Scotland) Act 1996 (DSA 1996) the Deer Commission of Scotland (DCS) was established as a non-departmental public body responsible with 'furthering the conservation, control and sustainable management of all species of wild deer in Scotland, and keeping under review all matters, including welfare, relating to wild deer' (Hunt 2003: 14). Their main responsibility is the sustainability, conservation and control of deer as part of the natural environment and heritage of Scotland. Under this charter it is the duty of the DCS to keep track of deer population and densities and how they effect the environment, agriculture and forestry and the interests of private owners of land occupied by deer such as game estates. The DSA 1996 sets out the DCS duties and also gives it a wide ranging set of powers to help it achieve its objectives. Its main tool is the power to apply measures set out in the Act, to reduce the deer population in any area where they are damaging agriculture, forestry, private property, themselves, or are causing a danger to public safety – normally through increased risk of traffic accidents involving deer.

Most game estates in Scotland belong to voluntary Deer Management Groups (DMG) that coordinate deer management based around a certain geographical area. There are more than 60 DMG in Scotland that are responsible for areas that range in size from 5,000 to 500,000 acres (DMG Newsletter One n.d.). The DMG groups comprise from three to as many as 30 properties, and are often based around a local sub-population of deer. DMG make annual counts of deer, agree to culling levels and share information to try and prevent and minimise damage done by deer. The DCS assists DMG by exercising regulatory functions such as entering into deer control agreements with property owners stating cull areas and numbers, authorising culls in certain situations, such as out of season or at night, and coordinating deer counts from DMG to help keep track of populations. The DCS also acts as a research body and actively promotes best practice methods for deer management on game estates.

Three main forms of culling are employed: sport, deer that are shot by game estate owners or their clients; protection, deer that are shot to protect agriculture and forestry; and control and management, the removal of sick or injured animals and reduction in overall population. Culling is normally carried out by stalking and shooting on foot so it is a time-consuming and costly business. On game estates, historically, the number of hinds culled yearly has been lower than the population increase. This has been because traditional game estate management has been focused on large mature stags and their capital and recreational value. High numbers of hinds was also thought to produce better numbers of stags, although this high-density form of deer management is now giving way on many estates with the introduction of widespread use of DMG, to more intensive culling to insure better habitat and healthier deer.

Fencing is also used in combination with culling as management strategy but has also been used in the past as an exclusive alternative. Fences are used on Scottish game estates to allow different land users to operate in close proximity. Protection of the now rare native Caledonian pine forest habitat from overgrazing by deer is a high priority, as is preventing them from getting onto high-risk public roads. Fences, however, are viewed by the DCS and other environmental management bodies, as only to be used where the full range of other measures for the management of deer have been examined and the negative impacts of fencing have been minimised. Negative impacts include effects on other land users, the disruption to traditional deer range and seasonal routes of travel, danger to other wildlife such as grouse and pheasants and loss of landscape and cultural values. There has been a move away from the use of fences in public policy and the use of higher culls as an alternative, particularly when the protection of important habitat and potential loss of rare bird life is concerned. Game estates, often operating with a fine bottom line, have often turned to fencing, because it was heavily subsidised by the Scottish Forestry Commission and culling is a very expensive business. This debate is an example of wider discussions currently underway in Scotland about rural land use (SNH 2004; DCS 2004a).

Red deer are a valued part of Scotland's past and present social and natural heritage and have to a large extent helped create the physical and social landscapes of the Highlands today. Their use as a private recreational hunting resource has had a major impact on the way that the Highlands have been transformed physically and developed socio-economically over time. I will now turn to the case of New Zealand. Focusing on the same animal, I will illustrate how hunting has been the single most defining feature effecting how red deer have helped to create some of New Zealand's physical and social landscapes.

Red deer in New Zealand – from pest to resource and back again

In the 100 years since red deer were introduced to New Zealand, they have been protected by law, hunted for trophies, culled in their thousands, become a valued wild resource and turned into a profitable farming industry. Now at the beginning of the twenty-first century the game estate industry is establishing itself and tourism hunting has an increasing importance as an alternative income source in the rural economy.

Within the global context of hunting, New Zealand is perhaps unique because of its ecological history. With no indigenous large mammals, the only native species being a tiny bat and seals living in the coastal zone, birds, insects and fish formed the basis of animal life within New Zealand's pre-human ecosystems. Dogs brought by the Maori, and later, pigs, deer, sheep, cattle and goats introduced with European settlement and trade in early nineteenth century, transformed this ecosystem dramatically. European settlers introduced as many species from 'home' as possible in an attempt to create a new landscape mirroring the one they had left behind. Hundreds of animal and plants species, including blackbirds, starlings, sparrows, cats, rabbits, hares, hedgehogs, trout and salmon were introduced in an attempt to 'civilise' the landscape and provide food and sport for the settlers, and also in a attempt to bring stability to what was seen by some (within the ecological thinking of the day) as a ecosystem out of balance (McDowell 1995).

The first red deer introductions took place in the South Island at Nelson in 1854. More deer were released into the area in 1861, and from 1863 deer were also released throughout the North Island. 1871 saw a major introduction with 17 deer from Invermark in Scotland being released in the province of Otago in the South Island. These deer differed from previous releases of farm bred stock from the United Kingdom as they were wild Highland deer. This herd came to be known by the early twentieth-century hunters as New Zealand's premier trophy herd. By 1883 the population had increased to the point that deer were being transplanted internally to different regions of the country to augment the importation of deer from the United Kingdom. Between 1854 and 1926 there were approximately 180 red deer releases around New Zealand by wealthy individuals or the local acclimatisation societies set up in the 1860s to promote the introduction of desirable animal and plant species.

The red deer herd was protected until 1870 with a New Zealand wide hunting ban, and after this date by strict local seasons, with an open season of only one month. By the end of the century deer hunting was open to all who joined an acclimatisation society and bought an annual licence.

Although deer were introduced to be hunted by all New Zealanders, partly as a reaction to the control of deer hunting by the wealthy and elite in the United Kingdom, it was wealthy locals and tourists who did most of the early hunting of red deer in New Zealand. The New Zealand government was quick to see the income potential of attracting well-to-do deerstalkers from the United Kingdom to hunt the thriving wild deer population and good trophy stags. By the early twentieth century there were top trophies coming out of New Zealand that rivaled or bettered anything produced back in the United Kingdom. The first three decades of the twentieth century are seen by many hunters as the heyday of recreational deerstalking in New Zealand with many world record trophies attained (Donne 1924; Logan and Harris 1967; McDowell 1995).

Altogether approximately 820 red deer were released over the 70-year period of deer introductions, encouraged by the government, particularly tourism departments. By 1919 there were an estimated 300,000 red deer and their numbers were increasing by 25 per cent a year (Yerex 2001). The deer population built up rapidly due to a lack of competition, predation and hunting pressure and soon damage to the environment and agricultural production from overgrazing by deer was outweighing any income gained from tourism. There was also a transformation in the way that New Zealanders viewed deer as more became aware of their country's unique native ecosystems and started to perceive themselves as New Zealanders rather than immigrants and colonists from the United Kingdom. New Zealand's natural history became something to be proud of and protect rather than something to alter and transform. Deerstalking was also perceived as a pastime of the rich and it did not gain enough popular support to attract the numbers of hunters needed to control population size or to form a strong pro-deer lobby.

1930 saw all protection removed from deer and from this date onwards there has been a concerted effort by various government departments to either eradicate completely or find some kind of method to control red deer in New Zealand. Between 1930 and 1956 approximately 670,000 red deer were killed by teams of government cullers from the deer control section of the Department of Internal Affairs (Yerex 2001). After 1956 this role was taken over by the New Zealand Forestry Service and the Noxious Animals Act (1956) defined red deer as a pest species to be destroyed where possible. Deer control policy changed, however, from outright extermination to a more scientific approach concentrating on areas that were deemed critical because of deer density or natural value. It was realised that the total eradication of deer was an impossible task so forest recovery involving the reduction of deer populations to a more sustainable level was seen as the way forward and deer were eventually reclassified as 'wild animals' under the Wild Animal Control Act (WACA 1977).

The commercial value of red deer had risen over time and by the late 1950s there was a strong market for skin and velvet, followed by a boom in the wild venison trade in the 1960s. The use of planes and later helicopters to retrieve deer carcasses from the ground shooters later gave way to the use of helicopters as shooting platforms. Helicopter 'gunship' teams comprising pilot and shooter armed with a high-powered semi -automatic rifle roamed riverbeds and the open 'tops' of mountains. The deer would be shot from the air; the helicopter would then land to pick up the carcasses or hover while they were slung underneath and then transport them back to the processing plant. The high returns available with the use of such technology meant that deer numbers were reduced significantly in some areas and commercial hunting took over as a *de facto* control method from less effective and uneconomical government driven ground shooting efforts.

In the 1970s it was realised that the farming of deer would eventually yield greater income in the face of declining deer numbers and the commercial hunting of deer turned to live capture to stock these new enterprises (Caughley 1983). By 2005 there were 2.2 million red deer on approximately 5,078 farms throughout New Zealand exporting venison and velvet to an international market worth approximately US$170 million a year to the New Zealand economy (Yerex 2001). Today wild deer occupy 44 per cent of the landmass and almost every major wilderness area and mountain chain in New Zealand (DOC 2005).

The early period of deer culling and venison recovery before the hunting process was mechanised through the use of transport technology such as helicopters, gave rise to one of New Zealand's strongest cultural stereotypes, the 'Good keen man'. 'Good keen men' went into the wilderness to earn a living culling deer, rejecting urban life and having to learn the skills necessary to survive in the bush. Today, this socio-cultural history of 'getting away from it' is reflected in the popularity of recreational hunting for red deer in New Zealand. From its upper-class beginnings in New Zealand, hunting became a popular 'everyman's' sport spawning a Deerstalkers Association (DSA), a lobby group for recreational hunting interests. There is open and free access all year round on public land administered by DOC, and the approximate 45,000 recreational hunters are seen as an important part of managing the estimated 250,000 wild deer (DOC 2005). Commercial hunting ventures are also given access to public land as helicopter shooting has proven the most efficient way of controlling deer numbers.

The WACA 1977 gives the Forestry's Service successor, the Department of Conservation (DOC), the legislative framework for the management of wild animals such as red deer in New Zealand. The WACA 1977 provides for:

> The control of wild animals generally, and for their eradication locally where necessary and practicable and for coordination of commercial and recreational hunters to ensure concerted action against the damaging effects of wild animals on vegetation, soils, water and wildlife.
>
> (DOC 1997: 23)

All red deer in New Zealand, whether they are in the wild, on a farm, or on a game estate are classified under this legislation as wild animals and DOC has the overall responsibility for their management. The main concern for DOC is that deer will escape captivity and increase the feral range and population of deer. DOC can prohibit certain geographical areas from having deer in captivity, can state where certain species are allowed to be held in captivity, and most importantly for game estates, provide regulations for perimeter fencing on game estates and farms.

Game estates started to appear in the early to mid-1980s by farmers introducing deer for hunting into marginal farming areas and through charging hunters for access to wild deer on their property. As a recognised pest species anyone can shoot deer on public and private land as long as they have the landowner's permission, whether it is the Crown, farmer or game estate owner. By the mid-1990s more up-market game estates were appearing that catered for international tourists mainly from the United States and Western Europe concentrating on hunting for top-class trophies (Yerex 2001). The New Zealand Association of Game Estates (NZAGE) emerged in 1997 as a body to provide support for the fledgling industries expansion. Faced with criticism, especially from recreational hunters because of perceived threats to free public access to hunting that commercialisation would bring and the view that it is not 'real' hunting, the association operates its own rigorous industry standards that its members must meet to become accredited. They currently have 16 member game estates, and there are another 10 properties that meet the requirements or are in the process of going through the process of accreditation (NZAGE pers. com. 2005). There are also many other game estates operating outside this framework, but no reliable figure exists for these unregulated establishments, but with the popularity of tourist hunting increasing, it may be significant.

The collective turnover in the industry is thought to be directly worth approximately USS10 million a year, although this figure does not take into account spending on accommodation, food and transport which would increase this figure substantially (NZAGE pers. com. 2005). To achieve consistent good quality heads for trophy hunters and a higher profit for the operator, stock on game estates are often sourced through the deer farming industry. These animals are bred for velvet harvest and so antler growth and mass are desirable qualities that are directly transferable into the game estate hunting industry as trophies. With the introduction of imported genetic material to increase the antler yield, New Zealand trophy red deer are now amongst the most sought after in the world (Yerex 2001). The real subsumption of nature via genetic manipulation now has a place in the New Zealand hunting tourism industry.

Tourist hunts for these trophy animals are undertaken on the property with the use of a guide who pinpoints the location of animals that are to be taken and then supervises the hunt. A quick search on the internet of web sites offering trophy hunting in New Zealand demonstrates the monetary value of this industry with prices reaching up to and sometimes beyond US$10,000 for a top trophy.

Discussion and conclusion

As the two different contextual settings have illustrated, the hunting of red deer affords an interesting example of how the hunting of game animals has shaped and transformed socio-economic, cultural and environmental characteristics of two quite different contexts. They also present a strong descriptive case study for the examination of nature–society relationships through the lens of a constructionist perspective. The same animal exists in two quite different social and physical settings, and the socio-economic and environmental transformations that hunting have produced in these case studies provide an excellent example of how societies 'construct' nature through social and material processes over time.

In the first case, red deer are an important naturalised part of Scotland's social and natural heritage. After years of mismanagement, they have been protected from overpopulation, and the damage that this causes to the wider environment, through the formation of regulatory bodies, such as the DCS and DMG, who undertake culling based on consultation with stakeholders and scientific methodology. On the other hand, red deer exist in New Zealand only through cultural and socio-economic processes that have transplanted them to an environment outside their natural range to which they have adapted into extremely successfully. This success has brought them important economic value as a resource within human socio-economic structures, such as farming, but has also left them to be classified as pests, and to be eradicated because of the danger that they pose to an environment that has evolved without ungulate browsing.

The hunting of red deer in the Highlands has traditionally been associated with the upper classes and the wealthy and their pursuit of private recreational hunting. The acquisition of deer forests physically transformed the Highlands, and led to the development of distinctive cultural traditions now associated with the hunt. This creation of both physical and social landscapes created through the hunting of red deer has also ensured that the wild deer became the external resource for an emerging industry in Scotland.

In comparison, in the New Zealand context, deer were originally 'harvested' from the wild and are now being used as the reproductive base for further use as hunting resources. Deer farmed for velvet and meat production are directly used as sources of trophy animals to be used on game estates. These deer, bred for meat, velvet harvest and ultimately trophy quality, are excellent examples of the 'real subsumption of nature' in use as an economic process and the material production of nature in general.

The increased commercialisation and economic value placed on recreational hunting has led to some tensions within the hunting community. Traditionally seen as a free 'resource' for all New Zealanders, game estates and guided tourist hunting are viewed by many recreational hunters as a direct threat to this status quo through direct economic value being placed on a 'wild' animal.

Red deer in both case studies exist within a multiple-layered intermeshed network that combines non-human natures (deer and the physical landscape) with human socio-economically created attributes, such as regulatory bodies, fencing, deer breeding, and hunting cultures and identities. The complex processes and broad structures of hunting in these two contexts provide an illustration of the way that hunting can provide an extremely useful field of enquiry when examining questions relating to the way that humans interact with, represent, and even create non-human nature.

References

Boyd, W., Prudham, W.S. and Schurman, R. (2001) 'Industrial dynamics and the problem of nature', *Society and Natural Resources*, 14(7): 555–70.

Castree, N. (2001) 'Marxism, capitalism and the production of nature', in N. Castree. and B. Braun (eds) *Social Nature*, Oxford: Blackwell, pp. 189–207.

Castree, N. (2005) *Nature: Key Ideas in Geography*, London: Routledge.

Caughley, G. (1983) *The Deer Wars: The Story of Deer in New Zealand*, Auckland: Heinemann.

Cronon, W. (1996) 'The trouble with wilderness; or, getting back to the wrong nature', *Environmental History*, 1: 7–55.

Deer Commission for Scotland (DCS) (2000) *Wild Deer in Scotland: A Long Term Vision*, Edinburgh: DCS.

Deer Commission for Scotland (DCS) (2004a) *Joint Policy Statement and Guidance on Deer Fencing*, Deer Commission for Scotland, Forestry Commission, Scottish Natural Heritage, Scottish Executive.

Deer Commission for Scotland (DCS) (2004b) *Annual Report*, Edinburgh: DCS.

Department of Conservation (DOC) (1997) *Issues and Options for Managing the Impacts of Deer on Native Forests and other Ecosystems*, Wellington: Department of Conservation.

Department of Conservation (DOC) (2005) *Update of the Specifications Governing the Keeping of Deer in Captivity in New Zealand for Deer Farming and for Safari Parks/Game Estates*, Wellington: Department of Conservation.

Donne, T.E (1924) *Red Deer Stalking in New Zealand*, Auckland: Halcyon Press

Gerber, J. (1997) 'Beyond dualism – the social construction of nature and the natural and social construction of human beings', *Progress in Human Geography*, 21(1): 1–17.

Higgins, P., Wightman, A. and McMillan, D. (2002) *Sporting Estates and Recreational Landuse in the Highlands and Islands of Scotland*, Economic and Social Research Council Report R000223163.

Hunt, J.F. (2003) *Impacts of Wild Deer in Scotland – How Fares the Public Interest?* Aberfeldy: WWF Scotland.

Inskipp, C. (2000) *Red Deer in Scotland*, N.p.: WWF and Scottish Natural Heritage, Queens Printer for Scotland.

Jarvie, G., Jackson, L. and Higgins, P. (1997) 'Scottish affairs, sporting estates and the aristocracy', *Scottish Affairs*, 19: 121–40.

Kong, L. and Yeoh, B.S.A (1996) 'Social constructions of nature in urban Singapore', *Southeast Asian Studies*, 34: 402–23.

Lever, C. (1985) *Naturalized Mammals of the World*, London: Longman.

Logan, P.C. and Harris, L.H. (1967) *Introduction and Establishment of Red Deer in New Zealand*, Wellington: New Zealand Forest Service.

McDowell, R.M. (1995) *Gamekeepers for the Nation*, Christchurch: Canterbury University Press.

NZAGE (2005) Representative interview, 14 September 2005.

Olwig, K. (1996) 'Nature, mapping the ghostly traces of a concept', in Earle, K., M. Mathewson and M. Kenzer (eds) *Concepts in Human Geography*, Lanham, MD: Rowman and Littlefield.

Orr, W. (1982) *Deer Forests, Landlords and Crofters*, Edinburgh: John Donald

Scottish Natural Heritage leaflet (2004) *Management: To Fence or Not to Fence*, Edinburgh: SNH.

Wightman, A. (2004) 'Hunting and hegemony in the highlands of Scotland: a study in the ideology of landscapes and landownership', *Noragric Working Paper No. 36.*

Wightman, A., Higgins, P., Jarvie, G. and Nicol, R. (2002) 'The cultural politics of hunting: sporting estates and recreational land use in the Islands and Highlands of Scotland', *Culture, Sport, Society*, 5(1): 53–70.

Yerex, D. (2001) *Deer: The New Zealand Story*, Christchurch: Canterbury University Press.

7 Shooting tigers as leisure in colonial India

Kevin Hannam

Introduction

Recent research has suggested that the dualistic oppositions between people and animals have to be transcended before a more sophisticated understanding of society can be reached (Wolch and Emel 1995). Within this framework the re-conceptualisation of animals in their own right has led to studies that demonstrate the importance of animals to human activities. In particular, both popular and scientific anthropomorphic representations are deconstructed as part of the investigation into the, 'continuing struggle between differentially empowered groups to define and represent the "true" meanings and values of wildlife and habitats' (Burgess 1993: 52). In this chapter, I seek to extend this recent area of research by focusing on the importance of tiger hunting in India as a leisure pursuit for the reproduction and maintenance of the British colonial State.

While hunting and killing tigers had been a sport of the earlier Mughal rulers of India, with the onset of colonialism it became a much more widespread leisure/tourism pursuit for the colonial elite (Pandian 2001). The blurring of the divide between leisure and tourism practices has recently been acknowledged (Crouch 1999) and although shooting tigers was primarily a domestic leisure or tourism practice, many Western visitors came to India to visit friends and relatives and engaged in the 'sport' of tiger shooting as part and parcel or their tour.

We need to note, however, that hunting and killing tigers as a leisure/tourism pursuit has been replaced by hunting and photographing tigers as a leisure/tourism pursuit even since the 1920s (Champion 1927). India currently has over 60 National Parks and over 300 wildlife sanctuaries. A specific feature of the management of the National Parks of India is the status accorded to one species, namely the tiger. This is enshrined in the Project Tiger scheme that was launched in 1973 on the recommendation of a special task force of the Indian Board of Wildlife. The main objective is to first ensure the maintenance of a viable population of tigers in India for scientific, economic, aesthetic, cultural and ecological values, i.e. for conservation. The second objective is to preserve, *for all times*, the areas of such biological importance as a national heritage for the benefit, education and enjoyment of the people, i.e. for tourism and recreation. In practice, however, the latter objective is clearly subordinated to the former. Whilst leading to the prohibition of hunting, Project Tiger also led to the prohibition of habitation and most productive activities in national parks designated as tiger sanctuaries. This has also led to conflicts with the local populations (Young *et al.* 2001; Hannam 2005). Over 20 Indian National Parks are additionally designated as tiger reserves. The designation of a National Park as a tiger reserve adds to its significance for potential tourists, but for the park management this is often seen as an added problem rather than an opportunity. However, the rules and regulations governing tourism in Indian national parks are on the whole already much stricter than those in parks elsewhere in the world: tourist behaviour in Indian national parks is highly regulated, with strict rules in place to protect wildlife (Hannam 2005). This is in marked contrast, as we shall see, to the colonial period when hunting and killing tigers for pleasure was the norm of the colonial elite.

Figure 7.1 Copy of frontispiece of 'The Book of the Tiger' by R.G Burton (1933) (London: Hutchinson)

Shikar

After the Indian Mutiny in 1857 India became part of the British Empire – the *Jewel in the Crown*, and was run by a relatively small number of highly trained civil servants, military officers, police officers and forest officers (Cohn 1983; 1987). These men wielded considerable individual power and would commonly spend their whole lives in India. As such they generated a particular way of life that outside office hours centred primarily on *shikar* (the Hindi term for hunting), in particular shooting tigers. As Mackenzie (1988: 180) has argued: 'The British and the tiger seemed in some ways to be locked in conflict for command of the Indian environment.' Indeed a vast number of memoirs were published on the subject of *shikar* from the early 1800s, right up until Indian independence.

Titles such as Silver Hackle's (1929) *Indian Jungle Lore and the Rifle*, Sanderson's (1878) *Thirteen Years among the Wild Beasts of India*, Stebbing's (1911) *Jungle By-ways in India*, Aitken's (1897) *A Naturalist on the Prowl or in the Jungle* and Glasfurd's (1928) *Musings of an Old Shikari* tend to give the impression that hunting was merely an escape from the routine pressures of administration. However, in this chapter I wish to argue that hunting in India was also essential to the reproduction of the British colonial State in India. More specifically I wish to argue that hunting tigers was emblematic of the exercise of colonial State power and reinforced both the claim to rule and the aura of British invincibility – the sense that the British colonial State was so powerful that it was useless to oppose it. In addition, through shooting tigers the British also enhanced the sense of the benevolent State: getting rid of the man-eating tiger fed into notion that the British were modernising and taming nature for the benefit of the Indian population (see Corbett 1944).

Moreover, whilst vast numbers of other animals were slaughtered for sport in India, shooting a tiger stands out in the memoirs as the aim of each and every sportsman, and was a prerequisite for becoming a fully accomplished man. Indeed, the individual huntsman, who was often in the jungle for a couple of months was often seen as having undergone a *rite de passage* – a series of transformations towards a 'wild' or 'savage' state, transgressing (and thus reinforcing) conventional social standards along the way (see Hell 1996). The spilling of blood was thus not regarded as a banal act. Indeed, relations between hunters and tigers followed a logic of institutionalised violence wherein the hunter appeared as the archetype of the fully accomplished hero and the animal itself became considered as an actor with considerable physical – particularly in the case of the 'man-eater' – and symbolic power. Indeed, the tiger was hunted because of its very potential to reverse the conventional power relations between the British elite and animals.

Thus, on the one hand, tiger *shikar* symbolised the right of the British officer and colleagues to take life or let live, in short to exercise, to a calculated degree, despotic power, the demonstrative exercise of power in itself. On the other hand, many of the methods of tiger hunting were informed by more subtle forms of disciplinary power (see Mann 1984). Through hunting, the forest officer, in particular, could claim sole responsibility for the management of the animal population and by default gain exclusive access and control of a large territorial area. Indeed, in the early nineteenth century: 'Game was everywhere plentiful and there was little limit or restriction imposed on what they shot, or where. In fact it was a matter of government policy to clear whole areas of game to open up fresh tracts for cultivation' (Elliott 1973: 23). However, by the late nineteenth century, who shot tigers, and for what reason, came to be carefully controlled by the State, not necessarily for altruistic reasons of conservation, but as a means to control the more remote rural areas of the Indian subcontinent. Governing India involved controlling the cultural and territorial distance between those who governed and those who were governed, and the tiger, as a symbol of both power and fear, came to occupy and inhabit the ambivalent realm in between. Moreover, shooting was a 'performative practice' which helped the British elite to also justify their rule to themselves; it boosted their own internal confidence (see Edensor 1998). As Osborne (1994) has noted, in order to secure rule the British had to develop a particular ethical competence. I shall demonstrate how the shooting of tigers fed into and helped to legitimise this ethic. It is thus this nexus between the killing of tigers, a certain type of masculinity and the ethical power of the colonial state that I wish to investigate.

Loneliness, danger and trophies

The reasons for hunting tigers given by members of the British ruling elite and their visitors in India were many and varied, but feelings of loneliness, danger and the desire for trophies were ostensibly important. Many expatriates claimed that hunting offered a kind of existential escape from the boredom and loneliness of everyday life. The forest officer, Best, for example, argued that:

> Throughout my service I killed a great many [tigers] and I was mad keen on their hunting, studying the phases of the moon in anticipation of hunting them at night, and I looked forward to each camping site as a possible place of going in their pursuit. All of which sounds bloodthirsty – which it was. My excuse being that tiger *shikar* was at certain seasons of the year my sole recreation in a very lonely existence.
>
> (Best 1935: 161–2)

Many other hunters argued that they hunted not out of boredom but because of the added danger involved in hunting tigers. This emphasis on danger fed into the late Victorian notion of manliness. Seeking out danger proved you were in some ways 'a real man'. Indeed, only the tiger was seen as providing an equal match and worthy adversary. Best (1931: 19) noted that 'For interest and danger, as well as for the trophy which may be secured, the hunting of the tiger is perhaps the most popular form of sport in India. All good men and true, hope when they come to the East, to take home at least one tiger skin.' It could be argued that the lure of tiger hunting was an essential element of the attractiveness of India as a destination for recruits to the public service of India: Glasfurd (1928) believed if the shooting of big-game became restricted then the 'most desirable type' of recruits would no longer be attracted.

Indeed, apart from the danger of hunting tigers, many members of the British elite were keen on hunting for the decorative trophies which could be collected and displayed often, it was claimed, for the enhancement of scientific knowledge. 'Silver Hackle' (1929: 67), like many others, argued that: 'A good sportsman does not kill for the mere pleasure of killing, and will always like to have something tangible to remind him of and make him live over again the hours spent in the pursuit of the different animals he has succeeded in bringing to bag, be they hours of which he has fond memories or the reverse.' In the same vein, Powell also noted that:

> The actual shooting of a tiger may sometimes be a comparatively unthrilling and even a regrettable ending to the chase. It is the associations with the animal that either make or mar its memory. ... It is the trouble one has taken, the energy expended, the jungle-craft employed, and the risks that make a tiger skin a treasured trophy. The actual shooting of the animal does not matter a hoot.
>
> (Powell 1957: 8)

Such trophy-ism, of course, can be viewed in the wider context of the Victorian predilection for collecting and displaying objects for scientific purposes and the birth of museums.

Masculinity and Britishness

However, beyond the rather glib assertions that hunting was either simply something to do to relieve the boredom, or to gain a decorative trophy, many other British hunters centred their arguments about their reasons for hunting around constructions of an aggressive masculine sense of British identity, which, conversely of course, excluded other competing definitions of both masculinity and nationhood. This late Victorian manliness emphasised the ideal of a virile, muscular and patriotic sense of endurance above all (see Mangan and Walvin 1987; Phillips 1997).

First, tiger shooting was seen as a quintessentially British thing to do:

> It has been said, that hunting instincts more or less pervade all nature, but Anglo-Saxons are the only true sportsmen in the world: and, in the case of English gentlemen, there is no doubt but that instinct and habit, alternatively cause and effect, do much in producing that activity, and energy of mind and body, that promptitude in danger, and passion for fair play which they carry with them wherever they wander.
>
> (Dunlop 1860: 2)

Hunting was thus seen as an instinctual part of the British national character. A character which was, *ceteris paribus*, caught up in an ethical tradition of British fair play and sportsmanship. Brown (1887: 278) similarly noted that: 'It has often been a matter of surprise to me that British sportsmen, with the means to gratify the love of sport that is inherent in most Englishmen, do not oftener go in for a *shikar* trip to the sunny land of the East.' Brown argued at length that hunting should be pursued in India both because it was much cheaper than in other parts of the British Empire and because servants, transport and medical aid were more easily procured in India. However, interestingly he concluded that hunting in India was much better than elsewhere because the British sportsman in India knows 'that, as one of dominant race, his wishes will be more likely to be forwarded by the native inhabitants, than they would be in any other part of the world' (Brown 1887: 278).

Whilst hunting, particularly of tigers, was seen as a natural, ethical, and indeed instinctual, British activity, on the other hand, it was also seen as a healthy and above all manly pursuit which would provide excellent training for the British subaltern. To perform the physical activity involved in hunting proved a man was capable of higher things; it was a test of muscularity, manliness and morality. This virile masculinity meant, of course, that women were generally not allowed to go hunting as it was deemed too dangerous but also because a hunting expedition was viewed as a good chance for men to bond together. However, this all male preserve was challenged by some female hunters – notably wives of forest officers (see, for example, Gardner 1895; Savory 1900; Baillie 1921; Smythies 1953).

However, the male hunter had to be, in particular, healthy, hence, Lieutenant Colonel Wood (1934: 8) noted that: 'One must be very fit to undertake serious big-game hunting. My advice to attain this is to avoid alcohol and over-eating; to lead a regular life, and acquire mental occupation and a good conscience. Avoid late nights, big dinners, meets, and stuffy clubrooms.' In the preface to his book, Stewart (1927: ix) meanwhile, felt certain that in peace time, there was '… no finer training than jungle shooting …'.

Such sentiments, though, linked the notion of manliness to that peculiar British ideal – the sense of fair play. As with any type of hunting there was a strict protocol attached to tiger hunting. For example, was it fair to sit up a tree and wait for a tiger, was it fair to shoot an animal at a water hole, was it fair to hunt with lights at night, was it fair to use a 12-bore shotgun? These questions were essential to the British ethic of fair play and correct form and separated the sportsman and gentleman from the lower classes, the rulers from the ruled. Even Indian princes or Maharajas were perceived as lacking in this notion of fair play, despite them being allowed to take part in hunting. Indeed, they were often deemed to have stepped too far and having gone on an orgy of killing rather than of 'fair' hunting. Mere 'natives' who hunted, meanwhile, were roundly condemned for poisoning and capturing tigers in nets and pits and then spearing them to death. These traditional Indian methods of hunting were generally deemed unsporting and cruel. The British, meanwhile, deployed two main methods of tiger hunting, each with a rather different sense of etiquette. As we shall see, there was considerable controversy over which was the best, or more masculine, method of hunting tigers. Moreover, each represented different elements of State power.

Beating versus tying up

The first of these methods involved beating a tiger towards a line of sportsmen sat on the back of elephants, one of whom would shoot the tiger as it attempted to escape. This communal and highly organised method of hunting was particularly expensive. Indeed, the forest officer, Stebbing, acknowledged that it was a genuinely elite form of sport:

> … it is the sport of kings and princes, bejeweled rajas, … Viceroys, and Lieutenant-Governors, the deputies of kings, and such minor fry as commissioners, moneyed globe-trotters, and suchlike. … to enjoy this form of sport in its pristine excellence not only requires a long purse, but added thereto, more than a nodding acquaintance with the great powers that be …
>
> (Stebbing 1911: 210)

Hence, on these hunts certain tigers were known as 'Viceroy's tigers': 'beasts that have been driven over time-honoured ground to a place where they are certain to come out for the brass hat to massacre', Best (1935: 185) noted. Such 'Viceroy's tigers' were inevitably larger than everyone else's as they were stretched during measurement. This was yet another mechanism for enforcing the strict administrative hierarchy in place in the British colonial State. The forest officer Benskin (1963: 118) noted that: 'Those chosen to run shoots for VIPs had to be experts in producing game at the right place, as well as sure and discreet longstops, so that the distinguished person got his tiger; tact, patience and diplomacy were essential qualifications.'

Originally, of course, this had been a Mughal method of hunting which displayed the monarch to his subjects at his most powerful; however, the British had taken it over and widely expanded it, thus continuing a highly visible and essentially despotic form of State territorialisation. Indeed, in the late nineteenth century the viceroys had become more and more aristocratic. This type of hunting thus represented a way in which the British were able to use the killing of tigers to build social bridges with the Indian aristocracy and gain added legitimacy for its rule. It represented the concern of the colonial State in India with the external form of its authority rather than with its internal strength.

Nevertheless, there was widespread disagreement about whether or not this method was the most sporting form of tiger hunting. Stebbing (1911: 210), for one, spoke up in favour of this method and argued that '… when tiger are afoot, and more especially when wounded tiger are afoot, it is hard to beat'. 'Silver Hackle' (1929: 40) similarly extolled the virtues of beating as a method, describing beating for tiger as almost an art, and noting that very few men were really proficient in it.

However, 'Silver Hackle' (1929: 39) generally preferred the alternative method of hunting for tigers in India, namely, tying up bait and waiting in the dark in solitude for the tiger to come along before shooting it. Tying up held a certain voyeurism for many hunters. Aitken (1897: 1–2), for example, noted quite proudly that he: '... always felt a strange pleasure in seeing without being seen. ... I cannot quite satisfactorily analyse this kind of enjoyment and am not sure it is very respectable, but it is very human. Stolen waters are sweet, and bread eaten in secret is pleasant.' Stebbing (1911: 238) in contrast, found this method boring, arguing that: 'This form of securing a much-coveted trophy is very monotonous in a way, since it means that you can do little yourself to assist matters.' Similarly, it was argued that tying out bait was not truly sporting in a British sense:

> I think that many people are too much guided by convention in the matter of sitting up; the general custom being to send the *shikari* out in advance to have the *machan* [tree platform] ready for the sportsman when he turns up at about four o'clock in the evening with the intention of sitting up until dark. It fits in with office hours and gives the minimum of discomfort but not, in my opinion, the maximum of results.
>
> (Best 1931: 62)

Occasionally, keen hunters would also venture to go after tigers on foot in order to prove their manliness: 'Shooting tigers on foot is the cream of sport; it requires knowledge of the locality, careful planning, crafty stalking, good shooting and, in addition, entails considerable hardship in enduring the heat. Occasionally risks have to be taken if the tiger is to be bagged', Best (1931: 69) argued.

Preparation and etiquette

More generally the question of sporting etiquette meant that the correct preparations had to be made for hunting. In this context comfort was the key. The hunter had to be seen to be maintaining his elite status in the field, often at considerable financial cost. Sanderson (1878: 182), the officer in charge of the government elephant catching operations, for example, argued that '... the sportsman should make himself and followers as comfortable as possible. ... Roughing it when there is no necessity – and there seldom is nowadays in India – is a mistake which only the inexperienced fall into.'

Hence, whilst there was a sense of adventure about *shikar*, there certainly wasn't a lack of comfort. We only need to note the amount taken on a two-month hunting trip for two officers:

> Two *shikaris*, one peon, one native blacksmith, one butler, two colassies or tent pitchers, one elephant with 'mahout' and two attendants, one horse each, eight bullock-carts, twenty-four pack bullocks with Bunjara owners, two field-officers' tents, one dhobie, ten dozen soda-water, eight dozen claret, six dozen beer, one dozen gin, one dozen brandy, crockery for breakfast, dinner

> etc. The best of stores comprising tea, coffee, chocolate, jam, sardines, bacon, hams, sausages, potted meats, etc., in fact all manner of 'Europe stores' that could add a little to the jungle fare obtained by our guns.
>
> (Brown 1887: 156–7)

One of the interesting things to note is that the above list makes no distinction between human beings, animals or provisions. Indeed, as we shall see, by and large, they were all treated as one and the same by the British hunting elite. Hunting in India, then, was a grand affair which only this elite could take part in and all others were excluded. Indians for example, but also 'other ranks', were excluded ostensibly because it was thought that they either wouldn't know how to hunt or that they would end up killing indiscriminately.

In the later colonial period other preparations also had to be undertaken though. An officer had to obtain the correct permits and get help from a variety of different sources. Interestingly, as the social distance between Indians and the British increased in the Edwardian period then hunting as sport became more and more regularised and codified, with permits being allocated. The key contact, Best (1935: 2–3) argues, was the district forest officer who allocated shooting blocks: '… the duty of the forest officer in connection with sport, is to allot blocks on application, to appoint a forest guard to accompany the permit holder's camp, and to see that the shooting rules and laws are obeyed by sportsmen.' Best (1935: 2–3) goes on to point out that: 'The old story of the shooting permits signed with red or black ink, according to the wishes of the forest officer as to whether the sportsman should be assisted or hindered has never been proved …'. Having obtained your permit you had to buy your maps from the Survey of India, Dehra-Dun, then, on arrival, call on the local officers of the district station and on the members of the nearest club, and finally, make friends with the nearest villagers to your block. However, the idea of permits still rankled with the old guard.

Exercising despotic and disciplinary power

It was the interaction with the rural inhabitants whilst out shooting that provided many officers with an alternative reason for hunting tigers, namely the enhancement of State disciplinary knowledge and the exercise of State disciplinary power. Major-General Wardrop (1923: 14) noted that *shikar*: '… helped enormously to keep up the prestige of the Government with the People'. It also enhanced the prestige of the government with the government itself. It gave the 'sportsman' the chance to really get to know India, and, as such, developed into a key power–knowledge relationship of British governmentality in India. As Miller and Rose have noted: 'Knowing an object in such a way that it can be governed is more than a purely speculative activity …' (Miller and Rose 1993: 79). It required a particular method of grasping the biopolitics of the population (Foucault 1981). In order to successfully manage the population and its environment the State would need to know its territory. It did this by creating mechanisms of classification while simultaneously compromising any instances of de-territorialised resistance.

Thus, 'Silver Hackle' (1929: 41) wrote of the importance of: 'On arriving at his shooting grounds, the sportsman should, with the help of the local inhabitants, get to learn as much of it as he can …'.

Indeed, it was argued that it was only really the hunter who knew a quarter of the Indian subcontinent. Stewart (1927: 254), for example, argued that: 'By searching out the country for good shooting spots you automatically learn the country, get to know the villagers and headmen, which is all part of your job …'. Getting information though was not always straightforward, however, as there was a degree of resistance to all the knowledge gathering that went with hunting. Captain Forsyth (1871: 290–1), for instance, argued that: 'A great many reasons, besides the simple one to which it is usually attributed, namely that "they are cursed niggers", combine to make the natives in most places very unwilling to give information.'.

Whilst hunting for tigers thus formed part of the disciplinary apparatus of the colonial State (see Bayly: 1993), there was also a strong element of despotic power in operation too. Many British officers would simply take what food they needed from villagers and force reluctant villagers to beat the jungle to flush out tigers with a considerable chance of being killed. Forsyth (1871: 290–1) noted that some hunters would harass 'the people in the matter of provisions … thrashing them all round if a tiger was not found for them when they arrived' in a district.

Conclusions

After the First World War stricter game laws were gradually introduced in all the provinces of India. On paper this appeared to give the tiger some protection. However, these laws were hardly enforced and there was little commitment to conservation. Tigers, in particular, were still seen by many as vermin: Eardley-Wilmot (1910: 89) even believed that the tiger's extinction was a duty of the colonial State: 'extinction appears to be a matter of time; for no Government would face the rare opportunity which would be afforded for misrepresentations by taking steps to protect so interesting a beast from extermination'. Eardley-Wilmot (1910: 89) did sound a note of remorse; however, this wasn't for the tiger – rather for the loss of sporting pleasure: 'Pity it is that he must disappear, and with him one of the greatest charms of forest life, and also a form of sport that has been not only enjoyable, but beneficial, to hundreds of exiles.' The object of the game laws that were passed seemed, at first, to be only to prolong the slaughter, through rationing. We even hear these overtones in the writings of such a noted *shikari* turned conservationist as Jim Corbett. He wrote that: '… in order to afford game animals the peace and protection which will enable them to live and reproduce their kind without damage to man, man should only be allowed to damage them under certain rules …' (cited in Booth 1990: 182).

Nevertheless, the end of the First World War was both a threshold for the British Empire and for the fortunes of the Indian tiger. Rather than being simply due to a new conservation ethos though, the upsurge in the tiger's fortunes was also due to the deaths of many of the keenest hunters on the fields of France:

> The war has greatly affected big game hunting in India. The 'old Contemptibles' are gone, and few of these who have taken their place during the great war have had the time or the necessary knowledge to pursue big game with much success; the result has been an undoubted increase in the numbers of some of the larger animals. The knowledge of where to go, and what organisation is required, was, in the old days, handed on through the messes or from friend to friend; a link in the chain has been broken by the war. …
>
> (Best 1931: xii–xiii)

By the 1930s, there had been a concerted move towards hunting with cameras. Wood pointed out that:

> Nowadays the feeling is not so much desire to kill as to take pictures of wild life which would be of interest to others. This is the right spirit, especially as game is decreasing year by year. Let my readers not think me a hypocrite, but the feeling is that, as one gets older, bloodlust gets less; one hates taking life, and feels a sense of remorse for all the animals and birds slain by rifle and gun.
>
> (Wood 1934: 9)

However, this movement away from the exercise of the despotic power to take the life of the tiger by colonial officials and their visitors, towards a documentation of the life of the tiger is emblematic of a further shift towards disciplinary techniques (see Foucault 1991). Indeed, a documentary database was generated in excess of any aesthetic or bureaucratic purpose. Numbers gradually became part of the illusion of control in which countable abstractions and photographic evidence, of both people and animals created a sense of a controllable reality for the colonial State (see Appadurai 1994). However, with the gains of the Independence movement, this illusion of control would soon be lost. With the post-independence ban on all tiger hunting and the clamp-down on poaching, the tiger would be saved from extinction – for the time being at least – with the establishment of national parks and protected areas and the development of nature-based tourism in India (see Hannam 2004).

References

Aitken, E. (1897) *A Naturalist on the Prowl or in the Jungle*, Calcutta: Thacker.

Appadurai, A. (1994) 'Number in the colonial imagination', in C. Breckenridge and P. van der Veer (eds) *Orientalism and the Postcolonial Predicament: Perspectives on South Asia*, Delhi: Oxford University Press, pp. 316–17.

Baillie, W. W. (Mrs) (1921) *Days and Nights of Shikar*, London: John Lane.

Bayly, C. (1993) 'Knowing the country: Empire and information in India', *Modern Asian Studies*, 27(1): 3–43.

Benskin, E. (1963) *Jungle Castaway*, London: Hale.

Best, J.W. (1931) *Indian Shikar Notes*, 3rd edn, Lahore: Pioneer.

Best, J.W. (1935) *Forest Life in India*, London: Murray.

Booth, M. (1990) *Carpet Sahib: A Life of Jim Corbett*, New Delhi: Oxford University Press.

Brown, J.M. (1887) *Shikar Sketches with Notes on Indian Field-Sports*, London: Hurst.

Burgess, J. (1993) 'Representing nature: conservation and the mass media', in F. Goldsmith and A. Warren (eds) *Conservation in Progress*, Chichester: Wiley.

Champion, F.W. (1927) *With a Camera in Tiger Land*, London: Chatto & Windus.

Cohn, B. (1983) 'Representing authority in Victorian India', in E. Hobsbawm and T. Ranger (eds) *The Invention of Tradition*, Cambridge: Cambridge University Press.

Cohn, B. (1987) 'The recruitment and training of British civil servants in India, 1600–1860', in B. Cohn, *An Anthropologist Among the Historians and Other Essays*, Delhi: Oxford University Press.

Corbett, J. (1944) *Man-eaters of Kumaon*, Oxford: Oxford University Press.

Crouch, D. (ed.) (1999) *Leisure/Tourism Geographies: Practices and Geographical Knowledge*, London: Routledge.

Dunlop, R.H.W. (1860) *Hunting in the Himalaya*, London.

Eardley-Wilmot, S. (1910) *Forest Life and Sport in India*, London: Edward Arnold.

Edensor, T. (1998) *Tourists at the Taj: Peformance and Meaning at a Symbolic Site*, London: Routledge.

Elliott, J.G. (1973) *Field Sports in India 1800–1947*, London: Gentry.

Forsyth, J. (1871) *The Highlands of Central India: Notes on their Forests and Wild Tribes, Natural History and Sports*, London: Chapman.

Foucault, M. (1981) *The History of Sexuality: An Introduction*, Harmondsworth: Pelican.

Foucault, M. (1991) *Discipline and Punish: The Birth of the Prison*, Harmondsworth: Pelican.

Gardner, A. (Mrs) (1895) *Rifle and Spear with the Rajpoots: Being the Narrative of a Winter's Travel and Sport in Northern India*, London: Chatto and Windus.

Glasfurd, A. (1928) *Musings of an Old Shikari: Reflections on Life and Sport in Jungle India*, London: John Lane.

Hannam, K. (2004) 'Tourism and forest management in India: the role of the State in limiting tourism development', *Tourism Geographies*, 6(3): 331–51.

Hannam, K. (2005) 'Tourism management issues in India's National Parks', *Current Issues in Tourism*, 8(2/3): 165–80.

Hell, B. (1996) 'Enraged hunters: the domain of the wild in north-western Europe', in P. Descola and G. Palsson (eds) *Nature and Society: Anthropological Perspectives*, London: Routledge, pp. 205–17.

Mackenzie, J. (1988) *The Empire of Nature: Hunting, Conservation and British Imperialism,*. Manchester: Manchester University Press.

Mangan, J. and Walvin, J. (eds) (1987) *Manliness and Morality: Middle-class Masculinity in Britain and America 1800–1940*, Manchester: Manchester University Press.

Mann, M. (1984) 'The autonomous power of the state', *Archives Européenees de Sociologie*, 25: 185–213.

Miller, P. and Rose, N. (1993) 'Governing economic life', in M. Gane and T. Johnson (eds) *Foucault's New Domains*, London: Routledge.

Osborne, T. (1994) 'Bureaucracy as vocation: governmentality and administration in nineteenth-century Britain', *Journal of Historical Sociology*, 7(3): 289–313.

Pandian A.S. (2001) 'Predatory care: the imperial hunt in Mughal and British India', *Journal of Historical Sociology*, 14(1): 79–107.

Phillips, R. (1997) *Mapping Men and Empire: A Geography of Adventure*, London: Routledge.

Powell, A.N. (1957) *Call of the Tiger*, London: Hale.

Sanderson, G.P. (1878) *Thirteen Years among the Wild Beasts of India: Their Haunts and Habits from Personal Observation; With an Account of the Modes of Capturing and Taming Elephants*, London: Allen.

Savory, I. (1900) *A Sportswoman in India: Personal Adventures and Experiences of Travel in Known and Unknown India*, London: Hutchinson & Co.; Philadelphia: J. B. Lippincott Company.

'Silver Hackle' (1929) *Indian Jungle Lore and the Rifle: Being Notes on Shikar and Wild Animal Life*, Calcutta: Thacker.

Smythies, O. (1953) *Tiger Lady: Adventures in the Indian Jungle*, London: Heinemann.

Stebbing, E.P. (1911) *Jungle By-ways in India: Leaves from the Note-book of a Sportsman and a Naturalist*, London: John Lane.

Stewart, A.E. (Col.) (1927) *Tiger and Other Game: The Practical Experiences of a Soldier Shikari in India,* London: Longmans.

Wardrop, A.E. (1923) *Days and Nights of Indian Big Game*, London: Macmillan & Co.

Wolch, J. and Emel, J. (1995) 'Bringing the animals back in', *Society and Space*, 13: 632–6.

Wood, H.S. (1934) *Shikar Memories: A Record of Sport and Observation in India and Burma*, London: Witherby.

Young, Z., Makoni, G. and Boehmer-Christiansen, S. (2001) 'Green aid in India and Zimbabwe – conserving whose community?', *Geoforum*, 32: 299–318.

Part III

Impacts of consumptive wildlife tourism

8 Conservation hunting concepts, Canada's Inuit, and polar bear hunting

Lee Foote and George Wenzel

Introduction

Travelers who undertake remote travel for hunting opportunities represent a specialized group of adventure tourists engaging in activities known as safari hunts, trophy hunting, hunting tourism or, more recently, conservation hunting (CH). In this chapter we provide a case study of polar bear (*Ursus maritimus*) hunting in Nunavut, Canada an introduction to CH origins and some contrasts between ecotourism and hunting. Conservation hunting can be considered a form of ecotourism partly because CH recognizes a broader reciprocity between hunters and local community members (Freeman *et al.* 2005).

Participants in CH tourism are highly motivated individuals willing to spend many thousands of dollars (usually in US currency) on major trips involving travel costs, permits, trophy fees, guide services, and taxidermy expenses. There are strong participatory motivations for the villagers, rural people, or indigenous communities that serve as hosts and service providers for CH. Most obvious is the range of employment opportunities arising from CH, but there are also important cultural exchanges, recognition and community pride in demonstrating subsistence and bushcraft skills required for CH that provide more global recognition and validation of very rural ways of life. This immersion into and participation in foreign hunting processes are alluring and meaningful to members of many modern, urban societies. Conservation hunting exchanges are not entirely seamless and when improperly managed can result in undesirable outcomes. The environmental footprint and animal harvest of conservation hunters is relatively light because of their low numbers, minimal infrastructure needs, high to extreme selectivity of harvest and because their activities are carefully scrutinized both by local people and governmental jurisdictions charged with regulating harvests to fit into biological management plans. Finally, CH tourism may contribute to the robustness (*sensu* Anderies *et al.* 2004) and sustainability of ecological/cultural systems. A 'robust' production system (e.g. controlled harvests of wild species) will typically perform less efficiently in wildlife production and income generation than will non-robust land uses such as exotic cattle on a tame grass pasture or commercial fishing. Robust systems are less likely to fail as rapidly as its non-robust counterpart when confronted with external disturbance, internal stresses or market failure. To increase the robustness of the grazing system, ranching of native ungulate species have been used as a replacement for cattle, an exotic species, on Kenyan ranches (Ehrlich and Ehrlich 1985). Robust systems are more sustainable in the long term and may provide decentralized benefits across a spectrum of social classes. Wildlife resources provide opportunities for community income through CH tourism, viewing tours, or subsistence use as in community-based natural resource management in South Africa, Botswana, Zimbabwe and Namibia (Child 2002). In the Northern Hemisphere, a similar suite of benefits come from Nunavut, Canada's polar bears and the brown bears (*Ursus arctos*) of Russia's Kamchatka peninsula. We use a more detailed description of polar bear hunting in Nunavut, Canada, to illustrate the ways polar bears are valued.

Case study: polar bears, conservation hunting and Inuit

The Canadian Arctic is home range for approximately one-half of the 25–30,000 polar bears living in the circumpolar world and most of Canada's population is found in the Nunavut Territory. Polar bears are the world's largest terrestrial carnivores and are among the most carefully managed of species (Fikkan *et al.* 1993) in the Northern Hemisphere.

No animal holds as significant a place in Canadian Inuit culture as the polar bear, *Nanuq* in Inuvialiuit. This prominence is evidenced by the fact that, with Inuit, polar bears are the other chief predator in the Arctic marine environment, sharing that environment on a virtually equal basis with humans, until the introduction of firearms. It is no surprise therefore, that *Nanuq* was a central figure in Inuit cosmology (see Boas 1888) and retains considerable symbolism for Inuit and non-Inuit, albeit often for different reasons, today. Inuit have hunted polar bears as an element of their overall set of ecological relations for millennia with this hunting (see Nelson 1969; Robbe 1994; Sandell and Sandell 1996) being conducted for spiritual-cultural reasons and to contribute to the traditional food economy.

Some 450 polar bears are hunted annually in Canada, far more than are hunted in Greenland or Alaska, whereas Norway and Russia, the other nations with significant polar bear populations, have banned all forms of polar bear hunting due to non-sustainable uses of bear populations. The majority of the harvest in Canada, about 325 animals, is taken by the Inuit of Nunavut. For Nunavummiut (Regional Inuit people), polar bear meat remains an important item in the modern Inuit diet and, with the advent of a fur trade with non-Inuit in the early twentieth century, polar bear hides became an economic resource. The sale of the skins is still a part of the contemporary economy.

Canada is also the only national jurisdiction in which trophy hunting of polar bears is sanctioned. Under the Agreement for the Conservation of Polar Bear (Lentfer 1974), Inuit are permitted to sell bears from their annual quota to non-aboriginal sport hunters. Typically, each year Nunavummiut make between 75 and 90 bears available to sport hunters from the United States, Western Europe, Japan and Latin America. With the average cost of a polar bear sport hunt now between CAN$30,000–35,000, this trophy hunt is now at least as important in economic terms as the traditional polar bear fur trade.

Polar bear trophy hunting may be one of the most taxing hunts anywhere in terms of the toll it exacts on hunters and their equipment. Conducted mainly between March, when temperatures average –25°C, and the end of May, when it may be as warm as +5°C, client hunters travel across the sea ice with an Inuit guide and his hunt assistant on hunts lasting up to 10 days, often covering 300 to 500km during a trip. By regulation, trophy hunting must be done using traditional means, which is by dogteam, and the conditions are frequently such that many hunters, in the face of this environment, replace high-tech parkas and boots with traditional Inuit caribou clothing.

The sport hunt and Inuit

The polar bear sport hunt is a complex topic and the background provided above is at best a modest introduction to a dynamic that has cultural, economic and social elements. It, for instance, by no means explicates the circumstances that led to the introduction of trophy hunting for polar bear on a broad scale to Nunavut Inuit communities (Wenzel and Bourgouin 2003). Nor does it speak in any full way to the cultural, institutional and socio-economic issues that have arisen among Inuit regarding sport hunting (Wenzel 2005) over the last decade, nor does it capture the evolution of polar bear hunting into CH aspects.

Nonetheless, it is important to address, even in a limited way, the benefits as understood by Inuit, of the sport hunt. An important part of this discussion will concern the financial benefits of the hunt, but there are also significant socio-cultural benefits that are often overlooked, if not lost altogether. To illuminate this discussion, we present data gathered in three Nunavut communities, Resolute Bay, Taloyoak and Clyde River (Figure 8.1), during 2001 and 2002 (see Table 8.1 for a summary).

The most obvious benefit Inuit derive via polar bear sport hunting is its financial contribution to the communities that host and stage trophy hunts. Such benefits most obviously accrue to the individuals who work as guides or are otherwise involved with trophy hunting, but there are also other returns, some economic and others less apparent, that transcend wages to individual Inuit.

A polar bear trophy hunt is by any standard expensive, easily costing the sport hunter up to CAN$35,000. This is the amount paid by client-hunters to the southern expediters that are the link between the sport hunt community and Inuit in Nunavut. In fact, the portion of the money that ultimately enters northern communities ranges from around 43 per cent (at Taloyoak) to about 60 per cent (Resolute Bay) of the price paid to wholesalers.

Even though a substantial proportion of the fee paid by a visitor-hunter goes to a southern expediter, hosting polar bear trophy hunting offers distinct economic benefit to northern communities and to the Inuit who participate in it as community outfitters, guides and hunt assistants. This is especially the case when the most recent information available (Government of Nunavut 1999) indicates that the

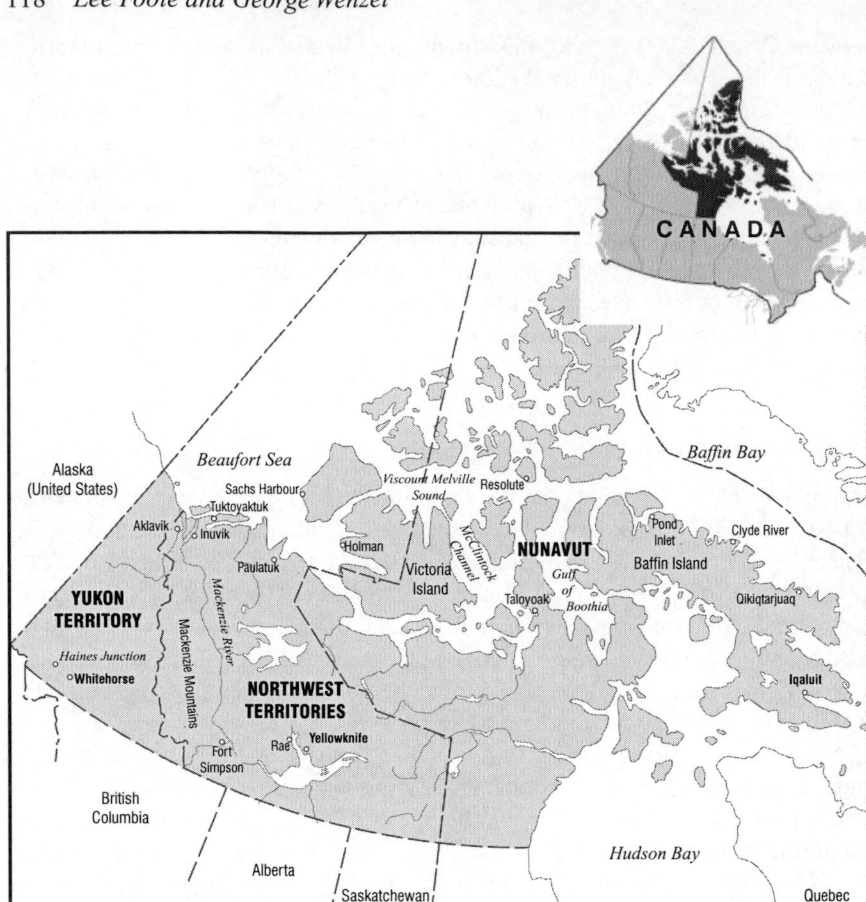

Figure 8.1 Map of Nunavut Territory showing communities where polar bear hunting was investigated

annual unemployment rate for Nunavut in that year was 21.5 per cent and, in the study communities it ranged from 10.2 per cent in Resolute Bay to 31.4 per cent at Clyde River.

As Table 8.1 shows, Inuit who work as guides and helpers receive considerable monetary benefit, earning at least as much (indeed, for guides, substantially more) than would have been the case if they were employed as minimum wage ($12.00/ hour) laborers in their communities. Roundly, the per-hunt salaries received by guides were $4,700 in Taloyoak, $5,100 in Clyde River and $9,000 in Resolute. Hunt helpers, not surprisingly, were less well remunerated, with helper salaries

Table 8.1 Polar bear sport hunt – economic attributes[1]

General features	*Clyde River*	*Resolute Bay*	*Taloyoak*
Annual polar bear quota	21	35	20
Annual sport hunts	10	20	10
Local outfitters	3 (private)	1 (private)	1 (community)
Wholesale hunt price[2]	$30,000	$34,500	$34,500
Local outfitter price[3]	$18,400	$19,000	$13,000
Local distribution			
Guides/helpers	10/10	5/9	5/9
Total guides' wages	$51,000	$180,000	$47,300
Total helpers' wages	$41,000	$100,000	$38,200
Gratuities	Avg. $1,800	Avg. $2,300	Avg. $1,500
Equipment capitalization[4]	$42,000	@$34,000	Unknown
Polar bear meat (kg)	@1,400	@3,000	@1,400
Polar bear meat $ value[5]	$14,000	$25,000	$2,000[6]

1 Not factored are fees to polar bear tag holders, additional charter or scheduled airline fares, local purchases of arts and handicrafts, and the cost of hunt consumables (food).
2 Total fee paid to southern broker by the individual hunter for his/her hunt (CAN$).
3 Contract fee between southern-based wholesaler and local outfitters.
4 These data refer to equipment purchased with sport hunt wages and are only partial.
5 Based on $8.50 per kg of imported meat (averaged across the communities).
6 As polar bear meat is generally used for dog fodder at Taloyoak, the value imputed to the meat entering the community is based on the price of imported dry dog food.

averaging $4,300 (range: Taloyoak – $3,800/Resolute – $5,000) for the duration of a hunt. Guides and hunt assistants almost always work at least two trophy trips per 8- to 12-week season and during 2001 the minimum number of hunts led by a Resolute guide was three, with one man leading six.

The pay received by guides and helpers takes on an even more impressive cast when it is broken down as an hourly wage. For the 240 hours that an Inuit spends guiding a sport hunter on a maximum duration hunt (24 hours per day for 10 days) from Resolute Bay, he or she (one Resolute guide is a woman – one of two accredited in Nunavut) earns $37.50/hour, while the helper counterpart receives $20.83/hour. It may seem strange to calculate guide or helper earnings as an hourly wage, but, when compared to the overall wage received by a full-time, minimum wage worker in Nunavut who works the same number of hours (the laborer earns $2,880 in 30 work days), the return is not only particularly impressive, but the opportunities available because of the sport hunt are critical in communities like Clyde River where unemployment and underemployment are chronic.

As important as the fact that sport hunt income is available in Nunavut's poor employment environment is who the recipients of these monies are. Almost all the Inuit from Clyde, Taloyoak and Resolute who guide are middle-aged Inuit who possess extraordinary traditional skills, able to 'age' a polar bear track, control a rambunctious team of 15 sled dogs, and respond to sudden changes in the weather or sea ice. All prefer work on the land to that available in their villages and, in the case of some, also lack sufficient command of English to hold high-paying wage employment. Indeed, many prefer work on the sport hunt not only because it offers high return, but also because they identify with being a hunter and consider their principal occupation to be hunting.

The sport hunt provides amounts of money substantial enough to meet 'village needs' – there are telephone bills to be paid and clothes to be bought – and to also invest in the capital and operational needs of full-time harvesting. Data on the amount of sport hunt income reinvested in hunting equipment by guides and helpers is limited, but it appears that a considerable portion of guiding income is spent in the other sectors of the Inuit economy, notably the harvesting of wild foods, with Resolute guides spending some 20 per cent of sport hunt income on equipment and those from Clyde almost 45 per cent (the apparent disparity between the two in harvesting investment relates principally to the larger incomes earned in Resolute (see Table 8.1, Equipment capitalization)).

This is of considerable significance not only for these 'investors', but also for their communities as Inuit consider the sharing of traditional foods an important cultural attribute (see Wenzel 1991, 1995). Interviews with 12 of the 20 men and women from the three communities who guided polar bear hunts in 2001 indicate that they spent at least 100 days in subsistence hunting the previous year. At Clyde, data on the harvest of the three most common traditional food species by Clyde guides that year was 22 caribou (approximately 1,200kg of meat), 198 ringed seals (4,950kg), and approximately 500 Arctic char (1,000kg). By way of comparison, the average Clyde hunter who did not guide others, harvested 1.5 caribou, 10.9 ringed seals and 44 Arctic char (Nunavut Wildlife Management Board 2004).

Additionally, communities retain virtually all the meat from sport-hunted polar bears. As this amount of food (some 200kg per bear) would almost certainly have been taken by subsistence hunters, counting the meat from trophy kills as a 'benefit' may seem a form of double counting. However, it must be remembered that the nearly 2,000kg received by the community via the sport hunt comes at essentially no cost as the 'expenses' are covered by the client-hunter.

Finally, there are 'psychic' benefits that certainly accrue to Inuit who guide or assist on sport hunts. These are not quantifiable in the sense of dollars or edible biomass, but they are quintessentially Inuktitut (the way of the Inuit), and essential to the well being of clients and to successful hunting. Perhaps the most important is that by working on the sport hunt, guides and assistants benefit from the simple fact of 'being on the land'. For older Inuit, time spent outside the communities is part of being truly Inuk. It is a time and place to exercise traditional skills that range from the actual tracking of a bear to reading the environment for hazards to handling dog teams. Furthermore, as many of the younger Inuit who function as hunt assistants may have had little, if any, opportunity to hunt polar bear themselves, working under the tutelage of an experienced hunter leads to a transfer of skills in the best milieu for learning about both polar bear and the land skills that formal schooling and life in the communities rarely affords.

Reciprocal benefits to the sport hunter

The European, American and other visitor-hunters who come to Nunavut also receive a variety of benefits. The most obvious is a polar bear hide that is destined for mounting. Insofar as hunters who have gone to Nunavut remark that a polar bear is found in only one in a thousand trophy rooms (Wenzel and Bourgouin 2003), its rarity must be counted as an obvious benefit. Also, the experience a visitor achieves with *Nanuq* and an environment as alien as any most will experience is also important. Numerous hunters, whether ultimately successful or not, have written about this as the central element of polar bear hunting.

The most trenchant justification for this form of CH, however, is the cross-cultural experience inherent in a polar bear hunt. It is, as already mentioned, one of the most physically demanding hunts in the world if only for the cold and starkness of the environment. But it is also an experience that once embarked on places the hunter in circumstances where he possesses relatively few skills, a condition made even more evident by the fact that most modern tools – from geographic position system locators to the best down clothing – fail as frequently as they work.

To hunt polar bear in Nunavut means that the visitor, perhaps more than is the case in hunts in other places, is near-completely reliant on local knowledge and local skills. Indeed, hunting in Nunavut may be one of the few ‘high end’ hunt experiences in which there is no southern professional hunter as a cultural buffer or any after-a-day’s-hunt amenities to escape the physical demands of riding a dog team at –30° or sleeping in an only slightly warmer snowhouse. From the instant a hunter leaves the community until he or she returns, he or she is literally in the hands of Inuit every moment.

Wenzel and Bourgouin (2003) note that, after the rarity of polar bear trophies, it is the experience of traveling and living with Inuit upon which those who have hunted in Nunavut most frequently comment. Traditional skills are not only still a part of modern Inuit life, but essential to it. Moreover, many come away realizing that the role of tradition among Inuit is not limited to the pursuit of a bear or a level of comfort with the environment, but also to the closeness that exists among Inuit and between them and the Arctic.

The origins of Conservation Hunting

Conservation Hunting has evolved out of simple safari-style hunting. The treatment of historical regulation of hunting is addressed elsewhere in this book (see chapter by Akama) and represents the conservation precursor to CH. Such restriction sentiments were transported to Africa during the early colonial period where a system of game laws, restrictions, and refuges was used to manage wildlife population (Adams 2004). A pivotal change was the restriction of the millennia-old custom of subsistence harvesting by indigenous people (Hutton 2005). Such denial of access was an imprecise form of management rife with resentment, the social strife resulting from recasting ancient patterns of subsistence use as illegal, then following such change with enforcement (Hutton 2005). By the mid-1800s some British colonies like India, Kenya, Botswana, and British Honduras had evolved elaborate systems of protected areas, commons and private holdings. By the mid-1800s, technological improvements in transportation, firearms and markets facilitated the efficient harvesting of large animals on a commercial scale and populations were diminished and extinction rates soared.

A global renaissance of interest in wildlife began to emerge in the early 1900s, likely as a response to messages of overharvest, scarcity and extirpations even in remote regions of the world. North Americans often believe that US President Theodore Roosevelt started the conservation movement in 1900–15 with strong political leadership, love of outdoors, and his sportsman's sentiments. Establishment of refuges and park-networks, seasonal limitations, stocking, fair chase and hunter self-limitation are activities that North Americans consider the basis of wildlife conservation, yet they are not original North American concepts. Most of Roosevelt's ideas on conservation were actually re-statements of earlier practices in Africa and India where British colonialists pioneered methods of preventing wildlife over-exploitation, thereby preserving sporting opportunities from disappearing as they had in their home countries (Adams 2004). The success of wildlife recovery in North America and Africa involved a system of using hunting fees to purchase protected parks and game reserves, and to enforce laws reducing excessive harvests in non-park areas. The changes in agricultural status, land ownership or park designation in North American, Africa and India were only lightly contested because of the widespread depopulation or subjugation of indigenous people through disenfranchisement, warfare and disease following colonial domination.

Maintenance of wildlife populations worked adequately until the mid-1900s but eventually it became evident that wildlife were being limited less by hunting than by habitat loss and alternative land uses (agriculture, human developments). Local people that may have subsisted on wildlife in an earlier era were now forbidden from subsistence harvests and came to see wildlife as a liability to human safety or crop production. A new paradigm was needed to reverse this value shift and to acknowledge the emerging self-determination and indigenous rights of local people so wildlife might be chosen as a reasonable land-use component. It was from this void that CH emerged.

Because the Arctic is an inaccessible region with few exportable resources, missionaries, fur trappers and whalers were the primary contacts with Canada's Inuit people prior to 1950. It was not until 1956 that restrictions were placed on the Inuit people of Arctic Canada to restrict the hunting methods for polar bears. Examples of restrictions included: no killing bears in dens, no killing sows with cubs, and designation of certain refuge areas. However, great flexibility was allowed for Inuit take and participation in sponsoring trophy hunting for polar bears by foreign hunters. Even though some people remain opposed to the killing of bears for any reason, no widespread organized opposition to polar bear hunting has been mounted on a moral basis thus far. The polar bear harvest is moderate with a total of about 600 worldwide killed annually and the hunt is virtually impossible to observe by anti-hunters or the media. Polar bear tag allotments are issued to Inuit hunters who have been harvesting bears annually for over 4,000 years, there is a strong record of sustainable use and, consequently, it is difficult for opposition groups to make claims that limited hunting is a substantive extinction threat to polar bears.

It is important to recall that polar bear tag allocation is through the hunter/trapper organizations and that Inuit hunters decide how many bears from their regional allotments will be harvested by native hunters and how many will be allocated for sale to foreign sport hunters – often around 15 per cent but with variations by community. Most sport hunters are from the United States, followed by substantial numbers from central European countries. Under this quota system, the same numbers of bears will be killed even if sport hunting were reduced or eliminated. The highly desirable sport hunting tags would simply be used by Inuit hunters for subsistence hunting to provide meat, recreational opportunities and furs for crafts. Polar bear harvest quotas are set after consultation with territorial biologists, Inuit traditional ecological knowledge, and with prudent oversight and recommendations from the Polar Bear Technical Working Group. Based on the 2004–05 observations of Inuit hunters seeing higher numbers of polar bears, the Nunavut territorial government increased its annual hunting quotas by 29 per cent – to 518 tags, an increase of 115 bears – despite the concerns of biologists that this was too many tags. Because the Inuit value the bears so highly and seek conservative harvests, there is substantial compromise and adjustment with input from authorities (Freeman 1986), and in this case a scaling back did occur after the season. Of late, climate change scenarios may pose significant range shifts, cumulative impacts and reductions in southern sub-populations due to reductions in sea ice. It is unknown if simultaneous range extensions are occurring at the northern sub-populations though.

Contrasting effects of bear-watching and polar bear hunting

Roxe (1998) described ecotourism as responsible travel that conserves the natural environment and sustains the well-being of local cultures. Most ecotourists on polar bear viewing trips arrive in the Arctic, more particularly in staging areas where polar bears concentrate, with expectations of safely observing the bears. Quite reasonably, tour members expect to have heated lodging and prepared meals. This experience requires substantial infrastructure to isolate them from the bears (e.g. elevated tour buses, gated observation decks and secure lodging). Such experiences are inherently social and group activities since viewing buggies are designed for 10–50 people and tour boats may accommodate over 100 people per voyage. A cursory review of internet advertisements showed costs for five-day polar bear viewing excursions to range from US$4,300 to US$5,500 (approximately $1,000 per day). Bear-viewing tours are described in advertisements by tour operators for Natural Habitats Worldwide Safaris (2006) in terms such as:

> Our day is spent viewing bears and wildlife in the company of our knowledgeable Expedition Leaders who will give us insight into the bears and their lives that cannot be learned elsewhere. This full day adventure is remarkably rewarding as our leaders and drivers know the best places to view the bears in secluded areas. We return in the late afternoon and have a short time to relax before dinner and evening presentation. On the other day (or on the final day in Churchill should weather and other factors not cooperate), we will schedule one of nature's greatest expeditions – a helicopter journey to the female bears' denning area! .... hopefully, have the rare opportunity to actually crawl inside an unoccupied polar bear den – a truly unique experience!
>
> (Natural Habitats Worldwide Safaris 2006)

The Dymond Lake Lodge (www.churchillwild.com) advertised the ease of the experience by specifying 'This trip requires very little to no walking. Buggies can sometimes offer a bumpy ride' and 'It's like watching your favorite nature channel ... without the television'.

The experience of polar bear viewing is simultaneously voyeuristic and vicarious in that participants pay to be in the proximity of bears for viewing while also paying to be isolated from the field conditions and the ways of life that constitute actually sharing the bears' environment. Problematically, there is little about polar bear viewing tours that reflect the living conditions of indigenous people or the relationship they have with polar bears. Tour operators recognize and respond to their clients' demand for comfort, safety and gourmet meals. Luxury accommodations set in one of the harshest climates on earth holds a curious attraction for foreign visitors. Most tourism companies in Churchill work to minimize their ecological impacts and to avoid wildlife disturbance yet perverse incentives exist to accommodate visitor desires, sometimes to the detriment of the resource. Isaacs (2000: 67) observed 'The rigors of a market system that caters to the resource-intensive preferences of modern consumers will make it difficult for low-impact ecotourism operators to prosper.' Amongst the advertisements for polar-bear watching ecotourism, we found no evidence or suggestion that any of the polar bear viewing ecotourism companies were Inuit-owned and because almost all tourist needs are met by non-Inuit ecotour companies, local people are likely to receive a reduced share of profits flowing from bear viewers. The provision of financial benefits to local indigenous people is one of the primary criteria defining ecotourism. Bear viewing may be more accurately characterized as simply tourism, though because there are no accepted standards or certifying organizations, the term 'eco-tour' will probably continue to be used in advertising this activity.

It is misleading to characterize hunting as consumptive use and viewing only tours as non-consumptive use. Both activities are likely to have demographic and survival costs to sub-populations of polar bears. Behavioral research has shown enhanced vigilance of polar bears in the presence of bear-watching tour buses, possibly increasing basal metabolic demands during this fasting period (Dyck and Badak 2004). Less scrupulous tour operators have disturbed bears (Anonymous

1998) in ways such as baiting bears into scenic or viewable settings (Herrero and Herrero 1997). The town of Churchill expands from a base population of 900 to approximately 2,500 during the bear-watching season thereby increasing the risk of human–bear conflicts. Bears may escape people if they choose to use other parts of the 150km of undisturbed coastline along the western Hudson Bay instead of the 10km accessible to bear-watching tours. Simply because there may be some costs to the bears does not mean bear watching is not a worthwhile activity. Of possible political and conservation benefit, Lemlin and Wiersma (2005) found that bear-watchers self-reported that they had gained appreciation and introspection that would lead to a greater commitment to aiding the future well-being of polar bears.

Like the non-extractive users of polar bears, hunters arrive by airplane in the Arctic with a variety of expectations somewhat specific to their endeavor. They seek an opportunity to pursue polar bears and generally expect to experience dramatic solitude accompanied by one or two guides and a dog team.

The economic, social and biological impacts differ greatly between the polar bear watch and the bear hunting parties. Nunavut law requires polar bear hunters to use only traditional means of travel while hunting, meaning a dog sled. Advertisements by polar bear outfitter Rick Herscher describe the difficulty of the endeavor:

> Polar Bear hunting is very rigorous. Only dedicated hunters in excellent physical condition should undertake this hunt. Hunters and guides camp on the ice pack and cover miles every day via dog sled. Hunters will ride, glass, and spend up to 12 hours a day searching for bears. Weather conditions are unpredictable and often severe. ... This is, by far, the toughest hunt on earth and not for the weak at heart! Hunting the arctic ice pack by dog team and sled you can experience temperatures to –40F with nothing between you and the elements but a canvas tent and the clothes on your back.
>
> (Herscher 2006: 1)

This does not suggest that some hunters would not use heated snow cabs for hunting polar bears if permitted, but the requirement of traditional travel helps ensure Inuit involvement and makes polar bear hunting a unique experience.

Some northern entrepreneurs in Alaska and British Columbia operate their lodges for both brown bear hunting and wildlife viewing ecotourism businesses. Various clienteles may be separated temporally into summer ecotourism, fall hunting and winter sports or wildlife viewing (Weaver *et al.* 1996).

The cultural exchange between regional hosting cultures and foreign hunters is a CH component easily recognized by participants yet remains difficult to measure. Visiting hunters often refer to experiencing a deep appreciation of their hosting cultures, or express their surprise at being changed in some meaningful way. Hunters often reciprocate by bringing gifts, financial support and donations, employment or educational opportunities, and world views difficult to access by remote cultures. Important bonds of shared hunting experiences may lead to intimate confirmation of the value and worth of those hunting and living in both subsistence and urbanized societies, thereby creating an important bridge across social, religious, geographic, linguistic and cultural gulfs.

Concluding remarks

At the core of CH are the three-way reciprocal benefits wherein: (1) hunters reap profound emotional and experiential benefits; (2) hosting communities find value, both tangible and intangible, in the process of supporting CH for species such as Canada's polar bear, Asia's Marco Polo sheep (*Ovis ammon*), North America's Dall sheep (*Ovis dalii*), African elephant (*Loxodonta africanus*), South America's jaguar (*Panthera onca*), and many other species; and (3) ecosystem robustness and sustainability are usually enhanced by increased value resulting in higher conservation priority given to the habitats of hunted species. Importantly, these species, like *Nanuq*, are accorded great intrinsic value and afforded protection resulting from CH concepts being incorporated into carefully regulated hunting protocols. As environmental ethicist Holmes Rolston III contends, our human-ness renders all values subjective and, without humans present, assignment of value may be impossible (Rolston 1981). Conservation hunting affects the way people value wildlife and wildlife habitat. Consequently, CH can indeed contribute to sustainability of hunted wildlife populations and their habitat by providing the incentive for local people living in close contact with wildlife species to become their stewards.

Acknowledgments

The authors thank the Canadian National Center of Excellence ArticNet programme University of Alberta for support, Ms C. Mason for maps, Canadian Circumpolar Institute. Dr N. Krogman provided a helpful review.

References

Adams, W.M. (2004) *Against Extinction: The Story of Conservation*, London: Earthscan.

Anderies, J.M., Janssen, M.A. and Ostrom, E. (2004) 'A framework to analyze the robustness of social-ecological systems from an institutional perspective', *Ecology and Society*, 9(1): 18. <http://www.ecologyandsociety.org/vol9/iss1/art18/> (accessed 20 April 2006).

Boas, F. (1888) 'The Central Eskimo', *Sixth Annual Report of the Bureau of American Ethnology for the Years 1884–1885*, Washington, DC: The Smithsonian Institution, pp. 339–669.

Child, B. (2002) 'Review of African wildlife and livelihoods: the promise and performance of community conservation', *Nature*, 415(6872): 581–2.

Dyck, M.G. and Badak, R.K. (2004) 'Vigilance behavior of polar bears (*Ursus maritimus*) in the context of wildlife-viewing activities at Churchill, Manitoba, Canada', *Biological Conservation*, 116: 343–50.

Ehrlich, A. and Ehrlich, P. (1985) 'Ecoscience: grazing systems from theory to practice', *Mother Earth News* (Vol. 94) July/August issue.

Fikkan, A., Osherenko G. and Arikainen, A. (1993) 'Polar bears: the importance of simplicity', in O. Young and G. Osherenko (eds) *Polar Politics: Creating International Environmental Regimes*, Ithaca: Cornell University Press, pp. 96–151.

Freeman, M.R. (1986) *Polar Bears and Whales: Contrasts in International Wildlife Regimes. Issues in the North, Vol. I, Occasional Publication No. 40*, Edmonton: Canadian Circumpolar Institute, University of Alberta.

Freeman, M.R., Hudson, R.J. and Foote, L. (2005) *Conservation Hunting: People and Wildlife in Canada's North*, Canadian Circumpolar Institute, Occasional Papers Series No. 56.

Government of Nunavut (1999) *Nunavut Community Labour Force Survey: Overall Results and Basic Tables*, Iqaluit: Nunavut Bureau of Statistics.

Herrero, J. and Herrero, S. (1997) *Visitor Safety in Polar Bear Viewing Activities in the Churchill Region of Manitoba, Canada*, BIOS Environmental Research and Planning Association Ltd for Manitoba Natural Resources and Parks Canada.

Herscher R. (2006) ***Polar bear hunting***. <http://www.polarbearhunting.net/> (accessed 15 June 2006).

Hutton, J. (2005) 'Exploitation and conservation: lessons from Southern Africa', in M. Freeman, R.J. Hudson and L. Foote (eds) *Conservation Hunting: People and Wildlife in Canada's North*, Canadian Circumpolar Institute, Occasional Papers Series No. 56, pp. 28–37.

Isaacs, J.C. (2000) 'The limited potential of ecotourism to contribute to wildlife conservation', *The Wildlife Society Bulletin*, 28(1): 61–9.

Lemlin, R.H. and Wiersma, E.C. (2005) *Interviews with the Polar Bear Viewer*, Published abstract, 11th Canadian Congress on Leisure Research, Nanimo, BC, Canada, 17–20 May, Available at <http://www.lin.ca/resource/html/cclr%2011/CCLR11-80.pdf> (accessed 7 July 2006).

Lentfer, J. (1974) 'Agreement on the conservation of polar bears', *Polar Record*, 17(108): 327–30.

Natural Habitats Worldwide Safaris (2006) <http://www.nathab.com/Polar%20Bears%20 Churchill%20Land%20Air%20tours//> (accessed 15 June 2006).

Nelson, R. (1969) *Hunters of the Northern Ice*, Chicago: University of Chicago Press.

Nunavut Wildlife Management Board (2004) *The Nunavut Wildlife Harvest Study*, Iqaluit: Nunavut Wildlife Management Board.

Robbe, P. (1994) *Les Inuit d'Ammassalik, Chasseurs de l'Arctique*, Mémoires du Muséum National D'Histoire Naturelle, Paris: Éditions du Museum.

Rolston, H. III (1981) *Environmental Ethics*, Buffalo: Prometheus Books.

Roxe, H. (1998) 'Interested in an ecotour? Tread carefully', *Time International*, 6 August: 6.

Sandell, H. and Sandell, B. (1996) 'Polar bear hunting and hunters in Ittoqqortoormiit/ Scoresbysund, NE Greenland', *Arctic Anthropology*, 33(2): 77–93.

Weaver, D., Glenn, C. and Rounds, R. (1996) 'Private ecotourism operations in Manitoba, Canada', *Journal of Sustainable Tourism*, 4(3): 135–46.

Wenzel, G.W. (1991) *Animal Rights, Human Rights: Ecology, Economy and Ideology in the Canadian Arctic*, Toronto: University of Toronto Press.

Wenzel, G.W. (1995) 'Ningiqtuq: Inuit resource sharing and generalized reciprocity in Clyde River, Nunavut', *Arctic Anthropology*, 32(2): 43–60.

Wenzel, G.W. (2005) 'Nunavut Inuit and polar bear: the cultural politics of the sport hunt. indigenous use and management of marine resources', in N. Kishigami and J. Savelle (eds) *Senri Ethnological Series* No. 67, Osaka: National Museum of Ethnology, pp. 363–88.

Wenzel, G.W. and Bourgouin, F. (2003) *Polar Bear Management in the Qikiqtaaluk and Kitikmeot Regions of Nunavut: Inuit, Outfitted Hunting and Conservation*, Unpublished Report to the Department of Sustainable Development, Government of Nunavut.

9 Environmental values of consumptive and non-consumptive marine tourists

Jackie Dawson and Brent Lovelock

Introduction

In 1971 Dr Seuss introduced the world to the eco-friendly Lorax and his environmentally destructive neighbour, the Once-ler. This book, ostensibly for children, tells the tale of two individuals who value the environment in very different ways. The Once-ler capitalises on the forest as an economic commodity while the Lorax rhythmically argues for its natural value: 'I'm the Lorax who speaks for the trees which you seem to be chopping as fast as you please', he says to the Once-ler as the Truffula forest slowly disappears (Dr Seuss 1971: 16). Opposing environmental orientations can be traced back to the utilitarian-conservation debates between John Muir (1838–1914) and Gifford Pinchot (1865–1946) in which Muir publicly crusaded for wilderness preservation, while Pinchot argued for the anthropocentric use of public lands (Nash 1967). More recently this separation of values has been confirmed by researchers such as Dunlap and Heffernan (1975), who have identified a non-consumptive/consumptive dichotomy – or more accurately, continuum. They argue that, like the Lorax and the ecocentric values expressed by Muir, people may be considered non-consumptive, suggesting their actions do not involve extracting anything from the environment. Conversely, individuals such as the Once-ler and the ideas expressed by Gifford Pinchot may be seen as having a consumptive outlook, meaning that their activities involve taking something from, or disrupting the natural environment.

In the field of outdoor aquatic recreation non-consumptive activities may include activities such as sea kayaking, sailing, canoeing, swimming, wildlife viewing and natural photography (Jackson 1989). Weaver (2001) makes a case for distinguishing these activities as non-consumptive, arguing that they must be environmentally sensitive – considering that vessels travelling across water leave no trace. The alternative categorisation, consumptive activities, include pastimes such as hunting and fishing (Vaske *et al.* 1982), which may have significant impacts upon ecosystems. However, there is much academic debate surrounding these distinctions. It has been argued that there is a clear distinction between the two hinging upon the primary goal of consumptive recreationists, whose goal it is to extract something from the environment (Vaske *et al.* 1982). This is debatable considering the fact that some fishers or hunters may be motivated to participate in order to experience nature, to enjoy the company of friends and family or simply to get away from a busy urban life. However, Vaske *et al.* (1982) believe the consumptive aspect of their experience to be critical, as shown in the way that consumptive users report lower satisfaction levels than non-consumptive users – except those hunters and fishers who are successful in killing their prey or bagging their catch.

A recent contribution to the debate as to what exactly defines and differentiates non-consumptive from consumptive tourism is seen in Holland *et al.* (1998) and Fennel (2000) who discuss fishing as a traditionally consumptive activity but consider the possibility of including 'catch and release' fishing as a form of ecotourism. While Holland *et al.* (1998) create a solid argument for billfish angling by relating the activity to the (admittedly loose) definitions of ecotourism, Fennel (2000) disputes the argument, suggesting that despite pro-environmental intentions or motivations, even 'catch and release' fishing should not be considered non-consumptive ecotourism for it is philosophically and fundamentally different in intent. Angling, regardless of intention imposes pain and extracts living things from their natural environment and is therefore consumptive. Arguably, this distinct difference is what separates non-consumptive from consumptive activities; it is critical to the ongoing debate, and is thus the dichotomy adhered to for the purposes of this chapter.

Environmental values

The non-consumptive/consumptive dichotomy raises some important issues regarding how environmental values are reflected in these seemingly polarised forms of tourism. Values can be defined as evolving and enduring beliefs (Rokeach 1973). They represent hypothetical constructs, which are manifested in humans through experience and communication (Pizam and Calantone 1987). Acknowledging the differences between non-consumptive and consumptive tourism it is reasonable to assume that the tourists involved in the different types of activities characterising these types of activities are likely to value the environment quite differently. This presumption has been supported by research conducted to date that suggests that non-consumptive tourists value the environment more than consumptive users. However, contradictory results have been reported (e.g. Van Lierre and Noe 1981; Jackson 1989; Theodori *et al.* 1998).

It is difficult to quantitatively measure intangible feelings and values, and some research results in the field are ambiguous. Consequently, many attempts have been made to refine the assessment techniques employed, including the development of a number of quantitative value scales, e.g. Rokeach's (1968) Value Survey; Kahle *et al.*'s (1986) List of Values; Pelletier *et al.*'s (1998) Motivations Towards the Environment Scale; Dunlap and Van Lierre's (1978) New Environmental Paradigm; and Dunlap *et al.*'s (2000) New Ecological Paradigm. While other methods of value assessment have been used, they are generally based on the above paradigms (e.g. Pinhey and Grimes 1979; Pizam and Calantone 1987; Zwick and Solan 1996). Each of the more commonly used values scales exhibits a number of advantages and disadvantages (see Table 9.1).

Table 9.1 Features of selected environment value assessment methodologies

	Advantages	*Disadvantages*
Value survey (Rokeach 1968)	– Widely used – Vast published results – Successful use – Determined validity – Other paradigms based on it	– Lack of focus – Long and time consuming – Ambiguous and inconclusive in assessing values – Arbitrary and subjective – Researchers question usefulness and meaning – Difficulties with data analysis
List of values (Kahle *et al.* 1986)	– Simple version of value scale – Short 9-point scale – Includes internal–external assessment – Published success	– Assumes values are situationally static – Generic – Not activity or site specific
Motivations towards the environment scale (Pelletier *et al.* 1998)	– Examines values and pro-environmental – Behaviour – Includes internal–external assessment – Unknown due to lack of thorough testing	– New – Not widely used
New environmental paradigm (Dunlap and Van Lierre 1978)	– Widely used – Vast published results – Determined validity	– Inconclusive/unsupportive – Ambiguous and generic – Measures overall world view – Not activity or site specific – Taps 'primitive beliefs'
New ecological paradigm (Dunlap *et al.* 2000)	– Similar to NEP – Updated version of NEO – Increased internal consistency	– Similar to NEP –Not widely tested – Only slightly increased internal consistency

Sources: Geisler *et al.* 1977; Albrecht *et al.* 1982; Braithwaite and Law,1985; Geller and Lasley 1985; Kahle *et al.* 1986; Blamey and Braithwaite 1987; Pizam and Calantone 1987; Homer and Kahle 1988; Jackson 1989; Noe and Snow 1990; Uysal *et al.* 1994; Madrigal 1995; Pelletier *et al.* 1998; Dunlap *et al.* 2000; Lalonde and Jackson 2002

The biggest failing of the value assessment techniques is that they assume environmental values are enduring and stable beliefs. For example, the scales may examine global beliefs or contextual beliefs or situational motivations to participate, but fail to assess all three areas. Recent research suggests that values are not stagnant and instead should be seen as transitional and situationally dependent (Gnoth 1997; Crick-Furman and Prentice 2000) and therefore should be assessed accordingly. Notably, how people depict an environment and what they value within it varies according to their immediate aims and objectives within a particular context (Crick-Furman and Prentice 2000). Other research complements this notion, suggesting that including site-specific environmental values may be more effective in determining true environmental concern than the more traditional worldview assessment. Further support for accessing environmental world-views (global values), resource specific environmental values (contextual values), and motivations to participate in activities (situational values) stems from the ideas driving Vallerand's (1997) Hierarchical Model of Intrinsic and Extrinsic Motivation. Vallerand suggests that, like values, motivation is transitional and exists within the individual at three hierarchical levels of generality: global, contextual and situational. According to Vallerand (1997), examining global, contextual and situational motivation allows researchers to consider motivation with heightened precision and refinement rather than a more generic examination.

The suggestion that the major flaw with current environmental value methodologies is that they assume environmental values are stable, prompted the authors to develop a modified environmental values scale, which takes into account the transitional nature of our attitudes and beliefs. This transitional nature of values (Crick-Furman and Prentice 2000) makes it impossible to construct a universal and flawless assessment scale. However, by combining the most effective aspects of the existing scales, values can be measured to their highest consistency (i.e. Dunlap *et al.*'s (2000), New Ecological Paradigm; Blamy and Braithwaite's (1997), Social Values modified from Rokeach's Values Survey; Vallerand's (1997), Hierarchical Model of Intrinsic and Extrinsic motivation; and Crick-Furman and Prentice's (2000), Multiple Values Research). The incorporation of existing methodologies made it logical to label the new method 'The Integrated Values Scale'. As discussed, there are various aspects to consider when accessing environmental values including: (1) the examination of environmental world-views; (2) the assessment of resource specific environmental values; and (3) evaluating tourists' motivation to participate in the environment. The Integrated Values Scale employed in this research examines all three. This new integrated approach of value examination is unique thus far in the field.

Field testing of the Integrated Values Scale

The Integrated Values Scale (IVS) was used in this research to investigate non-consumptive and consumptive marine tourists in the South Island of New Zealand. The specific non-consumptive and consumptive marine user groups targeted were sea kayaking and sea fishing tourists who utilised ocean waterways around the South Island of New Zealand in the winter and spring of 2002. Sea kayaking and sea fishing tour companies were contacted inviting their participation in the research project. In total 42 per cent (10) of the sea kayaking operators and 58 per cent (15) of the sea fishing operators agreed to participate. The companies were scattered throughout six regions of the South Island thus generating a geographically representative sample of operators (see Figure 9.1).

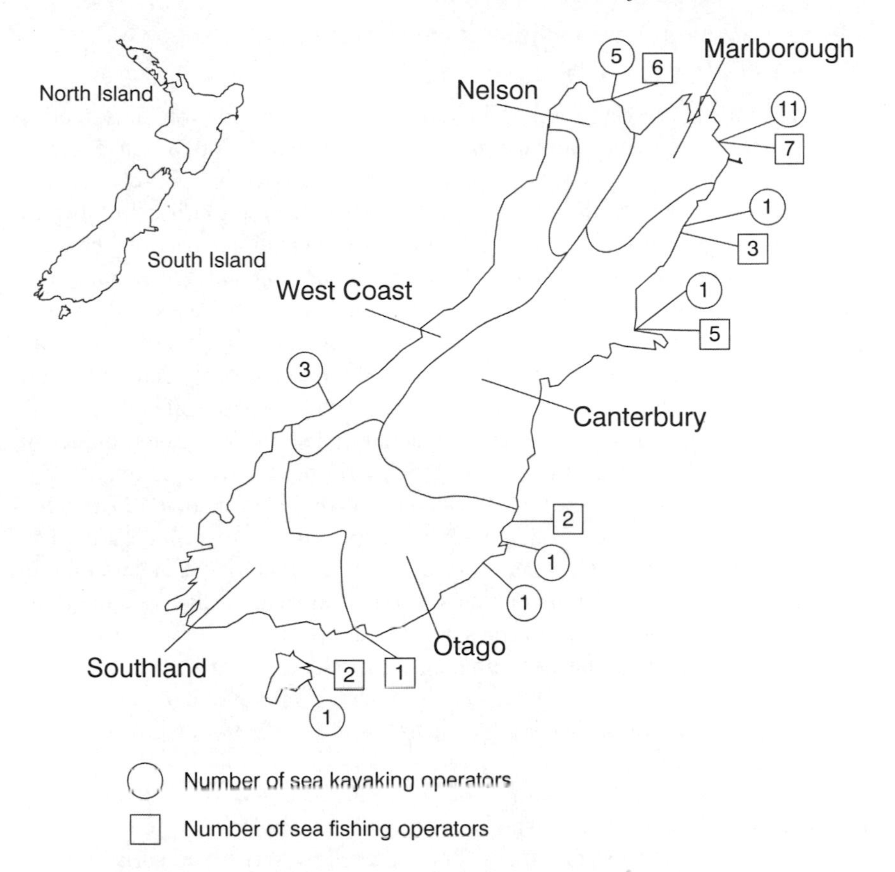

Figure 9.1 Study areas, South Island, New Zealand

The IVS was administered to marine tourists through a self-completion survey questionnaire, which was distributed by the tour operator or by the researcher. In total there were 585 surveys distributed of which 197 were returned, generating a response rate of 33.7 per cent (34.2 per cent sea kayaking and 32.7 per cent sea fishing).

Due to the distribution methodology, which involved targeting individuals participating in commercially guided sea kayaking and sea fishing activities, this method does not take into account the unknown number of tourists participating in these experiences independently from an operator. According to Hobson (1977), those participating independently generally own their own equipment and therefore participate more often. This cohort is a more 'specialised' or advanced group and may report different environmental values from their guided counterparts within the commercial tourism sector.

Findings: environmental values of consumptive and non-consumptive tourists

The purpose of this research was to investigate whether or not non-consumptive and consumptive tourists maintain different environmental values. In order to accomplish this it was important to assess some secondary elements. For example, the first step involved profiling sea kayaking and sea fishing tourists with regard to their demographics, trip characteristics, specialisation, past experience, participation rates, environmental values and environmental behaviours thus allowing for conclusions to be drawn as to whether or not sea kayaking and sea fishing tourism draw distinctly different individuals and could therefore be considered separate user groups. These secondary factors (demographics, specialisation, past experience and participation rates) were specifically chosen, as previous research suggests they are influential in distinguishing environmental value levels (e.g. Dunlap and Heffernan 1975; Hobson 1977; Hines *et al.* 1987; Schreyer, *et al.* 1984; Uysal *et al.* 1994). The second step involved examining the relationship between socio-demographic, experience and specialisation data and environmental values. How important are these characteristics in determining environmental values of marine tourists, compared to an analysis based upon the simple typology of consumptive versus non-consumptive?

Upon examination of the demographics, trip characteristics, specialisation, past experience, participation rates, environmental values and environmental behaviours, of sea kayaking and sea fishing tourists, a number of distinctions were made (see Table 9.2).

The typical sea kayaking tourist could be characterised as a young, highly educated, international female, who was generally unemployed (often a student, or on extended holiday) at the time of the research. In direct comparison, a sea fishing tourist tended to be an older, moderately educated, domestic male, who was generally employed full-time at the time of the research. Further differences evident in profiling the sea kayaking and sea fishing tourists were in their differing participation rates in environmentally 'sensitive' and environmentally 'insensitive' activities. Sea kayaking tourists were more likely to participate in activities which are arguably deemed to have less environmental impact (e.g. sailing, canoeing, swimming, wildlife viewing), while sea fishing tourists were more likely to participate in activities arguably deemed to have higher environmental impact (e.g. jet boating, 4-wheel driving, hunting). Sea kayaking and sea fishing tourists' trip characteristics also diversified the two groups. Sea kayak trips were longer, group sizes were smaller, trips were more formalised (guided), fewer operator employees guided formal trips, travel parties were smaller (solo or one other person versus larger groups), and respondents were generally provided with more environmental interpretation, than those participating on sea fishing trips. These combined differences make it reasonable to believe that the activities, sea kayaking and sea fishing, are quite different and do in fact draw distinctly separate groups of people. The higher recorded environmental value and behaviour scores by sea kayaking tourists versus sea fishing tourists provide a further distinction.

Table 9.2 Characteristics of sea kayaking and sea fishing tourists

Profile category	*Sea kayaking*	*Sea fishing*
Gender	Predominantly female	Predominantly male
Age	Younger	Older
Place of residence	International	Domestic
Formal education	Higher	Lower
Employment status	Often part-time	Full-time
Income	Equal	Equal
Nature of travel party	Smaller	Larger
Frequency of participation in outdoor recreational activities	Higher	Lower
Frequency of participation in environmentally 'sensitive' activities	Higher	Lower
Frequency of participation in environmentally 'insensitive' activities	Lower	Higher
Frequency of participation in respective activities	Lower	Higher
Frequency of participation in each other's activities	Equal	Equal
Ownership of sea kayaking/sea fishing equipment	Lower	Higher
Trip duration	Longer	Shorter
Group size	Smaller	Larger
Formality of trip	Less	More
Environmental value levels	Higher	Lower
Environmental behaviour levels	Higher	Lower

Past research has identified strong statistical relationships between particular demographic characteristics and environmental value segregation (e.g. Luzar *et al.* 1995; Jurowdki *et al.* 1995). Other characteristics that researchers suggest have an influence on environmental values are: past experience in outdoor recreational activities (Schreyer *et al.* 1984); frequency of participation in specific activities (Hobson 1977); and ownership of equipment needed for specific activities (Hobson 1977).

Each of these variables is linked to environmental values with different levels of statistical significance (high, moderate, low) (Table 9.3). For example, moderate relationships were discovered between gender and specialisation with environmental values (Hobson 1977; Luzar *et al.* 1995). Weak relationships emerged with place of residence, income and past experience in outdoor recreational activities (Schreyer *et al.* 1984; Hines *et al.* 1987; Harper 2001), and no relationship was seen between employment status and environmental values. The past research indicates that young, well-educated females who are committed to their activity via frequency in participation, or ownership of recreational equipment, are the most likely cohort to be concerned about the natural environment.

After profiling sea kayaking and sea fishing tourists, thus identifying that these recreationists can in fact be considered separate and distinct user groups, the extent to which statistically significant environmental value differences were related to tourist profiles (demographics and specialisation) and tourist typology

Table 9.3 Relationships between socio-demographics and environmental values – selected studies

Category	*Reference*	*Higher environmental values found in*
Gender	Uysal *et al.* (1994); Luzar *et al.* (1995); Harper (2001)	Females
Age	Dunlap and Heffernan (1975); Geisler (1977); Pinhey and Grimes (1979); Van Liere and Dunlap (1981) Hines *et al.* (1987); Samdahl and Robertson (1989); Luzar *et al.* (1995); Jurowdki *et al.* (1995)	Younger (i.e. <30)
Place of residence	Geisler (1977); Harper (2001)	
Level of education	Dunlap and Heffernan (1975); Geisler (1977); Pinhey and Grimes (1979); Van Liere and Dunlap (1981) Hines *et al.* (1987); Samdahl and Robertson (1989); Luzar *et al.* (1995), Jurowdki *et al.* (1995)	More educated
Income	Hines *et al.* (1987)	
Age of exposure to outdoor recreation	Schreyer *et al.* (1984)	Previous exposure to outdoor recreational activities
Frequency of participation in respective activity	Hobson (1977)	More participation in the respective activity
Ownership of equipment	Hobson (1977)	More specialised in the respective activity

(non-consumptive versus consumptive) was investigated (through independent t-tests and one-way analysis of variance (ANOVA) tests).

The findings of this research were consistent with the previously discovered value relationships. Table 9.4 summarises the strong, moderate and weak statistically significant relationships between values and demographic characteristics in previous research and those revealed in this study. In this research, statistically strong relationships were found between gender and age and environmental values. Moderate relationships were discovered with place of residence, and no relationships emerged between employment status, income levels, past experience in outdoor recreational activities or specialisation and environmental values. Finding no relationship between employment status and environmental values is reflective of past research (e.g. Hines *et al.* 1987). However, weak and moderate relationships have been found with income, past experience and specialisation (Hobson 1977; Hines *et al.* 1987). This difference is acknowledged, and could possibly be attributed to the previously mentioned research limitation, where participants were commercially guided by

Table 9.4 Statistically significant relationships between socio-demographic and environmental values

Category	*Statistical significance found in previous research*	*Statistical significance found in this research*
Gender	✓✓	✓✓✓
Age	✓✓✓	✓✓✓
Place of residence	✓	✓✓
Level of education	✓✓✓	✓✓
Employment status		
Income	✓	
Experience in outdoor recreational activities	✓	
Frequency of participation in respective activity	✓✓	(*)
Ownership of equipment	✓✓	(*)

Notes: ✓=weak statistical significance, ✓✓ = moderate statistical significance, ✓✓✓ = strong statistical significance, (*) = specialisation factors

tourism operators, but independent sea kayaking and sea fishing tourists were excluded.

The results of this research are congruent with those found previously. There is undoubtedly a relationship between the socio-demographic characteristics of sea kayaking and sea fishing tourists and their environmental values. Furthermore, the socio-demographic characteristics linked to environmental values, are those characteristics that separate the consumptive from the non-consumptive tourists. The highest environmental values are seen in young, educated females with previous exposure to the outdoors – this is also the predominant descriptor of the typical (non-consumptive) sea kayaking tourist in this research.

Conclusion

That non-consumptive sea kayaking and consumptive sea fishing tourists are two distinct user groups in terms of their socio-demographic characteristics and environmental values has been empirically demonstrated in this chapter. Examining the level to which these user groups value the environment creates a foundation for widespread understanding of tourists' relationship with the environment. In this research it was discovered that sea kayaking (non-consumptive) tourists value the environment more than sea fishers (consumptive), however, both groups indicated significant appreciation for the situational environments in which they recreate. The finding that non-consumptive marine tourists value the environment more than consumptive marine tourists is generally supported by environmental values research (e.g. Dunlap and Heffernan 1975; Jackson 1989; Zwick and Solan 1996), that has previously examined this dichotomy, but in terrestrial settings. Interestingly, however, recent work has suggested that this dichotomy may be too simplistic, and that further divisions occur within the consumptive group.

This Norwegian study revealed that various *types* of fishing and various *types* of hunting show different associations with environmental attitudes as measured by the NEP (Bjerke *et al.* 2006).

The incorporation of the new 'Integrated Values Scale' proved successful within this research. The scale uncovered statistically significant relationships between gender, age and education and environmental values. These characteristics are the same elements that Uysal *et al.* (1994), Jurowdki *et al.* (1995) and Luzar *et al.* (1995), among others, also found to be most influential. In contrast with other previously used scales, the IVS did not reveal any relationship between specialisation and environmental values.

The fact that these marine user groups both value the environment (to some extent) is promising, in view of the current growth of nature-based tourism. The demonstrable differences between non-consumptive sea kayaking and consumptive sea fishing tourists should be capitalised on in order to encourage pro-environmental values and behaviours in the future. This could be accomplished through specific user-group-targeted education programmes. For example, it was discovered that sea kayaking tourists maintain intrinsic and ecocentric motivations towards activity participation versus the extrinsic and anthropocentric orientations held by sea fishers. Catering to these orientations could be effective in enhancing environmental values and behaviours – for example through providing pre-trip environmental interpretation that addresses sea fishers' more extrinsic and anthropocentric orientations.

Similar outcomes could be achieved by utilising incentives. Offering consumptive wildlife tourists rewards for pro-environmental behaviours in activity-specific situations could be implemented. This could be applicable, for example, for sea fishing tourists who often engage in serious sport fishing competitions as well as friendly companionship competitions. These competitions could involve incentives whereby tangible rewards are presented to those consumptive sea fishers who adhere to environmental regulations. For example, sea fishers could be rewarded for sustainable practice with the presentation of tangible rewards such as fishing gifts or being featured in a fishing magazine. Discounts could be offered on fishing or hunting licences for individuals who engage in positive environmental practices or who are active members of environmental organisations.

For the future of consumptive wildlife tourism (as well as non-consumptive tourism) it is important to continue to examine issues such as the environmental orientation of user groups. Such research will be able to suggest ways of fostering pro-environmental values among consumptive wildlife tourists, thus contributing to sustainable tourism outcomes within this growing tourism sector.

References

Bjerke, T., Thrane, C. and Kleiven, J. (2006) 'Outdoor recreation interests and environmental attitudes in Norway', *Managing Leisure*, 11(2): 116–28.

Blamey, R.K. and Braithwaite, V.A. (1997) 'A social values segmentation of the potential ecotourism market', *Journal of Sustainable Tourism*, 5(1): 29–45.

Crick-Furman, D. and Prentice, R. (2000) 'Modelling tourists' multiple values', *Annals of Tourism Research*, 27(1): 69–92.

Dr Seuss (1971) *The Lorax*, New York: Random House.

Dunlap, R.E. and Heffernan, R.B. (1975) 'Outdoor recreation and environmental concern: an empirical examination', *Rural Sociology*, 40(1): 18–30.

Dunlap, R.E. and Van Lierre, K.D. (1978) 'The new environmental paradigm', *Journal of Environmental Education*, 9(4): 10–19.

Dunlap, R.E., Van Liere, K.D., Mertig, A.G. and Jones, R.E. (2000) 'Measuring endorsement of the new ecological paradigm: a revised NEP scale', *Journal of Social Issues*, 56(3): 425–42.

Fennel, D.A. (2000) 'Comment: ecotourism on trial – the case of billfish angling as Ecotourism', *Journal of Sustainable Tourism*, 8(4): 341–5.

Gnoth, J. (1997) 'Tourism motivation and expectation formation', *Annals of Tourism Research*, 21(2): 283–304.

Harper, C.L. (2001) *Environment and Society: Human Perspectives on Environmental Issues*, 2nd edn, Upper Saddle river, NJ: Prentice-Hall Inc.

Hines, J.M., Hungerford, H.R. and Tomera, A.N. (1987) 'Analysis and synthesis of research on responsible environmental behaviour: a meta analysis', *Environmental Education*, 18: 1–8.

Hobson, B. (1977) 'Leisure value systems and recreational specialization: the case of trout fishermen', *Journal of Leisure Research*, 9(3): 174–87.

Holland, S.M., Ditton, R.B. and Graefe, A.R. (1998) 'An ecotourism perspective on billfish fisheries' *Journal of Sustainable Tourism*, 6(2): 97–115.

Jackson, E.L. (1989) *Environmental Attitudes, Values, and Recreation: Mapping the Past, Charting the Future*, State College, PA: Venture Publishing.

Jurowdki, C., Uysal, M., Williams, D.R. and Noe, F.P. (1995) 'An examination of preferences and evaluations of visitors based on environmental attitudes: Biscayne Bay National Park', *Journal of Sustainable Tourism*, 3(2): 73–86.

Kahle, L.R., Beatty, S.E. and Homer, P.M. (1986) 'Alternative measurement approaches to consumer values: the list of values (lov) and values and lifestyles (vals)', *Journal of Consumer Research*, 13: 405–9.

Luzar, J.E., Diagne, A., Gann, C. and Henning, B.R. (1995) 'Evaluating nature based tourism using the new environmental paradigm', *Journal of Agriculture and Applied Economics*, 27(2): 544–55.

Nash, R. (1967) *Wilderness and the American Mind*, London: Yale University Press.

Noe, F. and Snow, R. (1990) 'The new environmental paradigm and further scale analysis', *Journal of Environmental Education*, 21: 20–6.

Pelletier, L., Green-Demers, I., Tuson, K.M., Noels, K. and Deaton, A.M. (1998) 'Why are you doing things for the environment? The motivation toward the environment scale (MTES)', *Journal of Applied Social Psychology*, 28(5): 437–68.

Pinhey, T.K. and Grimes, M.D. (1979) 'Outdoor recreation and environmental concern: a re-examination of the Dunlap–Heffernan thesis', *Leisure Sciences*, 2: 1–11.

Pizam, A. and Calantone, R. (1987) 'Beyond psychographics – values as determinants of tourist behaviour', *International Journal of Hospitality Management*, 6(3): 177–81.

Rokeach, M. (1973) *The Nature of Human Values*, New York: Free Press.

Schreyer, R., Lime, D.W. and Williams, D.R. (1984) 'Characterizing the influence of past experience on recreation behaviour', *Journal of Leisure Research*, 16(1): 34–50.

Theodori, G.L., Luloff, A.E. and Wilits, F.K. (1998) 'The association of outdoor recreation and environment concern: re-examining the Dunlap–Heffernan thesis', *Rural Sociology*, 63(1): 94–108.

Uysal, M., Jurowski, C, Noe, F.P. and McDonald, C.D. (1994) 'Environmental attitude by trip and visitor characteristics', *Tourism Management*, 15(4): 284–94.

Vallerand, R.J. (1997) 'Toward a hierarchical model of intrinsic and extrinsic motivation', *Advances in Experimental Social Psychology*, 29: 271–360.

Van Liere, K.D. and Noe, F.P. (1981) 'Outdoor recreation and environmental attitudes: further examination of the Dunlap–Hefferenan thesis', *Rural Sociology*, 46(3): 505–13.

Vaske, J.J., Donnelly, M.P., Hegerlein, T.A. and Shelby, B. (1982) 'Differences in reported satisfaction ratings by consumptive and non-consumptive recreationists', *Journal of Leisure Research*, 14(3): 195–206.

Weaver, D.B. (ed.) (2001) *The Encyclopedia of Ecotourism*, New York: CABI Publishing.

Zwick, R. and Solan, D. (1996) *Value Differences Between Consumptive and Non-consumptive Recreationists*, Proceedings of the 1996 Northeastern Recreation Research Symposium. New York: Northeastern Recreation Research, General Technical Report (NE 232): pp. 223–6.

10 The success and sustainability of consumptive wildlife tourism in Africa

Joseph E. Mbaiwa

Introduction

Safari hunting has a long history in Africa. It started in the 1800s with the arrival of European traders in the continent (Tlou 1985). However, safari hunting was not perceived as a tourism activity in Africa until the 1960s and 1970s when most African countries obtained independence from European colonial rule. Safari hunting has since become formalised as part of the tourism industry. It is commonly referred to as consumptive tourism of which sport or trophy hunting is the main tourist activity. Although safari hunting is one of the main tourism activities in Eastern and Southern African countries, studies have shown that global participation in safari hunting tourism is on the decline (MacKay and Campbell 2004). This decline is largely a result of the opposition to hunting by anti-hunting groups (Baker 1997; MacKay and Campbell 2004). Anti-hunting groups argue that the international killing of wild animals is not only immoral and abhorrent but it is also one of the means by which the extinction of animal species has been accelerated (Baker 1997). While this may be so, governments in Africa encourage safari hunting because of the assumption that it will contribute to economic development of their countries. Governments view tourism as a rapid means for national and regional development, bringing employment, exchange earnings, balance of payments advantages and important infrastructure developments benefiting both host populations and visitors (Glasson *et al.* 1995). Safari hunting is thus an important economic activity desirable for economic reasons in many countries.

Since the 1980s, issues of sustainability in natural resource use including safari hunting have become critical. As a result, African countries that encourage safari hunting find themselves in a dilemma as to whether hunting in their countries constitutes sustainable tourism. These countries face pressure from international conservation organisations like the International Conservation of Nature and Flora and Fauna (IUCN) and the United Nations Convention on International Trade in Endangered Species (CITES) to promote the sustainable harvesting of wildlife resources. The concern over the sustainable harvesting of wildlife resources is part of the global effort to halt the degradation of natural resources. The goal to achieve sustainable wildlife utilisation is guided by the concept of sustainable development, formalised by the World Commission on Environment and Development in 1987.

Sustainable development has become a popular environmental management concept, to the extent that many in tourism research are now advocating for sustainable tourism development. In relation to consumptive wildlife tourism or safari hunting, sustainable tourism development would imply that wildlife resources should be harvested to meet the needs of the present generations without jeopardising the wildlife resource needs of future generations (Mbaiwa 2004). The objective of this chapter, therefore, is to use the sustainable tourism development framework to examine the success and sustainability of consumptive wildlife tourism in Africa. The chapter specifically assesses the economic and environmental impacts of consumptive wildlife tourism in Africa. Examples are largely drawn from the countries of Botswana, South Africa, Namibia, Zimbabwe and Tanzania. The chapter also examines the challenges of safari hunting tourism and explores prospects for the sustainable use of wildlife resources.

Consumptive wildlife tourism in Africa

Consumptive wildlife tourism in Africa began in the 1960s and 1970s and is undertaken in areas outside national parks and game reserves. However, some game reserves like those in South Africa and Tanzania allow safari hunting. There are three main hunting areas where safari hunting is carried out in Africa: areas designated for Community-Based Natural Resource Management (CBNRM) projects; concession areas; and game farms.

The Community-Based Natural Resource Management Programme

In Eastern and Southern Africa, local community participation in consumptive wildlife tourism is carried out through the CBNRM programme. The CBNRM programme is a collective term used for a number of similar but unconnected programmes in different countries of Eastern and Southern Africa. The programme aims at addressing problems of land use conflicts, the lack of direct wildlife-related economic benefits to people living in wildlife areas, and local community participation in wildlife resource management (Mbaiwa 2004). The basic principle behind the CBNRM programme is that of reforming the conventional 'protectionist conservation philosophy' and 'top down' approaches to development, and it is based on common property theory which discourages open access resource management, and promotes resource use rights of the local communities (Kgathi *et al.* 2002). The CBNRM programme assumes that once rural communities participate in natural resource utilisation and derive economic benefits from it, this will cultivate the spirit of ownership and will ultimately lead them to use natural resources found in their local areas sustainably (Twyman 2000; Tsing *et al.* 1999; Mbaiwa 2004). The CBNRM programme is so far carried out in Botswana, Namibia, Zimbabwe, Kenya, Tanzania, Malawi, Mozambique and Zambia. The Zimbabwean version of the CBNRM programme is known as the Communal

Area Management Programme for Indigenous Resources (CAMPFIRE) and was the first to be established in Africa, in 1986. For purposes of illustrating how consumptive wildlife tourism is carried out by rural communities in Africa, the CAMPFIRE programme of Zimbabwe and Botswana's CBNRM programme are briefly described below.

CAMPFIRE and consumptive wildlife tourism in Zimbabwe

Child *et al.* (2003) argue that CAMPFIRE is a long-term programmatic approach to rural development that uses wildlife and other natural resources as a mechanism for promoting devolved rural institutions and improved governance and livelihoods. The cornerstone of CAMPFIRE is the devolution of the rights to benefit from, dispose of, and manage natural resources. The idea of CAMPFIRE came about as a result of resource degradation in rich biodiversity areas of Zimbabwe. This decline was blamed on rural communities for their failure to use natural resources found in their local environment sustainably. In the 1980s, international conservation agencies like the IUCN noted that one of the factors that cause resource decline is the exclusion of rural communities from resource management. The exclusion of rural communities made them antagonistic to conventional wildlife management approaches and this led to the unsustainable use of natural resources such as poaching of game animals. As ideas of public participation became popular in the 1980s, it became necessary that rural communities should be given a major role in resource utilisation and management, particularly of wildlife, in order to minimise resource decline. The CAMPFIRE programme was thus adopted to provide an opportunity for rural participation in wildlife resource management. The programme was to be achieved through consumptive wildlife tourism.

A review of the CAMPFIRE programme found that after almost two decades of implementation, the programme has been a success (Child *et al.* 2003). The review points out that CAMPFIRE has socio-economic and environmental benefits. Environmental benefits include the protection of an area of wild land roughly equivalent in extent to the Parks and Wildlife Estates of Zimbabwe (i.e. some 50,000 square kilometres). There is also an increase of wildlife population in areas reserved for safari hunting. In the 10 years since its inception, wildlife populations increased by about 50 per cent, with elephant doubling from 4,000 to 8,000 in CAMPFIRE Areas (Child *et al.* 2003). The increase in the elephant population in CAMPFIRE hunting areas disputes the popular belief by anti-hunting groups that safari hunting leads to a decline of wildlife species. Instead, safari hunting has become one of the land use options that can promote the sustainability of wildlife species in Zimbabwe. CAMPFIRE is also recognised for having reduced or contained veld fires in various districts (Child *et al.* 2003). CAMPFIRE has also led to the reduction of land use conflicts between agricultural production and wildlife management. Poaching has also been contained resulting in reduced levels of illegal wildlife off-take (Child *et al.* 2003). These environmental benefits indicate that CAMPFIRE is an effective strategy that promotes the sustainable wildlife harvesting for tourism purposes.

Social benefits of the CAMPFIRE programme include its acceptance countrywide as shown by a total of 23 districts that have established CAMPFIRE projects (Child *et al.* 2003). The membership of the CAMPFIRE Association has also increased to 52 of Zimbabwe's 57 Rural District Councils in the last 10 years. This shows that rural communities accept aims of CAMPFIRE which are primarily the conservation of wildlife resources and rural development through safari hunting tourism. Economic benefits from CAMPFIRE include an increase of revenues from safari hunting to US$2 million annually (Child *et al.* 2003). Between 1989 and 2001, CAMPFIRE generated direct income of over USD20 million, with an economic impact of US$100 million (Figure 10.1).

The revenue generated from CAMPFIRE is being re-invested in community projects such as building schools in rural areas and buying boreholes to provide water to both human beings and livestock. In addition, revenue has been devolved to participating communities at household levels (Table 10.1). Muir-Leresche *et al.* (2003) indicate that, by the late 1990s, an estimated 90,000 households (630,000 people) were benefiting from CAMPFIRE revenue. This also shows that safari hunting tourism benefits the poor rural economies in Africa.

Revenue obtained from safari hunting has also been reinvested by rural communities in the construction of lodges. Child *et al.* (2003) note that at least 12 high-end tourism lodges have been developed in communal areas with funds generated through safari hunting. These lodges provide employment opportunities to people in rural Zimbabwe that were not available before the CAMPFIRE programme was initiated in 1986. Thus consumptive wildlife tourism is not only beneficial to the safari hunters that visit Zimbabwe from developed countries, but it also has benefits to rural communities and encourages the sustainable use of environmental resources. Consumptive tourism is thus important to small rural economies in Zimbabwe and also promotes sustainability in wildlife use.

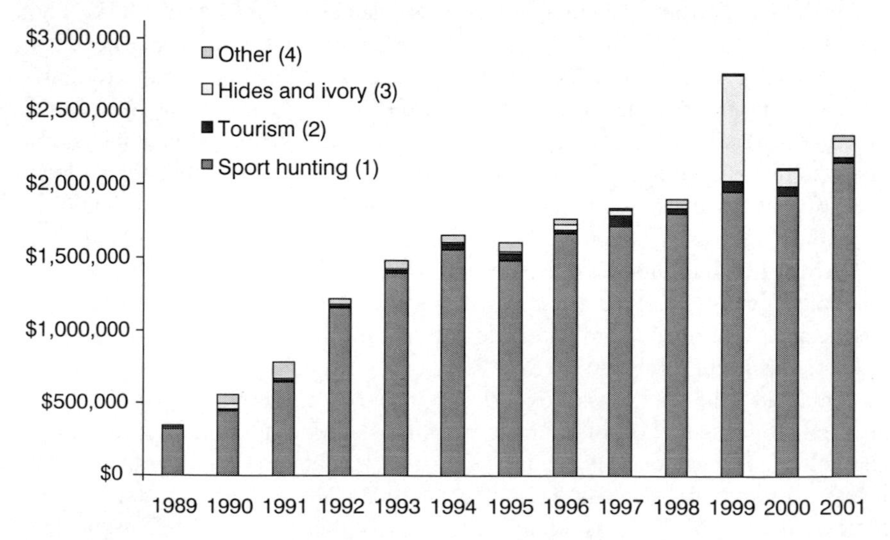

Figure 10.1 Direct Income to CAMPFIRE, 1989–2001 (Source: Child *et al.* 2003)

Table 10.1 Summary of CAMPFIRE Revenue and Ward and Household Beneficiaries

Year	*Total income US$*	*Number of districts*	*Number of wards*	*Number of households*
1989	348,811	3	15	7,861
1990	556,433	9	41	22,084
1991	776,021	11	57	52,456
1992	1,216,678	12	74	70,311
1993	1,483,873	12	98	90,475
1994	1,642,671	14	101	96,437
1995	1,591,567	14	111	98,964
1996	1,755,912	19	96	85,543
1997	1,837,438	17	98	93,605
1998	1,891,766	15	92	80,498
1999	2,753,958	16	112	95,726
2000	2,105,204	14	108	88,072
2001	2,328,452	14	94	76,683
Total	20,288,784			

Source: Child *et al.* (2003)

CBNRM and consumptive wildlife tourism in Botswana

Consumptive wildlife tourism in Botswana is carried out in areas around national parks and game reserves known as Wildlife Management Areas (WMAs). WMAs occupy about 22 per cent of Botswana's surface land area while national parks and game reserves occupy 17 per cent of the country's surface area. This means that a total of 39 per cent of Botswana's surface land area is set aside for wildlife protection (Botswana's land surface area is about 581,730 square kilometres, that is, about the size of Kenya, France or the state of Texas in the United States). The concept of Wildlife Management Areas (WMAs) arose from a need for conservation and controlled utilisation of wildlife outside national parks and game reserves, along with the desirability of creating buffer zones between parks and reserves and areas of more intensive land use. WMAs are further sub-divided into smaller land units known as Controlled Hunting Areas (CHAs). Botswana is divided into 163 CHAs which are zoned for various types of wildlife utilisation (both consumptive and non-consumptive uses). The Government leases CHAs to rural communities for CBNRM projects and to safari companies for safari hunting purposes. The CHAs that are directly leased to safari companies are also known as concession areas. The Department of Wildlife and National Parks (DWNP) uses CHAs as administrative blocks to determine wildlife quotas for safari hunting by rural communities involved in CBNRM and safari hunting companies in concession areas.

The number of CBNRM projects in Botswana has grown rapidly since 1996. By 2006, there were 91 CBNRM projects registered in Botswana. These projects cover approximately 150 villages in Botswana's 10 districts and serve a total of 135,000 people – 10 per cent of Botswana's population (Schuster 2007). In the Okavango Delta, 12 of them were involved in Joint Venture Agreements (JVA) with at least seven private safari companies (Table 10.2). Revenue from these projects was 16.3 million Botswana Pula (BWP) in 2006 (Schuster 2007). Arntzen *et al.* (2003) note that this is a lot of money considering the small size of villages involved in CBNRM in Botswana. They also note that employment generated by CBNRM projects is estimated to around 1,000–1,500 jobs with an average employment of 21 employees per project in 2001. Safari hunting by local communities thus serves as an alternative form of employment in wildlife regions. People employed by the CBNRM project at Sankoyo Village in the Okavango and by the joint venture safari hunting company have improved their shelter (homes), support siblings to meet the costs associated with school and provide support for their families (Arntzen *et al.* 2003). In this regard, safari hunting has a fairly substantial impact on livelihoods on the residents of Sankoyo and other communities in Botswana.

Different communities engaged in CBNRM are re-investing revenue from safari hunting into other income generating activities like photographic tourism. Mbaiwa (2004) indicates that in re-investing funds from consumptive tourism, the Sankoyo community established a 16-bed photographers lodge (Santawani Lodge), a cultural tourism centre (Shandrika) where tourists can view the cultural activities and way of life of the people of Sankoyo and a campsite (Kazikini) where tourists who do not want to stay in a lodge can camp. Santawnai Lodge, Kazikini and Shandrika generate income (Table 10.3) and employment for the people of Sankoyo. Santawani Lodge and Kazikini Campsite respectively employed 16 and 15 people in June 2004.

Arntzen *et al.* (2003) note that the Sankoyo CBNRM project is heavily dependent on wildlife resources, particularly the wildlife quota where over 70 per cent of its income is from the sale of the wildlife quota to safari hunters. This income subsequently ends up in the households in the form of dividends. Mbaiwa (2004) indicates that between 1996 and 2001, each household was paid BWP200, this sum increased to BWP250 in 2002, BWP300 in 2003 and BWP500 in 2004 (in 1996, all 34 households at Sankoyo received the dividends, households increased to 49 households in 2004). It is from this background that safari hunting

Table 10.2 Revenue generated from CBNRM projects, 2006

Activity	*Amount in Pula*	*Percentage of total*
Trophy (safari) hunting	11,900,000	72
Photographic and cultural tourism	3,100,000	20
Veld marketing	710,801	4
Craft production	600,000	4
Total	16,310,801	100

Source: Schuster (2007) 1 USD = 6 BWP (Sept. 2006)

Table 10.3 Income generated by Kazikini Campground and Santawani Lodge (in BWP)

Year	*Kazikini campground and restaurants*	*Santawani Lodge*
2001	27,926	48,204
2002	26,623	59,897
2003	159,746	188,536
Total	214,295	296,637

Source: Mbaiwa (2004) 1 USD = 6 BWP (Sept. 2006)

tourism has become a source of rural livelihood option to many communities living in wildlife areas of Botswana. Because of the economic benefits that rural communities derive from safari hunting, poaching is reported to have gone down in the last decade (Mbaiwa 2004; Arntzen *et al.* 2003).

Finally, the tourism industry in Botswana is the second largest economic sector after diamond mining. It contributes 5 per cent to Botswana Gross Domestic Product (Mbaiwa 2004). In analysing all aspects of consumptive wildlife tourism in Botswana, a study by the Botswana Wildlife Management Association (BWMA), an association of safari hunting companies in Botswana found that the turnover of the industry in 2000 is estimated at nearly BWP60 million. About 47 per cent of this revenue was generated from daily fees, 35 per cent from trophy income and the balance from levies, tips and curio sales (BWMA 2001). The BWMA indicates that the contribution to the local economy is 49.5 per cent, the national economy 24.8 per cent and other parties 24.8 per cent. As a result, BWMA notes that the per capita contribution of safari hunting in the rural districts where hunting is carried out in Botswana is worth more than 10 times the per capita contribution of the industry to Botswana as a whole, making sport hunting a vital component in rural economies. The sustainability of consumptive wildlife tourism in Botswana is assisted through regulation of wildlife harvesting. It is carried out over hunting seasons (April to September) in specific demarcated areas known as CHAs. The number of animals hunted in a particular year and in a concession area is also determined after annual wildlife statistics are provided by the DWNP (Mbaiwa 2004). Because of these limitations placed on wildlife harvesting in Botswana, the BWMA (2001: 6) notes 'analysis of quota trends and trophy quality indicates that the current off-take is probably within acceptable limits'.

The safari hunting tourism industry in Botswana remains one of the pillars of the tourism industry in the country. It continues to attract many safari hunters from developed countries. The BWMA indicates that the attraction for discerning hunters to come to Botswana is a combination of the standard of ethics with the hunt and the premium associated with hunting in un-fenced open country. The BWMA further notes that the attraction provided by dangerous game and charismatic species will continue in Botswana, and a premium will still be paid for hunting good quality trophies.

Game farming

Consumptive wildlife tourism in game ranches is increasingly becoming popular in Africa. According to the BWMA (2001), there is a dramatic conversion of cattle ranches into game ranches for safari hunting tourism purposes in Southern Africa. The BWMA indicates that in South Africa, Zimbabwe and Namibia, there is a close economic association between safari outfitters and game ranchers and, in many cases, the same individuals and companies are involved in both sectors. Game farming in Botswana is carried out in the Haenaveld Farms and along the Molopo River and in the Tuli Block Farms. Cattle ranches in these areas are being converted into game farms for both photographic and safari hunting tourism purposes. While game farming in Botswana, Namibia and Zimbabwe is rather small and at an infant stage, it is however more developed in South Africa. South Africa is the leading country in game farming and consumptive wildlife tourism not only in Africa but also in the entire world. Bezuidenhout (2003) states that 85 per cent of all Africa's trophy exports come from South Africa, confirming that consumptive wildlife tourism is more developed there than in the the other African countries.

Game farming in South Africa has a longer history than in other African states. Louw (2004) notes that game farming in South Africa dates back to 1945. The first game farm was in the Walboom district near Thabazimbi (Louw 2004) and game farming continued to grow throughout the 1960s. However, at that time there was no economic value attached to game farming except for free hunting trips organised by the farmer's friends and family to that particular ranch. Louw (2004) indicates that hunting was at the time considered a recreational activity to be enjoyed by those with time and the inclination. Safari hunting became an organised tourism activity around the 1970s, and since then South Africa's game farming industry rapidly grew to its present form. The end of Apartheid and the release of Nelson Mandela from prison in 1990 saw an end to economic sanctions and the acceptance of South Africa in the world economic market. These changes positively impacted on South Africa's economic sectors including consumptive wildlife tourism. The commercialisation of hunting in farms since the 1990s has led to an increase in game farms for tourism purposes. At present, South Africa boasts 5,000 game ranches and more than 4,000 ranches with a mixture of game and stock (Louw 2004). The increase in game farms in South Africa is a result of the fact that game farming has proved to be more profitable than livestock farming (Lunn 2004). As the South African game industry continues to grow, game numbers have also increased to the point that a market had to be found for 425,000 excess game animals during 2001 (Louw 2004). Table 10.4 shows hunting statistics for Eastern Cape, Northern Cape and the Limpopo Province which are the three leading safari hunting provinces in South Africa for the period 1 November 1999 to 31 October 2000.

In terms of revenue generation, in 1995, a total of 120,000 local hunters spent about SAR850 million on hunting. In 1997, the gross annual income of the private

Table 10.4 Hunting statistics in three leading hunting provinces of South Africa

	Eastern Cape	*Northern Cape*	*Limpopo Province*
Active hunting outfitters	91	77	334
Active professional hunters	187	215	463
Clients	1,002	452	941
Total animals	7,915	3,552	4,666
Animals/client	7.90	7.86	4.96
Clients/hunting outfitters	11.01	5.87	2.2
Clients/professional hunters	5.36	2.1	2.03
Total hunting days	9,223	3,729	9,900
Average length of hunt	9.2	8.25	10.52
Complaints investigated	3	11	67

Source: Louw (2004)

Table 10.5 Gross income of South Africa's game industry in 2000

Activity	*Income generated (in Rands*)*
Biltong hunters	450 million
Trophy hunters	153 million
Live game sales	180 million
Eco-tourism	40 million
Venison sales	20 million
Total	843 million

Source: Louw (2004) *1 US$ is equivalent to approx. SAR 7.00 (mid-2006)

game industry was estimated to be SAR1,000 million. Revenue generated from safari hunting in 2000 is shown in Table 10.5.

The above findings indicate that game farming and consumptive wildlife tourism positively contributes to the economic development of South Africa. Game farming in South Africa and in the rest of the African Continent particularly in Botswana, Zimbabwe, Namibia, Tanzania and Kenya is carried out based on modern scientific methods of farming. In this sense, game farming is carried out based on environmentally sustainable practices particularly in harvesting and breeding of wildlife species. Based on these observations, it is rather difficult to assume that safari hunting will lead to the degradation of wildlife species as anti-hunting groups argue.

Problems of consumptive wildlife tourism in Africa

Consumptive wildlife tourism in Africa has several problems that threaten its sustainability, these include the following.

Repatriation of revenue

Tourism development in Africa is characterised by the repatriation of revenue to developed countries. The dependency of the African tourism industry on tour operators (outfitters), tourists, imported food, airlines and travel agents from industrialised countries has resulted in substantial revenue leakages from African (Mbaiwa 2005). Most tourism companies in developing countries enjoy tax holidays. Where such companies should pay tax, it is often difficult for Africa governments to obtain the tax income because packages are bought and paid for outside Africa in developed countries. These companies never file tax returns in Africa since all the financial transactions are made outside the continent (South Africa may be an exception). In her study of Tanzania's safari hunting industry, Baker (1997: 276) observed that, 'most African hunting safaris are arranged by specialized companies called outfitters, many of which are based in the United States'. According to Baker, these companies sell hunting packages based on the desires of the client while in the United States. The outfitter makes most or all the logistical arrangements for a safari hunting trip in Tanzania including the acquisition of the necessary permits and the provision of a professional hunter to accompany the tourist hunter. As an illustration to her argument, Baker provides examples of a Houston, Texas based company operating in Tanzania and also a Colorado based company. The nature of operation of these two companies, and others, through having their headquarters in the United States where all their clientele originate and packages are sold, has led to substantial revenue leakages from Tanzania to the United States.

As in Tanzania, most African countries involved in tourism depend on safari companies from Western countries. In Botswana, safari hunting companies are also largely foreign owned with headquarters in developed countries. Some of the major safari hunting companies operating in Botswana include: Safari South, Johan Calitz Safaris, Landela Botswana, HCH Safaris and Rann Hunting Safaris. As is the case with outfitters operating in Tanzania, safari hunting activities organised by these companies are undertaken outside Botswana where their clients originate. According to Scott Wilson Consultants (2001), safari hunting in Botswana particularly in the Okavango Delta starts in the United States. These consultants note that safari hunters from all over the world attend the Safari Club International (SCI) convention held every January and organised in various cities in the United States. At this convention safari hunting companies operating in Botswana sell their hunts for up to 2–3 years in advance. The majority of the hunters that come for sport or safari hunting are Americans (followed by Spanish and Italians). Thus while hunting may be undertaken in wilderness areas of Botswana, all other arrangements are made in the United States, hence the revenue leakages.

Safari hunting companies operating in Botswana have small convenience offices in the gateway tourism towns of Maun and Kasane. These offices are meant to facilitate the movement of hunting clients upon arrival from the airport to and from hunting areas. The tourists' hunting packages are paid for in developed countries and include airfares, hunting permits, accommodation while in Botswana and food. The food offered to hunting tourists while in Botswana is generally imported, accommodation is in luxurious tents owned by these foreign safari hunting companies and the airlines tourists use are foreign-based (except for Air Botswana which has flights from Botswana to Johannesburg and Cape Town in South Africa). As a result of this arrangement, it has been difficult to retain much of the tourism revenue (i.e. both consumptive and photographic) in Botswana. Studies (e.g. BTDP 1999; Mbaiwa 2005) have shown that Botswana retains only 29 per cent of the tourism revenue while 71 per cent leaks out of the country.

The problems of consumptive wildlife tourism in Africa indicate that even if it is economically beneficial to host countries, particularly when considering employment opportunities and income that remains in rural communities, much of the funds leak to developed countries. This problem is likely to remain for some time into the future mainly because the majority of African countries (maybe with the exception of South Africa) still have limited capacity in terms of capital, entrepreneurship and marketing skills to effectively operate tourism businesses. This means that consumptive tourism in Africa will remain dependent on developed countries for its survival for a considerable number of years to come. This model of consumptive wildlife tourism thus poses questions of sustainability, particularly in terms of social equity. While the industry might appear environmentally sustainable, the distribution of economic benefits between African countries and their partners in developed countries is highly skewed towards the latter. The ideals of a fully sustainable tourism development in Africa thus are illusive and difficult to achieve.

The ban on elephant products by CITES

The global ban on elephant products by the Convention of International Trade in Endangered Species of Wild Fauna and Flora (CITES) has impact on the development of consumptive tourism in Africa. CITES regulates trade in endangered species through listing species on three 'appendices' which have restrictions attached to each of them. Elephant resources are currently listed under Appendix I. Appendix I species are those threatened with extinction as well as being actually or potentially affected by trade. As a result, export and import of Appendix I species is not allowed by signatory nations except under specialised conditions of non-commercial use, such as scientific research. While there is a global ban on elephant products, elephant hunting and sale of products is a very profitable business in Eastern and Southern Africa. For example, studies in Botswana by BWMA (2001), Mbaiwa (2004), and Arntzen *et al.* (2003) have shown that elephant hunting is very popular with safari hunters from industrialised countries and it is so far the most profitable safari hunting activity in the country. Its demand is further shown by the sale of

Figure 10.2 Elephants in the Okavango Delta, Botswana. Photo: J.E. Mbaiwa.

animals three years in advance to safari hunters from all over the world attending the SCI convention in the United States (Scott Wilson Consultants 2001). However, the CITES ban limits the ability of these destinations to fully maximise the economic benefits of the elephant hunting resource.

Although the ban on elephant products is acceptable in Eastern Africa, particularly in Kenya, it has been received with mixed feelings in Southern Africa especially in Botswana, South Africa, Namibia and Zimbabwe. East Africa supports the ban because elephant populations in the area are low. On the other hand, Southern Africa does not support the ban because elephant populations in the region are high. The trade embargo on elephant products has led to the rapid increase of the elephant population in Southern Africa. For example, Botswana's elephant population increased from 45,449 elephants in the 1980s to 79,480 by 1995 and to 122,000 animals by 2003 (Mbaiwa 2004). The DWNP in Botswana states that the country's elephant population is beyond the carrying capacity of rangelands and that elephants have proved to be environmentally destructive to the vegetation. In addition, elephants now roam from areas designated for wildlife conservation to nearby crop fields owned by subsistence farmers and cause some crop damage. This contributes to land use conflicts in the area. Although the ban on elephant products is meant to promote the conservation of the African elephant, it has also led to environmental and land use conflicts in Southern Africa. The indiscriminate ban on elephant products by CITES is thus failing to promote both environmental sustainability and the necessary economic benefits to Southern African countries. International trade and conservation agreements can only contribute to sustainable outcomes when they take into consideration local stakeholder interests and environmental conditions.

Conclusion

Consumptive wildlife tourism in Africa has the potential to contribute to sustainable tourism development. The CAMPFIRE programme in Zimbabwe and the CBNRM programme in Botswana show that consumptive wildlife tourism contributes to sustainable rural livelihoods and wildlife conservation. Benefits from consumptive wildlife tourism to local communities include the availability of game meat, creation of employment opportunities, income generation and access to land and wildlife resources. As such, an arbitrary ban on safari hunting as proposed by anti-hunting groups is likely to hurt small and remote economies of Africa. This suggests that a global campaign against safari hunting should not be applied indiscriminately to different parts of the world. While the ban might be necessary in Eastern Africa where populations of some game species (e.g. elephant) have declined in recent decades, it is not appropriate throughout Southern Africa. International conservation and trade agreements such as CITES hence need to take into consideration local and regional conditions in order to contribute towards a sustainable tourism industry.

Moreover, wildlife harvesting in Africa is regulated in order to promote its sustainability. For example, safari hunting in Botswana is regulated temporally and spatially, with hunting quotas every year being determined by wildlife professionals. Sustainable practices are also evident in the game farming industries of Eastern and Southern Africa, where principles of animal husbandry are employed to help ensure its sustainability. In South Africa, game farming has become a subject of scientific research with results published in academic journals and in magazines such as South Africa's *Farmer's Weekly*. In this way, scientific and sustainable methods of game farming are disseminated not only in South Africa but also to other African countries. Collectively, these approaches show that consumptive wildlife tourism is not necessarily an evil that should be avoided – as proposed by anti-hunting groups. Rather, the industry can be sustainable depending on the various control measures put in place by stakeholders involved in it.

References

Arntzen, J., Molokomme, K., Tshosa, O., Moleele, N., Mazambani, D. and Terry, B. (2003) *Review of CBNRM in Botswana*, Gaborone: Applied Research Unit.

Baker, J.E. (1997) 'Development of a model system for touristic hunting revenue collection and allocation', *Tourism Management*, 18(5): 273–86.

Bezuidenhout, R. (2003) 'Not an easy game', *Farmer's Weekly*, pp. 48–9.

Botswana Tourism Development Programme, BTDP (1999) *Tourism Economic Impact Assessment*, Gabarone: Department of Tourism.

BWMA (2001) *Economic Analysis of Commercial Consumptive Use of Wildlife in Botswana: Final Report*, Maun: BWMA.

Child, B., Jones, B., Mazambani, M., Mlalazi, A. and Moinuddin, H. (2003) *Final Evaluation Report: Zimbabwe Natural Resources Management Programme*, Harare: USAID.

Glasson, J., Godfrey, K. and Goodey, B. (1995) *Towards Visitor Impact Management: Visitor Impacts, Carrying Capacity and Management Responses in Europe's Historic Towns and Cities*, Aldershot: Avebury.

Kgathi, D.L., Mbaiwa, J.E. and Motsholapheko, M. (2002) *Local Institutions and Natural Resource Management in Ngamiland*, Maun: University of Botswana.

Louw, C. (2004) 'Scores in the game game', *Farmer's Weekly*, pp. 36–7.

Lunn, H. (2004) 'Does game tourism pay?', *Farmer's Weekly*, pp. 36–7.

MacKay, K.J. and Campbell, J.M. (2004) 'An examination of residents' support for hunting as a tourism product', *Tourism Management*, 25: 443–52.

Mbaiwa, J.E. (2004) 'The socio-economic impacts and challenges of a community-based safari hunting tourism in the Okavango Delta, Botswana', *Journal of Tourism Studies*, 15(2): 37–50.

Mbaiwa, J.E. (2005) 'Enclave tourism and its socio-economic impacts in the Okavango Delta, Botswana', *Tourism Management*, 26(2): 157–72.

Muir-Leresche, K., Bond, I., Chambati, W. and Khumalo, A. (2003) *An Analysis of CAMPFIRE Revenue Generation and Distribution: The First Decade (1989–2000)*, Harare: WWF-SARPO.

Schuster, B. (2007) *Proceedings of the 4th National CBNRM Conference in Botswana and the CBNRM Status Report*, 20–23 November 2006, IUCN Botswana, Gaborone, Botswana.

Scott Wilson Consultants (2001) *Integrated Programme for the Eradication of Tsetse and Tyrpanosomiasis from Ngamiland: Environmental Impact Assessment, Draft Final Report*, Edinburgh: Scott Wilson Resource Consultants.

Tlou, T. (1985) *History of Ngamiland: 1750–1996 The Formation of an African State*, Gaborone: Macmillan Publishing Company.

Tsing, A.L., Brosius, J.P. and Zerner, C. (1999) 'Assessing community-based natural resource management', *Ambio*, 28: 197–8.

Twyman, C. (2000) 'Participatory conservation? Community-based natural resource management in Botswana', *The Geographical Journal*, 166 (4): 323–35.

11 Trophy hunting and recreational angling in Namibia

An economic, social and environmental comparison

Jonathan I. Barnes and Marina Novelli

Introduction

Over the late twentieth and early twenty-first centuries, wildlife tourism has experienced significant growth, rooted in an ever-increasing demand for nature-based activities, usually linked to non-consumptive practices such as bird watching, photographic safaris, conservation holidays and rural activities in general. In this context, also, consumptive practices such as trophy hunting and fishing have become an important activity at some tourism destinations. Historically, these activities find their origin some 10,000 years ago when, prior to the agricultural revolution, hunting and gathering were the major economic activities devised by humans (Hummel 1994). In fact, while 'hunting animals for food and for sport has existed for thousands of years, the idea of visiting and observing wild animals for recreational purposes, as a tourist attraction, has been a more recent phenomenon' (Orams 2002: 282). Popular and fashionable within societies of the developed world, recreational 'safaris' (wildlife viewing, wildlife hunting and angling) aim at outdoor experiences characterised by the enjoyment of adventure, thrill of the chase, challenge of shooting, uniqueness of wildlife, landscapes and coastlines (Novelli and Humavindu 2005: 172), the contest of skills and general entertainment.

This chapter discusses both hunting and fishing and highlights some of the fundamental differences between these two activities. As Hummel (1994: 161) notes: '[b]oth hunters and fishers seek to find and capture live animals. Hunters, however, seldom have the option of being successful and allowing the animal to live', while '[a] fisher, however, can maintain the option of conquering his prey and allowing it to live (catch and release), assuming it is not injured'. Hummel further points out that sportsmanship in fishing seems to be less controversial than in hunting. This is since in fishing 'the fish is thought to have a "choice" whether or not to bite a bait or lure', while in hunting 'sportsmanship requires the quickest and most humane means of dispatching the object of the hunt'. Another interesting point is that although 'the thrill of fishing is agreed by many to derive from the sensations of struggle which are transmitted to the "hunter" via the sensitive tackle', few wildlife defenders would agree that fishing might be more painful than hunting for the animal involved. Hummel (1994: 162) raises further considerations on the uses of the natural environments, as 'the habitat approached by the fisher is not the object of intense incompatible competition for use as game lands. Farmers, hikers, nature watchers are threatened by activities of hunters. Fishers, however, utilise waters inhabited by swimmers, skiers, boaters, who, in fact, are a substantial threat to fisher success.'

Sport or trophy hunting has increasingly become part of the conservation argument and policy, being seen by some as a low-impact sustainable use approach, adding value to natural resources (Hofer 2002; Novelli and Humavindu 2005; Novelli *et al.* 2006). However, '[t]rophy hunting is a controversial and misunderstood activity for several reasons. First, trophy hunting is controversial on ethical, social and cultural levels. The practice of trophy hunting generates contradictory positions towards hunting in general. While some believe that the consumptive use of individual animals for the sake of the population, the species, or the ecosystem, is ethically acceptable, others vehemently oppose the killing of animals for personal satisfaction' (Hofer 2002: 14). Opposition to trophy hunting tends to be reinforced by the media, which often reports on illegal or unethical practices, making use of shocking illustrations and association with historical abuses.

While opposition to hunting is often vehement, angling practices are seen as less detrimental to the environment – especially if they involve catch and release. Some opposition to trophy hunting is due to doubts about its social equity and economic viability. However, an increasing number of studies indicate that, through trophy hunting, wildlife becomes economically important for the rural populations and increases their interest, concern and protective attitude towards the preservation of this new or newly recognised source of income (Baker 1997; Barnes 2001; Humavindu 2002; Barnes *et al.* 2002a; Novelli and Humavindu 2005). There are also indications that, through trophy hunting, government agencies are driven to implement adequate legislation, support protection strategies, conduct research and monitoring activities and to aim at the reallocation of revenues to management, protection and nature conservation (Hofer 2002; Novelli and Humavindu 2005).

Given the above setting, this chapter discusses the two main forms of consumptive wildlife tourism in Namibia, trophy hunting and recreational shore angling. Their economic value, impacts, contribution to development, and social and environmental characteristics are compared.

The African context

Wildlife-based tourism has become an important foreign exchange earner in many countries (Reynolds and Braithwaite 2001). African wildlife tourism sectors boomed in the mid-1960s, with increased interest in nature and wildlife conservation, travel affordability and accessibility to unspoiled and remote areas, among Western tourists. Reynold and Braithwaite describe a range of wildlife tourism experiences, with both non-consumptive and consumptive products and practices to be found, rooted in different and specific interests and historical backgrounds. They list hunting and fishing tours as 'consumptive use of wildlife in natural habitat, semi-captive or farmed condition' (2001: 33).

In a progressively more urbanised world, people now travel to reconnect with nature (Orams 2002) being increasingly stimulated by media documentaries and travel programmes. The range of opportunities for people to interact with wildlife grows, manifested in a spectrum of activities available to the public.

In relation to this, attention is often placed on the effects of visitors on the host environment. Baker (1997) suggests that, in the case of Africa, while there seems to be agreement on the necessity of preserving the continent's wildlife heritage for future generations, there is no consensus on the strategy. The conservation community and the public are split over the best methods for Africa, and also the best methods for individual communities in Africa.

In southern Africa, commercial utilisation of wildlife has been actively promoted and has taken place on private, communal and public land, involving a wide range of activities, such as: wildlife viewing tourism, safari hunting tourism, community wildlife use, game ranching, and intensive wildlife farming. Consumptive products have consisted of meat, hides, skins, ivory and live sales. The economic characteristics of wildlife use activities are varied, ranging from low-input, small-scale, labour-intensive subsistence use of low-density, free-ranging wildlife, to capital-intensive farming enterprises with captive breeding and rearing. The different activities vary widely in terms of efficiency of land use, capital, labour, management, transport costs and environmental compatibility (with tourism at the compatible extreme and intensive farming at the incompatible extreme). They also differ in terms of private profitability, economic rates of return and contribution to national income per unit of land (Barnes 1998).

Commercial wildlife use activities in southern Africa provide income for modern private sector entrepreneurs, but they also contribute to the livelihoods of historically marginalised rural communities in southern Africa. Here, they are often complementary to other household coping strategies, such as livestock keeping and crop production, and have contributed to development in communal areas (Ashley and Barnes 1996; Ashley and LaFranchi 1997; Barnes 2002).

Where land is designated specifically for wildlife and forest conservation, such as in national parks, game and forest reserves, wildlife use has mainly involved non-consumptive tourism. Indeed, non-consumptive wildlife tourism has generally emerged as by far the most economically important wildlife use in southern African countries (Novelli *et al.* 2006). Of lesser economic importance in southern Africa are the consumptive wildlife tourism uses, trophy hunting and recreational angling. Consumptive and non-consumptive tourism may seem mutually exclusive, necessitating a choice between one or the other. However, in southern Africa they are commonly practised side by side, and occupy settings with different resource arrays. There is growing evidence that they are not entirely mutually exclusive especially at the district and national levels (Barnes 1998, 2001).

The Namibian context

Namibia embraces 824,000 square km on the west coast of southern Africa, with a population of 1.8 million people. Its environment ranges from the extremely arid Namib desert in the west, along the coast, through arid karroid shrub lands and arid and semi-arid savannas, to semi-arid woodlands in the north east. The main types of land tenure are: state-owned communal (tribally occupied) land; privately owned (commercial farming) land; and state-owned (public) land (Figure 11.1). Land use is dominated by extensive use of natural rangeland with livestock and wildlife. Species rich, highly valuable wildlife communities occur in parts of the communal land, and local communities have limited custodial rights to use these. Communities are able to form communal conservancies for this purpose within the National Community-Based Natural Resources Management (CBNRM) Programme (Jones 1995). Private land contains large numbers of wildlife, dominated by plains game species. Here, landholders have limited custodial rights to use wildlife either individually or collectively through commercial conservancies. Use of wildlife in the country is primarily through tourism, with some consumptive use for meat. At least half of all tourism in Namibia is directed at nature-based pursuits, dominated by non-consumptive activities, which take place in protected areas, on communal land and on private land. These involve self-drive or guided camping safaris and luxury lodge experiences, some of which are promoted as ecotourism operations.

On the coast, the Namib Desert environment is extremely arid, and the waters are part of the Benguela marine system, which is characterised by cold but nutrient-rich up-welling, and abundant fish resources with relatively low species diversity. The marine fish stocks support important industrial fisheries (Molloy and Reinikainen 2003), involving demersal species, such as hake, pelagic species, such as sardines, horse mackerel, anchovy and tuna, and crustaceans, such as

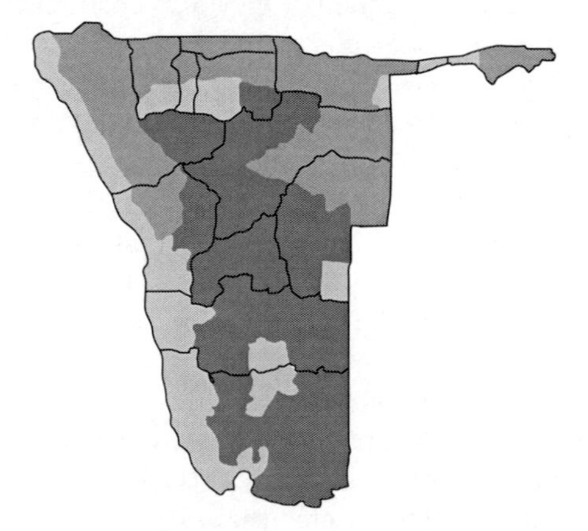

Figure 11.1 Land tenure in Namibia
Dark shading = private land, intermediate shading = communal land, pale shading = state land (Source; Mendelsohn *et al.* 2002).

lobster and crab. There is some commercial harvesting of seals, and an inshore line fishery, involving both commercial boats and recreational angling. The latter is described in detail below.

Landholders on private land and communities on communal land have been given custodial and use rights over their wildlife, and this has resulted in considerable investment in wildlife stocks in these areas. It has also resulted in the use of these stocks by landholders, for meat, consumptive tourism (trophy hunting), and non-consumptive tourism (wildlife viewing). On private land, owners have developed commercial wildlife use activities, and in several situations have joined together with groups of neighbours to form conservancies, which provide economies of scale in wildlife management (Barnes and de Jager 1996). On communal land, communities have formed management entities, termed conservancies, through which they are able to exercise custodial rights.

The introduction of wildlife-based tourism on private land has resulted in some conversion of land use from livestock to wildlife production. This has been partly due to higher financial incentives associated with wildlife, and partly due to the need to diversify income and reduce dependence on livestock, which is no longer subsidised. Among communities on communal land the introduction of wildlife-based tourism has not displaced livestock production significantly, but has tended to make use of new land, mostly unsuited to livestock. Wildlife tourism on Namibian communal land has thus emerged as largely complementary to traditional income earning activities. Is has provided significant new cash income for households, enhancing overall incomes, with little operating cost (Ashley and LaFranchi 1997).

Namibia trophy hunting context

This chapter draws on results of work by Barnes (1996a), Humavindu and Barnes (2003), Novelli and Humavindu (2005), Novelli *et al.* (2006), Samuelsson and Stage (2006) and Stage (2006), which describe the economic characteristics of trophy hunting. In Namibia, policy on wildlife explicitly encourages utilisation through tourism and consumptive harvesting. Wildlife's comparative advantage is mainly associated with its use for tourism. The hunting tourism industry involves guided visits for tourists who hunt trophy-quality game animals and retain the trophies. Trophy hunting clients are upper-income recreational hunters, mostly from Europe and the USA. Most trophy hunting is on private land where hunting bags comprise mainly plains game species. Smaller quotas, mostly involving high value species, are offered on communal land.

Trophy hunting is regulated both by government and private agents. The Namibia Professional Hunting Association (NAPHA) was founded in 1974 in order to promote Namibia as a hunting destination internationally and protect the right to hunt locally. The Association has an active working relationship with the Namibian Ministry of Environment and Tourism, and contributes to the realisation of legislation and to the implementation of regulations.

Figure 11.2 Trophy hunting – Dordabis Conservancy, Namibia (Source: Novelli 2003)

Namibian land-owners with investments in wildlife stocks can register with the government as hunting farms and then offer hunts. Similarly on communal land, either the state or community conservancies can offer hunts. Trophy hunting is only permitted in the company of a registered hunting guide.

Hunters can choose between predetermined hunting packages, containing varying numbers of animals from each species. Hunting bags on private land consists almost exclusively of plains game, including species such as gemsbok (*Oryx gazella*), springbok (*Antidorcas marsupialis*), kudu (*Tragelaphus strepsiceros*) warthog (*Phacochoerus africanus*), hartebeest (*Alcelaphus buselaphus caama*), mountain zebra (*Equus zebra hartmannae*), eland (*Taurotragus oryx*) and others. Hunting bags on communal land include plains game species but commonly also include high-value wildlife species, such as elephant (*Loxodonta africana*), leopard (*Panthera pardus*), buffalo (*Syncerus caffer*), lion (*Panthera leo*) and sable (*Hippotragus niger*). Hunters must obtain export permits in order to take trophies home.

Namibia recreational angling context

The marine environment supports a highly esteemed recreational fishery. Anglers mostly fish from the shore, from the beach, in the surf, using bait. Most frequently landed are kob (mostly silver kob, *Argyrosomus inodorus*, but also dusky kob, *A. coronus*), west coast steenbras (*Lithognathus aureti*), galjoen (*Dichistius capensis*) and blacktail (*Diplodus sargus*). To a lesser extent, sharks are targeted, including the copper shark (*Carcharhinus brachyurus*), the spotted gulley shark

(*Triakis megalopterus*) and the smoothhound (*Mustelus mustelus*). A small part of the recreational fishery involves inshore boat angling for a similar range of species, but also the pelagic snoek (*Thysites atun*). Catch and release is only practised to a limited extent, and mostly in the case of larger shark landings.

Access to angling on the Namibian coast is restricted to about one-quarter of the coastline, some 260 km, and most takes place in north-central stretches, between Walvis Bay and the Ugab river mouth. Land on the arid coastline is state controlled, and there are no resident communities, or private land holdings associated with the fishery. A set of regulations, under the Fisheries Act (Act 29 of 1992), came into force on 4 January 1993, implementing the 'sustainable conservation measures'. Angling is thus regulated to a limited extent by the state, which enforces restrictions on daily bag limits, fishing location and bait collection (MFMR 2005). Recently, anglers are required to purchase annual fishing licences, although there is no restriction on angler numbers. Angling is mostly unguided and practised by individuals or by groups. Individual anglers originate from coastal Namibia, inland Namibia and South Africa.

The marine recreational fishery is described in detail by Kirchner (1998); Holtzhausen (1999); Kirchner and Beyer (1999); Holtzhausen *et al.* (2001) and Holtzhausen and Kirchner (2001). The fish resource targeted by anglers is also utilised by an inshore commercial line fishery, and evidence suggests that, overall, off-takes have been unsustainable. This chapter draws on results of work done by Kirchner *et al.* (2000); Zeybrandt and Barnes (2001); Barnes *et al.* (2002b, 2004) and Kirchner and Stage (2005) on the economic characteristics of the coastal angling fishery.

Economic aspects of trophy hunting and recreational angling

Table 11.1 shows comparative data on the trophy hunting and coastal recreational angling sectors. The data for hunters are derived from analyses of hunting licence and trophy export permit records, as well as of results from a postal survey of hunters, by Humavindu and Barnes (2003); Samuelsson and Stage (2006) and Stage (2006). The data for anglers are based on analyses of a roving creel survey,

Table 11.1 Comparative average characteristics for the trophy hunting and coastal angling sectors in Namibia, 2005

Measure	*Units*	*Trophy hunting*	*Coastal angling*
Number of hunters/anglers	No./annum	3,640	8,270
– Foreign from overseas	%	75%	<3%?
– Foreign from Africa	%	22%	43%
– Domestic from Namibia	%	<3%?	54%
Number of hunting/fishing days	No./annum	51,000	173,000
Average length of trip	No. days	14	21
Total number trophy animals/fish taken	No./annum	13,300	464,100
Number trophy animals/fish taken per trip	No./trip	4	56
Price elasticity of demand for trip	–	not known	inelastic

and two surveys of angler expenditures undertaken by Kirchner and Beyer (1999); Kirchner *et al.* (2000) and Zeybrandt and Barnes (2001). Estimates of the price elasticity of demand for coastal angling trips were made by Zeybrandt and Barnes (2001) and Barnes *et al.* (2002b, 2004), but no elasticity estimates are available for hunting.

The number of anglers is more than twice the number of trophy hunters, and the number of angling days per annum is more than three times the number of trophy hunting days. Coastal angling trips tend to be longer than hunting trips, and anglers take many more fish per trip than hunters take trophies. Coastal angling takes nearly 460,000 fish per annum while the annual harvest of game animals is some 13,000. Of interest is the composition of the hunting and angling populations. Trophy hunters are nearly all foreign, and three-quarters are from overseas. On the other hand, coastal anglers are nearly all from Africa, and more than half of them are domestic tourists resident in Namibia.

Zeybrandt and Barnes (2001) and Barnes *et al.* (2002b, 2004) estimated the price elasticity of demand for angling trips. They found this demand to be inelastic, which suggests that, on average, anglers are willing to pay more than they actually do for a trip. Anglers largely use government-run campsites, for which prices are arbitrarily set, and at the time of the study angler numbers were unrestricted and unlicensed (government has since introduced a payment for licences system, which aims to capture some of this willingness to pay). No price elasticity estimates for hunting trips are available, but one might expect the price elasticity to be neutral, given that hunting is hosted by the private sector, prices tend to be market-related. Such an expectation is supported by findings for non-consumptive tourism in Botswana, where the price elasticity of demand for use of government campsites was inelastic, while that for use of private lodges was not (Barnes 1996b).

Table 11.2 draws on and synthesises data from Humavindu and Barnes (2003); Samuelsson and Stage (2006) and Stage (2006), for hunting and Kirchner *et al.* (2000); Zeybrandt and Barnes (2001); Barnes *et al.* (2002, 2004) and Kirchner and Stage (2005) for angling. Samuelsson and Stage (2006) and Kirchner and Stage (2005) made use of a social accounting matrix (SAM) for the Namibian economy (Lange *et al.* 2004) to measure the economic impact of direct expenditures for hunting and angling on the broader economy.

Table 11.2 shows some interesting differences in the financial and economic characteristics of trophy hunting and coastal angling. Hunters pay nearly nine times more for a trip than anglers. The aggregate expenditure (gross output) for the trophy hunting sector is some four times larger than that for the coastal angling sector. But in terms of contribution to the gross national product (GNP), trophy hunting adds some 12 times more than coastal angling. This is in terms of the direct contribution (that of the sector alone) as well as the indirect contribution (when the effect of the income multiplier in the broader economy is taken into account). Thus, for hunting, each (Namibian) dollar of expenditure generates some $0.47 in direct GNP, and a further $0.43 in indirect GNP via the income

Table 11.2 Comparative financial and economic characteristics for the trophy hunting and coastal angling sectors in Namibia, 2005

Measure	*Units*	*Trophy hunting*	*Coastal angling*
Hunter/angler expenditure per trip	N$/trip	54,120	6,270
Aggregate hunting/angling expenditure	N$/annum	202,349,200	51,648,300
Aggregate direct value added to GNP	N$/annum	95,104,100	7,833,900
– As % of wildlife-based tourism GNP	%	9	0.7
– As % of total tourism sector GNP	%	6	0.5
Aggregate indirect value added to GNP	N$/annum	86,179,900	7,050,500
– Income to communal land communities	%	14	None
– Income to low income employees	%	26	Not known
– Income to high income employees	%	5	Not known
– Income to commercial agriculture	%	5	None
– Income to other sectors	%	29	Not known
– Income to government	%	21	Not known
Total impact of hunting/angling on GNP	N$/annum	181,284,000	14,884,400
Aggregate Namibian consumer surplus	N$/annum	negligible	29,539,400
Total economic value of hunting/angling	N$/annum	181,284,000	44,423,700
Economic value per hunter/angler	N$	49,750	4,240

(Note: at time of writing US$1= N$7.4)

multiplier. For angling, each dollar of expenditure generates only $0.15 in direct GNP, and a further $0.14 indirectly via the multiplier.

The reason why the GNP contribution, relative to output, is so much lower for angling than it is for hunting, is because a large portion of anglers are Namibian, while almost no hunters are. If there was no angling, Namibian anglers would be expected to spend similar amounts on other recreational pursuits in the country, i.e. their contributions to GNP would happen anyway and cannot be attributed to the presence of angling opportunities. However, foreign anglers and hunters would likely not come to Namibia if there was no angling or hunting, so that their GNP contributions would be lost and can be attributed to the presence of angling and hunting opportunities (Samuelsson and Stage 2006; Storey and Allen 1993).

On the other hand, both hunters and anglers enjoy what is termed a consumer surplus. This means that some hunters and anglers pay less than they would be willing to pay for their experience. The consumer surplus of foreign hunters and anglers does not benefit Namibia, while that enjoyed by Namibian residents does. In Table 11.2, consumer surpluses enjoyed by Namibians are added to the GNP values to get the total economic values for hunting and angling. The estimated consumer surplus for Namibian anglers is some N$29.5 million.

Thus, the total economic value (GNP contributions plus any Namibian consumer surpluses) for trophy hunting is some four times more than that for coastal angling. Because the number of anglers per annum is more than twice that of hunters, the economic value generated per hunter is some nine times higher than that generated per angler.

According to Table 11.2, the direct GNP contribution of hunting tourism and coastal angling constitute some 6 per cent and 0.5 per cent, respectively, of the total GNP contribution of the tourism sector in Namibia. Nature-based tourism is estimated to make up some two-thirds of total tourism sector so this means that trophy hunting and coastal angling contribute some 9 per cent and 0.7 per cent, respectively, to direct nature-based tourism GNP. Thus, by far the bulk of the nature-based tourism sector constitutes non-consumptive tourism activities. Novelli *et al.* (2006) showed that while non-consumptive wildlife-based tourism earns the most income, hunting tourism occupies an important and complementary niche in Namibia. The same applies to coastal angling, so that both trophy hunting and angling occupy specific niches and do not displace other tourism activities or potential.

Samuelsson and Stage (2006) used the social accounting matrix to analyse to whom the total income (GNP) generated through trophy hunting accrues. Some 21 per cent is captured by government, and some 40 per cent accrues to low income earners and communal land communities. Hunting thus contributes significantly to poverty reduction and to the treasury. No such analysis exists for coastal angling, but since communal land is not involved here, it might be surmised that the impact of coastal angling on poverty alleviation would be less. Much of the economic value of coastal angling in Namibia takes the form of consumer surplus, enjoyed by middle-class anglers.

Social and environmental aspects of trophy hunting and recreational angling

The comparison suggests that trophy hunting contributes more to rural development than coastal angling. This is partly because of the setting. Some trophy hunting activities take place on communal land, where rural communities are able to benefit at least to some extent through conservancies. Here, rural communities also benefit through the empowerment, institutional development and capacity building that accompanies Community Based Natural Resource Management (CBNRM). No rural communities exist on the arid coast and the contribution of angling to low income households is restricted to wage payments within formal sector linkages. Trophy hunting takes place through guiding outfitters, which themselves directly create jobs and build capacity. Angling is a non-guided activity carried out by individuals, and it does not provide such benefits.

The trophy hunting industry is run though the private sector on private and communally controlled land. The landholders involved also benefit from the activities, and tend also to invest in the wildlife resources on their land. Resource production and use are thus linked in mutually reinforcing ways. With coastal angling the state facilitates a *de facto* open access fishery and the resource is not actively managed or owned. Trophy hunting off-takes are markedly selective and small, while angling catches (despite some catch restrictions) tend to be non-selective and larger, and the practice of catch and release is not prevalent. The numbers of trophy hunters are partially restricted through quota and licensing mechanisms, while the numbers of anglers is not. Generally trophy hunting is recognised as having had a positive conservation effect (Barnes 1996a; Novelli *et al*. 2006). In contrast, there is evidence that the line-fish resources which support angling have been over-utilised (Kirchner 1998; Holtzhausen, 1999; Holtzhausen *et al*. 2001; Holtzhausen and Kirchner 2001). Furthermore, the open access and unguided nature of coastal angling has tended to result in environmental problems due to littering and destructive off-road driving.

As noted above, there is a strong and apparently growing international animal rights lobby, which considers recreational hunting unethical and would like to see it ended (Novelli *et al*. 2006). The angling sector does not appear to suffer from the same opposition, perhaps because there is less public empathy for the resource it uses.

Conclusions

Comparison between the two main forms of consumptive tourism in Namibia, trophy hunting and coastal angling, shows that trophy hunting is more economically efficient than coastal angling. It also appears to be more socially and environmentally acceptable than coastal angling. Hunting tourism involves smaller numbers of tourists than angling tourism, but it generates significantly more income for the country. It also contributes more to poverty reduction and to development than angling. Hunting tourism appears to be more environmentally sustainable than angling tourism, with NAPHA claiming that a 38-year period of ethical, reasonably priced trophy hunting of the highest standard in Namibia has revealed that sustainable utilisation of wildlife resources has been a major factor in protecting game populations. Even depleted game species, which were formerly present in areas of Namibia, have been re-introduced through effective game management based on the principle of conservation through selective hunting. 'NAPHA is therefore convinced that man's oldest cultural heritage, namely hunting, carried out through sustainable game utilisation, is an effective tool to ensure the survival of wildlife and the well-being of local communities' (NAPHA 2005: n.p.).

The reasons for these differences are partly situational, but primarily related to property rights and institutional factors. Coastal angling takes place away from communal lands and makes use of a more or less openly accessible resource, while hunting takes place on private or communal land and makes use of an at least partially owned resource. Trophy hunting tourism by law involves only guided hunts, and it targets high income foreign clients. On private and communal land, there is a self-reinforcing link between investment in the wildlife resource and its use through hunting tourism. In coastal angling, there is no such link, instead the central government administers the use of a largely unmanaged resource by mostly unguided individuals. Investment and management of the resource is negligible and limited to application of limited catch restrictions and (recently) issuance of angling licences.

Both trophy hunting and coastal angling have important contributions to make to Namibian tourism, and Namibian development. They occupy niches, which are complementary in tourism, i.e. they do not displace other non-consumptive tourism activities but add to them. But the question arises as to whether there are policy interventions which might make coastal angling contribute more to the economy, poverty reduction and sustainable development. A start has been made since the data described above were collected, in that anglers now have to purchase licences. This allows capture of at least some of the consumer surplus, which results from non-market pricing in the system. These revenues can be reinvested in management of the system or invested in national development. Policies which promote more guided angling rather than individual use, could significantly enhance the economic contribution of the angling sector, enhance its contribution to poverty reduction, and make it more environmentally sustainable.

Acknowledgements

Some of the work of J.I. Barnes leading to this paper has been funded by the Swedish government through Sida. We thank Michael Humavindu and Jesper Stage for assistance with data collation and comments.

References

Ashley, C. and Barnes, J.I. (1996) *Wildlife Use for Economic Gain: The Potential for Wildlife to Contribute to Development in Namibia*, DEA Research Discussion Paper, No. 12. Windhoek, Namibia: Directorate of Environmental Affairs, Ministry of Environment and Tourism.

Ashley, C. and LaFranchi, C. (1997) *Livelihood Strategies of Rural Households in Caprivi: Implications for Conservancies and Natural Resource Management*, DEA Research Discussion Paper, No. 20. Windhoek, Namibia: Directorate of Environmental Affairs, Ministry of Environment and Tourism.

Baker, J.E. (1997) 'Development of a model system for touristic hunting revenue collection and allocation', *Tourism Management*, 18(5): 273–86.

Barnes, J.I. (1996a). 'Trophy hunting in Namibia', in P. Tarr (ed.) *Namibia Environment*, Vol. 1. Windhoek, Namibia: Directorate of Environmental Affairs, Ministry of Environment and Tourism, pp. 100–3.

Barnes, J.I. (1996b) 'Economic characteristics of the demand for wildlife viewing tourism in Botswana', *Development Southern Africa*, 13(3): 377–97.

Barnes, J.I. (1998) *Wildlife Economics: A Study of Direct Use Values in Botswana's Wildlife Sector*, PhD Thesis, University College, University of London.

Barnes, J.I. (2001) 'Economic returns and allocation of resources in the wildlife sector of Botswana', *South African Journal of Wildlife Research*, 31(3&4): 141–53.

Barnes, J.I. (2002) 'The economic returns to wildlife management in southern Africa', in D. Pearce, C. Pearce and C. Palmer (eds) *Valuing the Environment in Developing Countries: Case Studies*, Cheltenham: Edward Elgar, pp. 274–88.

Barnes, J.I. and de Jager, J.L.V. (1996) 'Economic and financial incentives for wildlife use on private land in Namibia and the implications for policy', *South African Journal of Wildlife Research*, 26(2): 37–46.

Barnes, J.I., MacGregor, J. and Weaver, L.C. (2002a) 'Economic efficiency and incentives for change within Namibia's community wildlife use initiatives', *World Development*, 30(4): 667–81.

Barnes, J.I., Zeybrandt, F., Kirchner, C.H. and Sakko, A.L. (2002b) *The Economic Value of Namibia's Recreational Shore Fishery: A Review*, No. 50, August 2002. Windhoek, Namibia: Directorate of Environmental Affairs, Ministry of Environment and Tourism.

Barnes, J.I., Zeybrandt, F., Kirchner, C.H., Sakko, A.L. and MacGregor, J. (2004) 'Economic valuation of the recreational shore fishery: a comparison of techniques', in U.R. Sumaila, S.I. Steinshamn, M.D. Skogen and D. Boyer (eds), *Ecological, Economic and Social Aspects of Namibian Fisheries*, Delft: Eburon Academic Publishers, pp. 215–30.

Hofer, D. (2002) *The Lion's Share of the Hunt. Trophy Hunting and Conservation: A Review of the Eurasian Tourist Hunting Market and Trophy Trade under CITES*, Brussels: TRAFFIC Europe Regional Report.

Holtzhausen, J.A. (1999) *Population Dynamics and Life History of Westcoast Steenbras (*Lithognathus aureti (Sparidae)*), and Management Options for the Sustainable Exploitation of the Steenbras Resource in Namibian Waters*, PhD Thesis, University of Port Elizabeth, South Africa.

Holtzhausen, J.A. and Kirchner, C.H. (2001) 'An assessment of the current status and potential yield of Namibia's northern West Coast steenbras *Lithognathus aureti* population', *South African Journal of Marine Science*, 23: 157–68.

Holtzhausen, J.A., Kirchner, C.H. and Voges, S.F. (2001) 'Observations on the linefish resources of Namibia, 1990–2000, with special reference to west coast steenbras and silver kob', *South African Journal of Marine Science*, 23: 135–44.

Humavindu, M.N. (2002) *Trophy Hunting in the Namibian Economy: An Assessment*, DEA Working Paper, Windhoek, Namibia: Directorate of Environmental Affairs, Ministry of Environment and Tourism.

Humavindu, M.N. and Barnes, J.I. (2003) 'Trophy hunting in the Namibian economy: an assessment', *South African Journal of Wildlife Research*, 33(2): 65–70.

Hummel, R. (1994) *Hunting and Fishing for Sport: Commerce, Controversy, Popular Culture*, Bowling Green: Bowling Green State University Popular Press.

Jones, B.T.B. (1995) *Wildlife Management, Utilisation and Tourism in Communal Areas: Benefits to Communities and Improved Resource Management*, Research Discussion Paper No. 5. Windhoek, Namibia: Directorate of Environmental Affairs, Ministry of Environment and Tourism.

Kirchner, C.H. (1998) *Population Dynamics and Stock Assessment of the Exploited Silver Kob* (Argyrosomus inodorus) *Stock in Namibian Waters*, PhD Thesis, University of Port Elizabeth, South Africa.

Kirchner, C.H. and Beyer, J. (1999) 'Estimation of total catch of silver kob *Argyrosomus inodorus* by recreational shore-anglers in Namibia, using a roving-roving creel survey', *South African Journal of Marine Science*, 21: 191–9.

Kirchner, C.H. and Stage, J. (2005) *An Economic Comparison of the Commercial and Recreational Line Fisheries in Namibia*, DEA Research Discussion Paper No. 71, Windhoek, Namibia: Directorate of Environmental Affairs, Ministry of Environment and Tourism.

Kirchner, C.H., Sakko, A.L. and Barnes, J.I. (2000) 'An economic valuation of the Namibian recreational shore-angling fishery', *South African Journal of Marine Science*, 22: 17–25.

Lange, G., Schade, K., Ashipala, J. and Haimbodi, N. (2004) *A Social Accounting Matrix for Namibia 2002: A Tool for Analyzing Economic Growth, Income Distribution and Poverty*, NEPRU Working Paper 97. Windhoek, Namibia: Namibia Economic Policy Research Unit.

Ministry of Fisheries and Marine Resources (MFMR) (2005) <http://mfrm.gov.na> (accessed 7 December 2005).

Molloy, F. and Reinikainen, T. (eds) (2003) *Namibia's Marine Environment*, Windhoek, Namibia: Directorate of Environmental Affairs, Ministry of Environment and Tourism.

NAPHA (2005) *Namibia Professional Hunting Association* <http://www.natron.net/napha/> (accessed 7 December 2005).

Novelli, M. and Humavindu, M.N. (2005) 'Wildlife tourism. Wildlife use vs local gain: trophy hunting in Namibia', in M. Novelli (ed.) *Niche Tourism: Contemporary Issues, Trends and Cases*, Oxford: Elsevier, pp. 171–82.

Novelli, M., Barnes, J.I. and Humavindu, M.N. (2006) 'The other side of the ecotourism coin: consumptive tourism in Southern Africa', *Journal of Ecotourism*, 5(1&2): 1–18.

Orams, M.B. (2002) 'Feeding wildlife as a tourist attraction: a review of issues and impacts', *Tourism Management*, 23: 281–93.

Reynolds, P.C. and Braithwaite, D. (2001) 'Towards a conceptual framework for wildlife tourism', *Tourism Management*, 22: 31–42.

Samuelsson, E. and Stage, J. (2006) *The Size and Distribution of the Economic Impacts of Namibian Hunting Tourism*, Research Discussion Paper No. 74. Windhoek, Namibia: Directorate of Environmental Affairs, Ministry of Environment and Tourism.

Stage, J. (2006) *The Willingness to Pay for Hunting in Namibia: Are the Prices Right?* Unpublished paper. Windhoek, Namibia: Directorate of Environmental Affairs, Ministry of Environment and Tourism.

Storey, D.A. and Allen, P.G. (1993) 'Economic impact of marine recreational fishing in Massachusetts', *North American Journal of Fisheries Management*, 13: 698–708.

Zeybrandt, F. and Barnes, J.I. (2001) 'Economic characteristics of demand in Namibia's marine recreational shore fishery', *South African Journal of Marine Science*, 23: 145–56.

12 Welfare foundations for efficient management of wildlife and fish resources for recreational use in Sweden

Leif Mattsson, Mattias Boman, Göran Ericsson, Anton Paulrud, Thomas Laitila, Bengt Kriström and Runar Brännlund

Introduction

In Sweden, hunting and fishing have always been of importance for many individuals. While the importance in old times was primarily assignable to food and other products obtained for a livelihood, hunting and fishing are nowadays leisure activities where the recreational aspect is very significant. Wildlife and fish populations have varied over time, depending on not only the harvesting through hunting and fishing, but also environmental changes associated with industrial and other activities. For example, the effects of forestry on wildlife habitats are considerable, and so are the effects of certain wildlife species on forestry (Persson 2003). Fish resources are affected by hydropower exploitation, and commercial fishing affects the possibilities for recreational fishing (Paulrud 2004).

Consequently, it is necessary to manage the Swedish wildlife and fish resources efficiently, so that hunting and fishing can maintain or improve their functions in a welfare context. This chapter examines the above resource-use options from a welfare economics perspective. Below, we begin with a short description of institutional settings for the use and management of wildlife and fish resources, followed by a demographic overview of the hunters and fishers, where we provide a background to the extent of the two activities. We then enter into problems associated with the management of fish and wildlife resources, and discuss what role welfare economics can play in solving the problems. To add some empirical substance to the theoretical foundations of this discussion, we also summarise some research results of relevance to current management issues. An important aim of this chapter is to look into the near future regarding research requirements relevant to hunting and fishing, and to discuss the potential for research-supported management of wildlife and fish resources for increased welfare.

Some institutional settings

Wildlife and fish are a public concern in Sweden (Naturvårdsverket 2005). Generally speaking, society is responsible for supervising the use and management of fish and wildlife resources nationally, regionally and locally, in order to ensure sustainability. The Swedish Parliament creates laws and assigns the right to declare directives to the government, which further designates power to authorities like the Swedish Environmental Protection Agency, the National Board of Forestry, the Swedish Board of Fisheries, and the Swedish Board of Agriculture. In their turn, these central authorities often assign power and tasks to regional bodies such as counties and regional boards of forestry and agriculture.

The Swedish Association for Hunting and Wildlife Management has for many decades been assigned by delegation from the government to direct hunting and wildlife management in practice, although without being a formal authority (Jägareförbundet 2005). In Sweden, the right to hunt comes with property ownership. Accordingly, the landowners are responsible for use and management of wildlife according to the hunting statues, and if they choose to lease out their hunting rights, management becomes a responsibility shared with the leaseholders. Since 1985 all new hunters are required to take a hunting education programme, that addresses ecological knowledge, rules and regulations (e.g. concerning hunting seasons and the use of rifles and shotguns), safety, and practical shooting proficiency.

In recent decades fishing cooperatives have been of great importance for the regional and local management of fish resources. In 1994 it was decided to provide the cooperatives with legal rights to decide on fishing regulations like fishing periods, size limits, fishing gear etc. Leisure use of fish resources takes place as subsistence fishing as well as sport-fishing (i.e. recreational fishing). The latter implies fishing with rod, hook and line primarily for recreational purposes, and the catch is intended for use in the household. Subsistence fishing is normally carried out with multi-catch equipment (for example net), but the catch is also (primarily) consumed within the household. There are many regulations with respect to leisure fishing, one of the most important being that leisure fishing is forbidden closer than 100 metres to stationary commercial fishing equipment.

Who are the hunters and fishers?

Almost 300,000 Swedes pay the mandatory annual hunting fee (SEK 200), and can thus be assumed to be active hunters. This corresponds to 3.3 per cent of the Swedish population. Hunting in Sweden is in several ways an organised and collective activity emphasising the social interaction (Thelander 1992; Heberlein 2005). A majority of the hunters are members of at least one hunting team. Most hunters are also members of at least one of the two national hunter organisations. The proportions of female and young hunters (aged 18–25 years) are around 5 per cent, respectively, and are slowly increasing (Naturvårdsverket 2005).

Today, hunting is not just an activity for countryside people. Half of the hunters (49 per cent) live in communities with more than 2,000 inhabitants, while less than a third (29 per cent) did so in 1969. Hunting is for many a life-long activity, although the age of initiation has increased along with the major wave of urbanisation in the late 1970s and early 1980s. Of all Swedes aged 16–65 years,

13 per cent live in a household with at least one hunter. Seven out of 10 Swedes say that they have at least one close friend who hunts and the same proportion of Swedes use game meat in the household at least once a year (Ericsson *et al.* 2004). According to a national survey in 1980 (Persson 1981), repeated in 2001 (Ericsson and Heberlein 2002), a great majority of Swedes continue to be positive to or accepting of hunting.

Similar to hunting, fishing is nowadays an activity not only for countryside people but urban people too, and it is for many a life-long form of recreation. A difference compared to hunting is that fishing is, to a lesser, extent practised collectively and is thus less characterised by social interaction. Of the 6.3 million people aged 16–74 years living in Sweden, 1.2 million engage in recreational fishing at least once a year, which is far more people than those practising subsistence fishing (0.4 million people of whom 0.2 million also do recreational fishing). The total number of days spent leisure fishing (recreational and subsistence fishing) is estimated to be 22 million and the associated spend amounts to SEK 2.3 billion (Fiskeriverket 2006). About 40 per cent of Swedes live in households with at least one member fishing, and in some areas in the northern part of Sweden this number is as high as 80–90 per cent. In contrast, only one out of 10 live in households where they regularly eat fish caught by some household member (Ericsson *et al.* 2005). These statistics suggest that fishing is being practised much more for recreational purposes than for provisioning.

Nevertheless, the extent of recreational fishing implies a certain importance from an economic point of view. Half of the population in Sweden agrees partly or totally with the statement that recreational fishing can be developed further in order to create new job opportunities through tourism. At the same time there are many people less inclined to allow tourism fishing full control over the fish resources (Ericsson *et al.* 2005).

Management problems

Society has a general interest in balancing positive and negative impacts from wildlife and fish, i.e. to reduce the negative impacts while achieving a high output of goods and services. Important management issues arise in a dynamic situation where the composition of wildlife and fish populations change over time simultaneously with changes in the population of, and demands from, hunters, fishers, landowners and a broader public. What are the preferences among these groups regarding the extent of harvesting of different species, and what does this imply for the size and composition of the populations of the species? An answer to this question requires consideration of all affected groups, as well as consideration of the biological effects. This calls for an interdisciplinary approach, which may produce management schemes that are welfare improving.

Apart from the goods and services derived from hunting, Swedish wildlife has a negative impact on aspects of the environment and on a sector like forestry. Included in the definition of wildlife management is, thus, not only the preservation of wildlife, but also the balancing of both private and public interests. In the Swedish forest ecosystems, moose, wild boar, roe deer, red deer, fallow deer and hares have negative impacts, e.g. in terms of damage to forest stands through browsing. Such impacts are important for setting harvest levels for big game like moose and red deer. Landowners and to some extent hunters face an impact that may affect forestry profits or prevent them from meeting goals set by the Swedish Forestry Act. Lately, impact from wildlife was ranked highest by private landowners as the main obstacle for forest production (Blennow and Sallnäs 2002). At the same time, game is also ranked high as an asset in the forest ecosystem. During the last decade populations of wolves, brown bears and lynx have rebounded strongly. In the wake of their increase conflicts arise from interactions with animal husbandry, hunters and the everyday life of people in large carnivore areas (Ericsson and Heberlein 2003). On the other hand, large carnivores are viewed positively by a considerable proportion of Swedes (Boman and Bostedt 1999), and an increased predation pressure may at least locally mitigate the damages of other species to forestry.

Diminishing fish stocks are a major problem for current Swedish fisheries management. As mentioned earlier, statistics suggest that sport-fishing is an important recreational activity for Swedes. However, there are also other important and competing uses of the water resources. For example, hydropower exploitation remains critically important for Sweden's electricity supply, but has a significant influence on fish stocks and habitat. Furthermore, commercial fishing provides employment in certain areas of Sweden, but naturally adds to the pressure on fish stocks. Continued growth in the popularity of sport-fishing as a tourist activity is adding to an increased demand on the fish resources. Between the group of non-tourist sport-fishers, the group of tourist sport-fishers and the group of producers of tourist fishing (tourist entrepreneurs, guides etc.) large interactions occur, both positive and negative. While the group of non-tourist sport-fishers is experiencing some negative impacts, e.g. increased costs due to commercialisation, there are also positive effects, such as an increased catch due to better management attributable to commercialisation.

Consequently, while for both hunting and fishing there are management problems inside each of the sectors, in terms of effective use of wildlife and fish resources, the management problems are, however, wider than that. The basic resource for hunting, i.e. the stock of wildlife, has negative (so called external) effects on other sectors, e.g. forestry, while in the case of fishing the negative effects are primarily going in the other direction, i.e. the basic resource in terms of fish stock is negatively affected by other sectors, e.g. hydropower.

A welfare economics view

Social sciences play an important role in resource management in at least two ways. The first, and perhaps most obvious, is in providing a rational basis for management objectives, or goals. The second, and maybe less obvious, is in the implementation process.

An ecosystem produces a number of different and interdependent goods. The utilisation of one good will affect possibilities of using other goods. In some cases the utilisation of a specific good in the system will have positive effects on other goods. In many cases the reverse is true, however, as when forest land is used both to produce timber and to provide habitat for game.

If we allow one of the parties (e.g. forest owners) to make the decision without consideration of the other party (e.g. hunters), it is likely that the outcome will be socially inefficient. The very essence of the problem suggests that it is unhelpful to let each actor define his/her goals and act accordingly, assuming there is a social goal that balances the welfare of each stakeholder. If there is no such goal, or if the welfare of only one group is targeted, the economic analysis is by and large uninteresting. If we want to strike a balance between different users, we first need to understand what the essential problem is and then propose instruments that can produce a socially efficient outcome.

In general terms, a natural resource conflict is often related to a so-called externality. Let us for a moment use the parable of spectators at a football game. A spectator may have as an objective, or goal, 'an undisturbed and good view of the game'. When rising to get a better view the spectator might make it more difficult for another person who has a similar objective. In this case, a spectator inflicts a negative externality on another spectator. How to internalise negative (or positive) external effects is a core subject in environmental and natural resource economics (Baumol and Oates 1988).

To solve the externality problem, we may use different instruments. One option is to use so-called incentive-based instruments and impose a tax on the fellow rising from the chair. Alternatively, we could organise a market and distribute 'rising rights', so that each spectator can buy the rights from each other to stand up. In both cases, the negative externality has a price (a tax or the price of a permit) that provides incentives to the individual. Another option is to impose a regulation that forces each spectator to sit down during the game (otherwise he/she would be subject to, let us say, some unspeakable penalty). This scheme would be equivalent to distributing zero 'rising rights'. Of course, enforcing each of these policies is not easy in practice (which is perhaps why we still have people blocking our views on our favourite football game). Without going into details, a long-standing argument, originating from ideas presented by the Nobel Laureate in economics Ronald Coase (1960), is that externality problems can be solved by voluntary agreements. Thus, each spectator can negotiate with the relevant neighbour to come up with a socially efficient solution. Observe that this assumes that property rights are well-defined, for example, that a spectator who blocks another spectator's view has the right to stand up. Conversely, spectators may have the right to an unblocked view, which is not equivalent to a ban on standing up. The reason is that the spectators can agree on a solution (e.g. 'you can stand up, if you pay my ticket').

Externalities are pervasive in natural resource management, and economics provides useful instruments to attack them. An important question is then, what is the socially optimal level of an externality? In many cases, it can be argued that the socially optimal level is not zero. It is seldom optimal to force everyone to sit down during a football game, for example. Likewise, it is not optimal to have no wildlife and only forestry, or no hydropower and only sport-fishing – or vice versa. To obtain information about the appropriate level of an externality, we need to know the value that people place on unpriced wildlife and water resources, to which we turn next.

A selection of contemporary research results

The importance of hunting as a leisure activity in Sweden, and increasing forest damage caused by wildlife, made the need for improved scientific knowledge relevant to the management of wildlife quite important from the 1980s. For example, lack of knowledge on the economic value of hunting was very obvious. A valuation study was therefore conducted, focused on hunting in the country during the hunting year 1986/87 (Mattsson 1989, 1990a. 1990b). The methodological approach was contingent valuation (Mitchell and Carson 1989) by means of a mail survey. The survey questionnaire was sent to a random sample of 2,500 hunters throughout Sweden, and 68 per cent responded.

According to this research the total hunting value of all game in the country amounted to SEK 2,405 million (recalculated into year 2005 monetary value), or about SEK 8,000 per hunter, two-thirds of which was attributable to recreation and one-third to meat. The moose hunting value was SEK 1,461 million (61 per cent of the total hunting value), while the value of hunting other game – roe deer, hare, game birds and other species – was SEK 944 million (39 per cent of the total hunting value). The proportions of recreation value and meat value varied a lot across different game species. For example, while the recreation value and the meat value of moose hunting amounted to 60 per cent and 40 per cent, respectively, the corresponding proportions for hare hunting were 90 per cent and 10 per cent.

Besides moose being the most valuable game species and one causing severe costs through forest damage, moose is also a species that responds relatively promptly to management actions. The moose was therefore analysed especially with regard to the hunting value given different population densities. Not surprisingly, the results showed an increasing moose hunting value – but at a decreasing rate – as the moose population density is increased. However, this decreasing marginal value was more pronounced in the northern part of Sweden than in the southern part, primarily because of a moose population density in the north that gave more moose per hunter than was the case in the south. In other words, the supply of moose hunting was ‘more sufficient’ in the northern part of the country than in the south.

The population of moose – as well as its value for hunting – is much dependent on the hunting policy. Based on economic as well as biological data, an analysis was made dealing with the economic benefits of the selective moose hunting practised in Sweden (where the hunting is primarily focused on low- and non-reproductive animals) as compared to a random (i.e. non-regulated) moose hunting (Ericsson *et al.* 2000). Present values of moose hunting produced by the two alternatives were calculated for a period of 10 years using different interest rates. Using an interest rate of 4 per cent, for example, the hunting value of the selective moose hunting exceeded the average hunting value of a random moose hunt by SEK 300 million, or SEK 1,280 for the average moose hunter. In determining whether the selective moose hunting policy is profitable or not, costs of supporting such a policy should also be considered. A most considerable part of the costs probably lies in providing the moose hunters with information, e.g. on what is to be gained in the long run by a selective hunting today.

In other countries there has also been a demand for scientific knowledge on the value of hunting. For example, in Norway, Sødal (1989) made a study on moose hunting during the hunting year 1987/88, i.e. one year after the Swedish study by Mattsson (1989) mentioned above. Sødal also used the contingent valuation method, with a survey design very similar to the one in the Swedish study. In the Norwegian study the moose hunting value per hunter and day spent hunting moose turned out to be 80 per cent of the corresponding value for Sweden (recalculated with consideration to difference in currency and hunting year). This difference in hunting value may partly be explained by the difference between the countries in moose population density, and similar to Sweden the Norwegian results showed a decreasing marginal value of an increasing moose population density. Furthermore, concerning the value of small game hunting, Young *et al.* (1987) made a contingent valuation study in Idaho, USA, during the hunting year 1982/83 (four years before the Swedish study). Also in this case the value per hunter and day happened to be 80 per cent of the corresponding value for Sweden (recalculated with consideration to difference in currency and hunting year). Here, however, the difference in value may result from differences between countries regarding the classification into 'small game' and 'big game', respectively. There are studies from many countries on the value of game for hunting as well as other resources for recreational use – for overviews, see Navrud (1992) and Wibe (1994). Among these studies there are of course considerable differences in hunting values to be found, depending on differences in game species compositions, demand for and supply of hunting possibilities, etc.

In recent years, the need for compatible research also on recreational fishing has become obvious. In 2005 a study was commissioned by the Swedish Board of Fisheries and carried out by Statistics Sweden (Fiskeriverket 2006). A survey questionnaire was sent to a random sample of 8,000 Swedes, and 62 per cent responded. It was found that the total value of all leisure fishing in Sweden was SEK 3,290 million. Creating possibilities to double the catch would result in an increase of the total value by SEK 168 million and an increase of the total number of fishing days by 3.5 million. For some fishers, however, there seems to be a saturation in the sense that they would decrease the number of fishing days if the catch per day increases.

Several other studies of recreational fishing have been conducted, showing the marginal value of more fish to catch. In studies of recreational fishing in the Swedish mountain region (Paulrud and Laitila 2004; Laitila and Paulrud (2006); Laitila *et al.* 2005) the estimated values of catching one extra large fish (brown trout or grayling over 40 cm long) ranged from SEK 50 to SEK 330. The largest estimate was obtained for a site in northern Sweden, usually reached by helicopter. The smallest estimate was obtained for a site in mid-Sweden, reachable by car from the Stockholm metropolitan area within a few hours. One possible explanation to the variation in the estimates is a difference in demographic composition of fishers at the different sites. Another, and perhaps more interesting, explanation is that the surroundings of the fishing site have effect on the valuation of the catch. Such an effect is partly indicated in Laitila *et al.* (2005), where the deconstruction of a dam was observed to have effect on the valuation of catch.

Bag limit is in some studies found to be relatively important while less so in other studies. Paulrud and Laitila (2004) report the estimated value of an extra fish in the bag limit to SEK 40. In the study by Laitila *et al.* (2005), the estimated value of an extra fish in the bag limit was higher, up to SEK 100. A small value of an increased bag limit suggests a managerial tool that protects recreational fishing sites from overfishing, while maintaining the value of recreational fishing. Another result with potentials for managerial use was found in both Paulrud and Laitila (2004) and Laitila *et al.* (2005). In both studies the difference in valuation of species, brown trout and grayling, was found to be insignificant. These results were obtained for fishing sites in the mountain region.

There are of course valuation studies made on recreational fishing in other countries too. In Scandinavia, Norway is known as a country where the fish resource is very important. Accordingly, several studies have been conducted with focus on the value of recreational fishing. Pioneering Norwegian studies in this field was made by Strand (e.g. 1981), followed by Navrud (1984) and Scancke (1984). Results from these studies, which concerned three different rivers in Norway, show some clear trends: the recreation value of fishing Atlantic salmon and sea trout is higher than that of fishing brown trout; a river with a large average size of the fish has a higher value for recreational fishing than a river with small average size of the fish; the larger stock of fish (i.e. the more fish there are in terms of fish population density) the larger recreation value of the fishing. These results, especially the latter two, are not very surprising – more and bigger fish result in a higher recreation value than the opposite. Nevertheless, the two most central factors in management of fish resources for recreational fishing become clear, although there are limitations because of biological reasons when it comes to combining the two.

Continued efforts

In the field of hunting, an important background for new research is the dynamics of harvesting and game populations. The hunting outcome of different species in Sweden during the past 45 years, which is also related to the population dynamics of the species, is illustrated in Figure 12.1.

Many game species are harvested in tens or hundreds of thousands. Among these, the harvests of moose, hare and game birds have decreased since the study 1986/87 by Mattsson (1990a, 1990b) mentioned earlier, while the opposite is true

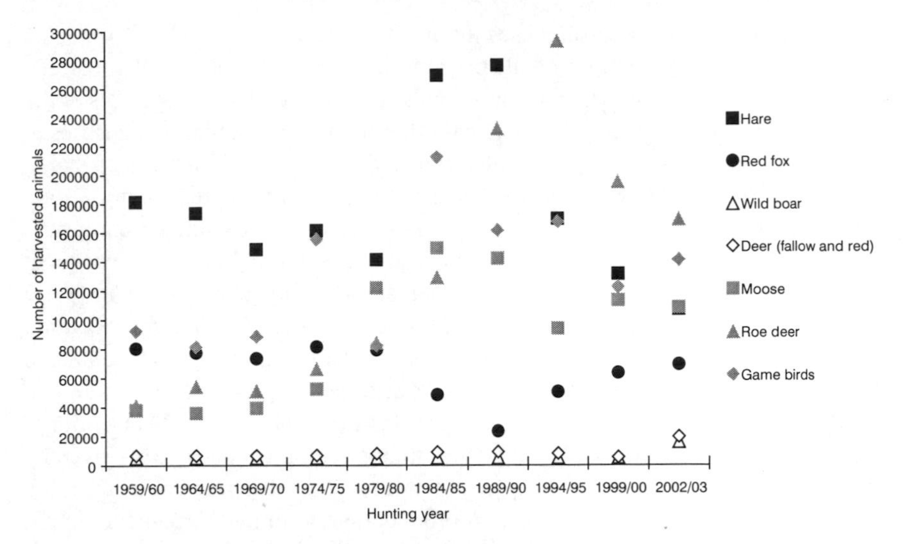

Figure 12.1 Number of harvested animals for selected species or groups of species (Source: Skogsstyrelsen 1999, 2002, 2004)

for red fox, fallow and red deer. Wild boar was not hunted at all in the middle of the 1980s, but presently more than 10,000 are harvested annually. The hunting of bear and lynx is increasing, but on a small scale (not displayed in Figure 12.1). These changes in hunting suggest that changes may also have taken place in attitudes and demand regarding hunting and wildlife management. In new research it is thus important to consider preferences regarding species composition, sex, age and spatial structure, substitution possibilities between game species, and risk and uncertainty. These preferences can then be statistically linked to the demographic characteristics of the hunters. Intertemporal comparisons can be made between the results of the original and repeated survey, which is also of methodological interest for assessing the temporal reliability of the contingent valuation method (Whitehead and Hoban 1999).

For hunters, game is of course considered as a good. For forest owners, game may be a good as well as a 'bad'. Many forest owners are hunters, and those who are not hunters themselves may get income from hunting by leasing out the hunting rights. On the other hand, game may affect forestry negatively through browsing damage. A recently started project will analyse the attitudes, preferences and valuations held by main groups with an interest in Swedish game resources: hunters who are not forest owners, forest owners who are not hunters and forest owners who are also hunters. A basis for the analysis will be a repetition and extension of selected parts of the valuation study by Mattsson (1990a, 1990b). The research will cover important game species in different parts of Sweden, and examine how the preferences and valuations among hunters regarding these species have changed over the past two decades. Special attention will be given to moose in the analysis. Moose is in many respects the most important game species in Sweden, but it also causes the most severe problems to forestry. This is why the extension of the original survey in the new research will also include forest owners, and the survey instruments will be designed to allow analyses of the hunting value and costs for forestry given different moose population densities.

In the field of sport-fishing, the valuation studies made so far have been static, while the interaction between sport-fishers and fish populations are dynamic. Good fishing conditions during one period attract sport-fishers who, without appropriate regulations, can reduce the stock of fish, thus attracting fewer fishers in subsequent periods. If the stock can recover it may again attract more sport-fishers. In future research, these kinds of dynamic interactions need to be considered for valuation of fishing sites. A socially optimised value of a sport-fishing site might be obtained through active management, with different regulations over periods depending on the dynamic behaviour of the sport-fishers. However, sport-fishers' behaviour is only one side of the coin. On the other side are the ecological constraints given by the potential for fish to grow and reproduce. Although sport-fishers value catching a large fish more than a smaller fish, reproduction and growth arguments might place a zero bag limit for smaller fish while not for larger fish. In a recently started project, one aim is to combine ecological and economic models for the purpose of handling the dynamic components involved.

Data sets collected in earlier studies are to be combined into an analysis of segments of sport-fishers. Such an analysis may shed light on the causes of the differences in results obtained. For instance, why is an extra fish valued to SEK 330 in one area of study but only SEK 50 in another? As mentioned earlier, one possible explanation is a difference in demographic composition of sport-fishers visiting the different sites. If this would be an important cause, it opens up an opportunity to use benefit transfer methods in these kinds of studies. That is, results obtained for one sport-fishing site can be extrapolated to be used for valuation of other sites. Information on composition of segments is still needed, however. On the other hand, if difference in demographic composition of sport-fishers is not an important cause, it indicates that valuation of sport-fishing site characteristics is dependent on other factors. Aspects like these are important to consider in upcoming research.

Potentials for the management of fish and wildlife

The new research efforts presented above are all included in a large Swedish research programme entitled 'Adaptive Management of Fish and Wildlife Populations'. This programme is characterised by, among other things, interdisciplinarity and the ambition to produce results valuable not only for the scientific community but indeed also for different stake-holders.

We believe that there may be ways to manage wildlife and fish populations in Sweden such that the benefits to hunters and fishers are increased, with non-increasing costs, e.g. to landowners. Put differently, the management would result in an increasing value of wildlife and fish populations from a societal point of view, without increasing negative externalities from these populations. If such a management is successful, it will be beneficial not only for hunters, fishers and landowners, but will also increase the potential for higher profits in the hunting and fishing tourism industry, by creating a more valuable 'product' to be sold on the market. The ambition is to evaluate current management against alternatives, based on e.g. renewed valuation efforts.

The ultimate goal of the research is to devise management strategies that increase the aggregate welfare of hunters, fishers and the parties carrying the costs of the wildlife and fish populations. It should, however, be emphasised that we have just embarked on the journey towards such management strategies. With respect to wildlife and fish, we are beginning to find answers to two of the basic economic questions: what to produce, and for whom? The third basic question – how to produce? – still presents a great challenge. Besides knowledge on hunters, fishers and landowners, the answers to these questions involve issues such as the geographical pattern of harvests and populations, population growth, age and sex distribution, food resources, etc. Thus, insights from both natural and social sciences are required, and the goal can therefore not be successfully reached without a strong interdisciplinary approach.

References

Baumol, W.J. and Oates, W.A. (1988) *The Theory of Environmental Policy*, Cambridge: Cambridge University Press.

Blennow, K. and Sallnäs, O. (2002) 'Risk perception among non-industrial private forest owners', *Scandinavian Journal of Forest Research*, 17: 472–9.

Boman, M. and Bostedt, G. (1999) 'Valuing the wolf in Sweden: are benefits contingent on the supply?', in M. Boman, R. Brännlund and B. Kriström (eds) *Topics in Environmental Economics*, Dordrecht: Kluwer Academic Publishers, pp. 157–74.

Coase, R.H. (1960) 'The problem of social cost', *Journal of Law and Economics*, 3: 1–44.

Ericsson, G. and Heberlein. T. (2002) *Fyra av fem svenskar stödjer jakt*, SLU Kontakt, Fakta Skog, 2, 4p.

Ericsson, G. and Heberlein, T. (2003) 'Attitudes of hunters, locals and the general public in Sweden now that the wolves are back?', *Biological Conservation*, 111(2): 149–59.

Ericsson, G., Boman, M. and Mattsson, L. (2000) Selective versus random moose harvesting – does it pay to be a prudent predator?, *Journal of Bioeconomics*, 2: 117–32.

Ericsson, G., Eriksson, T., Laitila, T., Sandström, C., Willebrand, T. and Öhlund, G. (2005) *Delrapport om jakt och fiske – Omfattning, betydelse och förvaltning*, FjällMistra-rapport nr. 14, Umeå.

Ericsson, G., Heberlein, T., Karlsson, J., Bjärvall, A. and Lundvall, A. (2004) 'Support for hunting as a means of wolf population control in Sweden', *Wildlife Biology*, 10: 269–76.

Fiskeriverket (2006) *Resurs- och miljööversikt: Fiskbestånd och miljö i hav och sötvatten*, Göteborg: Fiskeriverket.

Heberlein, T.A. (2005) 'In my opinion: wildlife caretaking vs. wildlife management – a short lesson in Swedish', *Wildlife Society Bulletin*, 33(1): 378–80.

Jägareförbundet (2005) <http://www.jagareforbundet.se> (accessed on 30 October 2005).

Laitila, T. and Paulrud, A. (2006) 'A multi-attribute extension of discrete-choice contingent valuation for valuation of angling site characteristics', *Journal of Leisure Research*, 2: 133–42.

Laitila, T., Jonsson, A. and Paulrud, A. (2005) *Regleringsdammen vid Storsjö-Kapell: Sportfiskarnas värdering av ett återställande till naturligt fjällfiske*, Mimeo, Umeå: Department of Forest Economics, Swedish University of Agricultural Sciences.

Mattsson, L. (1989) *Viltets jaktvärde – En ekonomisk analys*, Working Report 86, Umeå: Department of Forest Economics, Swedish University of Agricultural Sciences.

Mattsson, L. (1990a) 'Hunting in Sweden – extent, economic values and structural problems', *Scandinavian Journal of Forest Research*, 5: 563–73.

Mattsson, L. (1990b) 'Moose management and the economic value of hunting – towards bioeconomic analysis', *Scandinavian Journal of Forest Research*, 5: 575–81.

Mitchell, R.C. and Carson R.T. (1989) *Using Surveys to Value Public Goods – The Contingent Valuation Method*, Washington, DC: Resources for the Future.

Naturvårdsverket (2005) <http://www.naturvardsverket.se> (accessed on 31 August 2005).

Navrud, S. (1984) *Økonomisk verdsetting av fritidsfisket i Hallingdalselva i Gol kommune.* M.Sc. thesis. Department of Economics, Agricultural University of Norway, Ås.

Navrud, S. (ed.) (1992) *Pricing the European Environment*, Oslo: Scandinavian University Press.

Paulrud, A. (2004) *Economic Valuation of Sport-fishing in Sweden: Empirical Findings and Methodological Developments*, Doctoral thesis. Acta Universitatis Agriculturae Sueciae, Silvestria 323, Swedish University of Agricultural Sciences, Umeå.

Paulrud, A. and Laitila, T. (2004) 'Valuation of management policies for sport fishing on Sweden's Kaitum River', *Journal of Environmental Planning and Management*, 47: 863–79.

Persson, I.-L. (2003) *Moose Population Density and Habitat Productivity as Drivers of Ecosystem Processes in Northern Boreal Forests*, Doctoral thesis. Acta Universitatis Agriculturae Sueciae, Silvestria 272, Swedish University of Agricultural Sciences, Umeå.

Persson, R. (1981) *Jakt och jägare. Initiering och utövandemönster. En kartläggning av jakten och jägarna i Skåne*, Doctoral thesis. Department of Sociology, Lund University, Lund.

Scancke, E. (1984) *Fisket i Tinnelva.* M.Sc. thesis. Department of Economics, Oslo University, Oslo.

Skogsstyrelsen (1999) 'Skogsstyrelsen arsbok1999', Jonkoping: Skogsstyrelsen.

Skogsstyrelsen (2002) 'Skogsstyrelsen arsbok 2002', Jonkoping: Skogsstyrelsen.

Skogsstyrelsen (2004) 'Skogsstyrelsen arsbok 2004', Jonkoping: Skogsstyrelsen.

Sødal, D.P. (1989) *Okonomisk verdisetting av elgjakt*, Doctoral thesis. Department of Forest Economics, Agricultural University of Norway, Ås.

Strand, J. (1981) *Beregning av samfunnsøkonomisk verdi av fisket i Gaula-vassdraget*, Mimeo, Department of Economics, Oslo University, Oslo.

Thelander, B. (1992) 'The way we hunt in Sweden', in R. Bergström, H. Huldt and U. Nilsson (eds) *Swedish Game – Biology and Management*, Spånga: Svenska Jägareförbundet, pp. 50–63.

Whitehead, J.C. and Hoban, T.J. (1999) 'Testing for temporal reliability in contingent valuation with time for changes in factors affecting demand', *Land Economics*, 75(3): 453–65.

Wibe, S. (1994) *Non Wood Benefits in Forestry – Survey of Valuation Studies*, Working Report 199. Umeå: Department of Forest Economics, Swedish University of Agricultural Sciences.

Young, J.S., Donnelly, D.M., Sorg, C.F., Loomis, J.B. and Nelson, L.J. (1987) *Economic Value of Upland Game Hunting in Idaho*, Resource Bulletin RM-15. Fort Collins: Rocky Mountain Forest and RangeExperiment Station, USDA.

13 What happens in a Swedish rural community when the local moose hunt meets hunting tourism?

Yvonne Gunnarsdotter

Introduction

Hunting is a common leisure activity in Sweden, with moose hunting as the most popular form. Participation, however, is slightly declining with the exception of women. Currently, approximately 300,000 people hunt, of which 5 per cent are women (www.naturvardsverket.se). This can be compared with two growing sports: golf with 600,000 players (www.sgf.golf.se) and horse riding with 200,000 riders (www2.ridsport.se). Hunting tourism is a relatively new phenomenon with less than 300 enterprises offering this activity – mostly to Swedes but also to international hunters.

In this chapter the two phenomena of local hunting and hunting tourism are investigated. First the background and meaning of moose hunting in Sweden is introduced. Then hunting tourism is discussed, first from the landowners' and foreign hunters' perspective, and then from the perspective of the local hunters. Both local moose hunting teams and hunting tourism contribute in different ways to viable rural communities. The hunting teams help to maintain the sense of community and sense of place that the inhabitants develop over time. Hunting tourism supports the local economy by providing other sources of income than farming. These cultural and the economic processes are necessary for a viable community but they sometimes work counter to each other. The chapter addresses ways to handle the tensions that sometimes appear.

The empirical material in this study comes from fieldwork undertaken in Locknevi situated in southern Sweden (Figure 13.1). In the south hunting is less popular than in the north, which is more sparsely populated and thus offers more game. Locknevi, however, is typical of rural hunting locations. It is a parish with 500 inhabitants spread out over five villages, where most people commute for work in the nearest towns some 30km away. Fieldwork there was carried out through participant observation and interviews between 1999 and 2003, hunting being one case in a broader field study.

Figure 13.1 Location of Locknevi in Sweden.

Moose hunting in Sweden

Every year many of those who have moved away from Locknevi return with a son or another relative to take part in the moose hunt in the first week of October. Hunting is sometimes described as holy, and it engages men from different backgrounds and also a few women. Historically the peasants of the area had no or few rights to hunt, though poaching occurred, and what hunting that did occur focused on small game. Today the most common game species are moose and roe deer – both species being rare before 1950. The 30 moose hunting teams in Locknevi comprise groups of up to 20 landowners and their relatives (roe deer are usually hunted by smaller teams). All hunts are strictly regulated in terms of firearms, safety and game rules. These regulations are often controlled by hunter's associations which also arrange the obligatory hunting training classes.

Moose hunting teams increased in the 1960s, the main reason being growth in game numbers. Another reason was the decline of agriculture causing the community of work to be replaced by a community of leisure. In order to understand why the introduction of hunting tourism could cause problems, the meaning of hunting according to the local hunters is captured in six key terms: ritual, nature, animals, egalitarian friendship, maleness and place. Many hunters seem to perceive the meaning of hunting as a wholeness or fusion of hunter-forest-game-place-history.

Rituals in hunting contribute to the unity of the team and indirectly also to maintaining a sense of community. Two examples of almost universal rituals in hunting are the distribution of meat and the trophy. According to Johansson

(2000), historically the goal of hunting has not been the private individual consumption of meat. Sharing meat is an historic practice still relevant in the moose hunt of today. Since the emergence of the modern moose hunt in the 1950s the meat has been distributed among the hunters, irrespective of their social status. But over the last decade many of the teams distribute meat to the landowners too, even if they do not take part in the hunt. The trophy, usually the antlers, has a strong symbolic value, however, and is always given to the person who shot the animal.

The game is obviously a critical aspect of the hunt, with hunters telling anecdotes about how clever the game is. Their supposed cleverness and their great abilities to survive in the forest are challenges for hunters and many of them spend much of their leisure time in the forest studying the behaviour of the animals. When an animal is killed the norm prescribes that it should be treated with respect. A real hunter should be able to slaughter an animal in a neat and tidy way. This is sometimes difficult for those who are not used to rural life.

Irrespective of ownership, profession and other signs of status outside the hunting situation, the team is built on reciprocal social relations (Ekman 1991). The social space of hunting is informal and characterised of what anthropologists call 'joking relationships'. Older hunters especially may value the informal get together higher than the shooting. The egalitarian and reciprocal relations in a team have a clear boundary with the outside world. The hunters often tell jokes about people from outside, and guests have to stand some mild provocations.

Both agriculture and hunting are changing in terms of the male norm. Female hunters are getting more common and the few in Locknevi feel accepted. However, when a woman, who moved to Locknevi 25 years ago, invited another woman as a guest during a hunt some members of the team reacted in a negative way. The link to the real hunters (male, preferably landowners born in Locknevi) became too weak as she crossed an invisible border.

The sense of place is also a critical component of the hunter-forest-game-place-history synthesis. When passing a place during a hunt someone often tells an anecdote about what has happened there before. The stands where the hunters wait for game are named after persons or a characteristic situation. Through this continuous denomination the place becomes a part of the hunt instead of an object for the hunt. Abram (1996) suggests that a well-known landscape communicates with us and makes us remember by addressing all our senses. Telling anecdotes about what happened at different places when passing them is a way to mediate the memory of a landscape.

This synthesis of hunter-forest-game-place-history is considered to be impossible to recreate in tourism activities. Moreover, the introduction of outsiders through commercial hunting tourism activities may disturb this unity. The next section examines the history of hunting tourism within Locknevi and discusses the processes of social change that threaten the wholeness of the hunting experience as perceived by locals.

Hunting tourism

In the 1980s one landowner in Locknevi (Kjell – all names in the text are fictitious) started to lease hunting permits for moose and roe deer to German and Danish hunters. Kjell is considered an entrepreneur, always first to try something new, and hunting tourism is a small branch besides his forestry interests. He lives in a neighbouring parish on family property and the property he owns in Locknevi also used to belong to his family. Kjell does not hunt himself. If there are a few hunters coming for a weekend he gives them a map showing where they can hunt. The hunters stay in cottages and they pay per day or per animal they shoot. Kjell keeps the meat but the hunters get the trophy, in this case the antlers. The bigger groups staying for a week are taken care of by a Danish hunting leader, Jens, who is familiar with the land and who hunts together with the group.

By the time Kjell started his business the moose population had increased rapidly in the whole country and hunting tourism became an opportunity for landowners. Today about 260 enterprises in Sweden are concerned with hunting tourism, approximately half of them situated in northern Sweden, which is dominated by big forests and where the game is more common (Jaktturismnäringen i Sverige 2003). Many of these enterprises are engaged in other activities like farming, forestry or other forms of tourism. Hunting tourism is still a small branch, but with a potential to grow. A hunting tourist is defined as 'a person who temporarily leaves his daily surroundings (household, working place) to hunt' (Jaktturismnäringen i Sverige 2003). Most of them are Swedish, but foreign hunters are also eager to come. Compared to countries with a more developed hunting tourism, like Scotland and Poland, the Swedish prices are rather low. There is some hesitancy about inviting foreign hunters, though, both because of the more complicated arrangements required, and because of the sceptical attitudes that local inhabitants sometimes show (Jaktturismnäringen i Sverige 2003).

In Locknevi, a few years after Kjell's initiative, another landowner, a farmer, also started leasing hunting permits on a short-term basis. Like Kjell he does not hunt himself and he leaves all arrangements to the Danish hunting leader Jens. During the 1990s some other landowners started with hunting tourism on a different scale. All of them are hunters and sometimes they hunt together with the tourists. Jens now leases hunting rights on the whole or part of seven properties with long-term arrangements and in turn leases hunting permits for parties for a week. In Denmark Jens is a truck driver but he spends five weeks every year in Locknevi. He cannot take part in all hunting since there are several groups on different places at the same time.

An even more small-scale form of hunting tourism is performed by one of the moose hunting teams in Locknevi which invites Danish paying guests who join the team during the first moose hunting week. Apart from their own land this team leases a property from a company. They use the money from the Danish hunters to pay the fee for the lease. Other teams have discussed or tried the same kind of arrangement. An even more extensive form of hunting tourism is when some landowners invite foreign guests to hunt in exchange for a hunt at their home location.

The landowners' and foreign hunters' perspective

Economic development is the motive behind hunting tourism. Over the years Jens has become familiar with the landowners and hunters in Locknevi. He has heard about criticism of hunting tourism, however. he has never experienced any of this criticism personally. He has had only positive experiences except for one occasion when a landowner asked him to suggest a price for taking care of the deer hunt. Jens suspects that the landowner used him to trigger the price for the Swedish team leasing the hunt on his land. Jens is aware of the fact that he pays more for leasing than what is common and that 'his' hunters pay more than Swedish hunters do.

In contrast, Kjell, the local entrepreneur, is rather used to being criticised for the projects he carries out. He suspects that if he had been a hunter himself, locals would have been more accepting of his hunting tourist business. Being excluded from the community of hunters who live in or have moved from Locknevi, he is aware of the importance of personal relations. He states that those who criticise him do not know him. Some of the hunters in Locknevi, however, do know him from school and make friendly jokes about him.

Kjell is sympathetic to public criticism of the rising prices for leasing hunting permits. When he started his business, the local newspaper wrote an article with a headline about how Kjell had thrown out those who used to lease the hunting permits on his property and replaced them with Danish and German hunters. Kjell explains that they were friends who had hunted for free, but since he is known as a businessman people think that business is all he cares for. In the same article the regional hunting adviser thought that landowners ought not to engage in hunting tourism. Kjell describes the situation in the 1980s:

> I was a member of the regional board of LRF [Swedish Farmers Association] and I thought this [hunting tourism] could be a niche of agriculture and forestry … I wanted to discuss this with LRF. The chairman got offended and said it was not a question for LRF. They [LRF] thought this should be settled by the hunters' association, but they are the opposite party, not the landowners. At that time many people didn't consider hunting as something connected with agriculture and forestry. They did it as a hobby and not as a part of the business. That is possible if you have a good economy, but the properties were expensive. If I had been a hunter I would have been appreciated, but I don't hunt – I only get money. They are envious.
>
> (Kjell, landowner and businessman)

In the beginning of the 1990s LRF changed its attitude and started to support hunting tourism as a way to make more money from the land. Kjell is proud to be recognised as an entrepreneur, but at the same time the competition with other landowners has lowered his profit compared to the 1980s. He is very concerned about rights of ownership and the right to control his own property. He does not manage his forest in the new environmentally friendly way, where one should leave seed pines and fallen trees after the felling to increase biological diversity: 'Sometimes the interests of nature conservation do not correspond to mine. I might not want the lumber and I might violate the law to keep it tidy', Kjell explains.

Kjell has renovated the houses on his property in Locknevi and uses them to accommodate the hunting tourists. The neighbours appreciate that he looks after the buildings but they do not approve of him refusing to join the local moose management association. 'Those who don't join are put on the black list', Kjell says. One reason for him not to join is that he has properties in different areas with different management associations, and he wants the same rules for all hunters he administers. He wants control over his business and his property and sometimes this is more important than maximum profit. For example, he hesitates to take tenants because of the risk that they could be the 'wrong' kind of people.

Two of the other landowners who followed in Kjell's footsteps have since stopped their tourist business. They felt that engaging in hunting tourism had had a negative influence on their own hunting experience. However, they think it is up to the owners themselves to decide what to do with their own property. One of them had paying guests in the team but he got fed up with the comments from the others about how much 'your bloody Germans' got of the meat. The other team members also complained that the Germans never took care of a dead animal. But some landowners have more positive experiences of hunting together with the visitors:

> It's always nice when people you know come and visit. We usually spend a few days at his place, so I have been to Denmark many times. The Danes and the Germans are good at shooting.
>
> (Yngve, hunter/landowner living in Locknevi)

Thanks to Jens, who organises most of the hunting tourism, there is a certain amount of continuity in the hunting tourism in Locknevi. He brings knowledge about the place and the people. The Danish hunters that return are building up their own relation to the place and there are similarities between the Danish and the local hunters. But even though the Danish hunters also value the experience of being in the forest, the excitement when they come across an animal and the joy of the male friendship, hunting is still different to that of the hunters in Locknevi. For most of the Danes it does not matter if they are in Locknevi or somewhere in Poland, according to Jens. They do not have the opportunity to become familiar with the game as individuals as some of the local hunters have. The sense of camaraderie is probably not so strong in the foreign teams as they often consist of new members every year. Those who return lack strong bonds to each other and each other's relatives. Kjell captures their situation: 'They buy a hunting experience and everything else is up to me to arrange'.

However, the foreign hunters that are paying guests in a team still have a stronger relation to a place compared with those involved in the more large-scale forms of hunting tourism practised in other destinations. Bo, a hunter from Locknevi who was a paying guest in Kenya, says that he had 'much more fun than the dollar tourists in the lodges'. He believes Swedish hunting tourism to be a more genuine experience that people are willing to pay for.

The Locknevi team which has six Danish guests every year is one of the more traditional teams with several elders and where the meat is distributed to the local hunters and not to the landowners. At the same time they have introduced elements of this modern (touristic) form of hunt. The Danes are treated both as guests and customers. There is no doubt that they are welcome and the friendship seems to be mutual. They rarely take part in the conversation during the breaks though, partly because of language difficulties and partly because they have a different role from the other team members. They do not have to drive the game and the hunting leader tells them where to go and gives them a lift to and from the stand. They are not allowed to take any meat, but they get the trophy.

A guest should not leave empty-handed, and often the members of the hunting team have commented about the poor outcome for the Danes. 'There are far too few moose shot and it's bad for the Danes who have travelled so far', Karl-Gunnar says. A customer should have value for the money invested, and when one of the Danes shoots a moose one of the local hunters comments that it is good 'so that they will return next year'. This polite treatment could be interpreted both as the host's care and the salesman's service. The team has not discussed how to behave in relation to the Danish hunters and since the situation is new they have improvised and ended up with this mixed approach. Basically the Danes are guests but everybody is conscious of the fact that they are also useful. But exploiting someone is not socially acceptable and the following quotation reveals ambivalence about the activity and an eagerness for mutual benefit.

> Well, I might have exploited them a little, but they find it fantastic just to come here and sit down in silence. In Denmark you can't find a place to hunt without hearing the traffic or other sounds in the background. So they pay for the sense of community and for the silence … Maybe someone thought it was a bit strange in the beginning. Since they paid more than we, they were supposed to have some advantages … it has worked out good, really well … Other hunters might complain that we bring Danes and Germans to Locknevi.
>
> (Karl-Gunnar, hunter and landowner in Locknevi)

The local hunters' perspective

Many of the hunters and other inhabitants in Locknevi are sceptical about hunting tourism, but there is no open conflict. Their criticism can be summed up from cultural and social perspectives. The critique based on cultural arguments concerns the meaning of hunting and the way it changes through money and the presence of strangers. Both phenomena affect the wholeness of hunter-forest-game-place-history that many of the hunters feel. The critique mirrors what many of them think is the ethos of hunting: the mix of excitement, being in the forest and the spirit of community that has developed in a certain place over time. The social aspects of the critique concern the changing identity of being an inhabitant in Locknevi. This is caused partly by the fact that those who do not own land have difficulties to get access to a hunting team when the prices are rising. Also, local social relations are affected when it becomes more important to own land.

Cultural aspects – money

When people in Locknevi talk about hunting tourism they often compare how much a hunting permit costs for the Swedes, the Danes/Germans and for the Danish hunting leader Jens:

> Money has ruined the hunt … when they brought the Danes, the Germans and the money. The game does not belong to the property – that's the way it is. A moose walks 50 km. He [Kjell] can sell the meat for 2,500 [SEK] and the trophy for 5,000, that's ridiculous. They have the right to do it, but it takes away the joy for the neighbouring teams.
>
> (Ronny, hunter living in Locknevi all his life)

Even those who understand why people engage in hunting tourism can be critical towards the activity. 'If I didn't hunt myself I might do the same, but from a hunting point of view it's a damned thing', says Conny who is a hunter in Locknevi. Most people who do not own land themselves understand that landowners have a need to utilise the available resources of the property. To what extent hunting tourism is accepted is a matter of scale and if the landowner hunts himself. Few are critical of small-scale tourism when the hunting team invites paying guests. Magdalena, a hunter living in Locknevi, points out the importance for rural people to utilise the resources themselves: 'Otherwise people come from the city and build fishing camps and other things'.

What happens is that money dissolves the relations between the hunter, the forest, the game and the place. Setting a price results in the instrumental values of resources surpassing intrinsic values. A price demands a measurable object and to create that the relations between the components in a system are altered (Evernden 1987). Tourism is a phenomenon where instrumental values are produced from what used to be intrinsic values. According to Urry (2002) tourism both consumes and produces places. The consumption is made mainly through the 'Tourist Gaze' which objectifies 'the Other' including both people living in the place and the place itself. With money the tourist buys the right to use different objects, like the right to shoot certain animals. Also a place's production is made both through the gaze and with money: what the tourist looks at or buys becomes a tourist goal. This transformation of place to landscape, has arisen through the process of modernisation, with the separation of man from environment.

Svensson (1997) highlights the conflict between landowners' production and the tourists' consumption of landscape, using examples from today's medieval role play around Swedish castles. These actors exist in different landscapes with different interpretations of history. Instead of creating a situation in which modernity is contrasted against tradition we could learn how to handle the culturally complex cultivated landscape that is the result of the landscapes of both consumption and production. For the forest owners in Locknevi it is important to find socially acceptable compromises between forest production, the local moose hunt and hunting tourism.

One example of the difficulties of putting a price on values is when the authorities of New Zealand tried to value a nature reserve of religious importance for the Maori (Vadnijal and O'Connor 1994). The inhabitants were not able to value the place as an object since they perceived that it existed together with them: 'There are things, dimensions in life that are beyond money' (1994: 379). In a similar way, the local hunters' enchantment with the forests in Locknevi is meeting a process of disenchantment, a process where the world is reduced to measurable components (Berman 1981). The question is whose perspective should be considered, especially when visitors who are able to pay have the power to interpret the situation, regardless of how local inhabitants value a place.

Cultural aspects – strangers

People in Locknevi do not consider themselves as xenophobic, and they stress that they do not dislike the foreign hunters. They instead direct their critique towards the landowners who are responsible for the Danes and Germans following the rules (and who also brought the strangers). Those who have met the Danish hunting leader and have hunted together with the foreigners have a positive attitude towards the individuals, though they sometimes make jokes about them. But as a group the foreign hunters represent 'the Other', a strange body which does not belong. This view is revealed in expressions like 'now the Danes are invading' (Katrin, hunter who moved from Locknevi), or 'big hordes of Danes' (Kerstin, who lives in Locknevi and does not hunt). This kind of critique refers to the Danes (who are now in the majority) not hunting the same way as the local hunters do. As an example they are accused of not following the rules about which kind of moose they are allowed to shoot, which is important since the animals cross the borders between properties: 'They shoot everything on four legs', says the hunter Kristian.

The hunting tourists in Locknevi are viewed as a strange phenomenon, referred to in terms of 'culturally different', 'cultural clashes', 'unwritten laws', 'strangers' and 'proper behaviour'. The Danish hunters are probably not aware of the symbolic actions that make them look different. An ideal (but not always the praxis) is that alcohol and hunting do not mix well: 'We got fed up when we had collected two sacks of empty bottles of beer and liquor', says Katrin who used to let a cottage to Danish hunters.

Another kind of criticism concerns the way the foreign hunters treat the game and the dogs. Folke, a hunter in Locknevi, complains: 'They shoot all kinds and throw it on a car roof so that the blood is dripping along the doors'. They are also criticised for the way they dress: 'It looks a bit stupid when the Danes are coming dressed in camouflage clothes and lifting their legs high when walking on the roads', says Irene, who does not hunt. The tourists are also accused of being trophy hunters.

There are differences in the way the Danish and the German hunters are viewed. The Danes understand Swedish and appear more similar, but in contrast to the Germans they are many and they often hunt without a Swedish guide. The German hunters are considered more careful both with alcohol and shooting. On the other hand, hunting is considered an upper-class activity in Germany, which goes against the local ideal of an egalitarian community. The combination of the hunters being foreign and having paid make some of the inhabitants feel excluded from the forest. This is expressed by Kerstin, who is a tenant and who does not hunt: 'It feels like I'm intruding though I have all rights in the world to walk there'.

Some people suppose that when Swedes hunt abroad they behave differently compared to local hunters and that they are viewed with the same wariness as foreigners in Locknevi. Axel, who moved back to Locknevi, explains the local standpoint as a suspicion towards everything that is different: 'There is dissociation from everything foreign, and I mean foreign from Locknevi's perspective. Everything from Vimmerby [the nearest town] to Brussels'.

Doubt towards everything foreign is, according to Urry (2000), typical of local communities and expresses a will to maintain barriers. In the struggle to maintain a community, be it a hunting team, a parish or a nation, a need exists for a border between those who belong to the community and those who do not, between 'we' and 'they' (Cohen 1985). The flexibility that many inhabitants show in other contexts is difficult to uphold when 'they' are so many that they become a category of their own.

A stranger is someone impossible to classify, neither friend nor enemy (Bauman 1990), and is treated differently in a village than in a city. There is no place for a stranger in a small-scale community and he/she is quickly classified into either friend or enemy. In the abstract systems characterising urban settlements the opposite of a friend is no longer an enemy, and not a stranger in Bauman's sense, but just someone you do not know, says Giddens (1990). In contrast to local communities it is normal to meet a stranger, otherwise it would not be a city (Asplund 1991: 52).

Locknevi is a community with concrete relations between the inhabitants, even though they are also embedded in abstract systems guaranteeing social infrastructure and the welfare of individuals. Criticism directed towards the hunting tourists does not have to be xenophobic but mirrors ambivalence about how to classify this 'strange body'. As individuals the foreign hunters are rather easily classified as friends. This explains why it is easier to accept small-scale hunting tourism. Big groups coming for a week and people hunting without any contact with the local hunters are more difficult to include in the perception of hunting as an activity among equals.

Social aspects – exclusion

Hunting tourism is causing higher prices for hunting permits and some relatives and friends of the landowners cannot afford to hunt any longer. This highlights hierarchies and class perspectives. Erik reflects on this: ‘The sad thing about it [hunting tourism] is that the ordinary blue collar workers with limited economy cannot afford it. It triggers the prices and everything’. The space for an informal exchange economy diminishes, which affects mostly people with low income.

> Now I lease the hunting and I also hunt myself. He [who hunts on his property] doesn’t pay anything, but he is an electrician and helps me in the house. Many Swedes cannot accept money [from friends and relatives] and then the foreigners come and offer big sums.
>
> (Hans, hunter living in Locknevi)

When hunting tourism becomes an alternative it is tempting to put a price on one’s own hunt. One landowner says he experiences a conflict between hunting himself and letting hunting permits. He acknowledges the fact that the hours he hunts are expensive hours compared to the income he gives up.

For some of the inhabitants without land, hunting forms an important local network and a meeting place. For the first and second generations of people who have moved from Locknevi hunting is a link to the community, irrespective of whether they own land or not. Hunting is not so common among the youth and one reason is considered to be the high costs. If these groups (the young and those who have moved) give up hunting it will affect their identity in relation to Locknevi since there are rather few networks based on local identity, except the local voluntary associations.

Social aspects – changed relations

When the landowners accept payment for the hunt it affects not only the access to hunting, but also the social relations in the community. If a person accepts money, a risk exists that he/she exploits someone and is considered discourteous. This is reflected in the earlier quotation where a landowner talks about payment from the foreign hunters. The logic of the market economy is not evident to some elderly people in Locknevi, who argue that the price some of the foreign hunters pay ‘does not correspond to reality’. For them the price is not a point where the supply meets the demand, as in market pricing relations. Instead, a price should reflect egalitarian relations where people agree in consensus about what is reasonable for them. Neither the landowners nor anybody else is expected to take advantage of their position and ask for ‘too much’ or offer ‘too much’ money. This is based on norms like ‘good neighbours do not ask for money from each other’ and ‘balance between right and obligations should be maintained'. Violating these norms can cause conflicts and the norms have to be renegotiated.

One expression of social tensions is disputes about borders. One hunter in Locknevi says that hunting tourism has affected the good neighbourship between his own team and the neighbour who leases all hunting permits to the Danish hunting leader. A common and seemingly eternal dispute connected to hunting is when hunters cross a property border to follow a track from an animal. So far there has been no open conflict, but hunters tell stories about animals they have found on their property, that have died from a wound, without anybody telling the owner. This kind of behaviour is often attributed to the foreign hunters. But generally the local hunters strive for good relations towards the hunting tourists:

> We have decent relations towards them … The Danes have crossed our borders on some occasions. They are not familiar with the territory and that is nothing to make a fuss about. You shouldn't create bad relations. I think we have an agreement that if something happens we should contact each other. If an animal goes in [to another property] and lies down, you are allowed to shoot it.
>
> (Sune, hunter who has returned to Locknevi)

Everybody agrees that there is a great responsibility for the landowners to ensure that the foreign hunters stick to the rules. It is also expected that the landowners act according to the norms. One of the landowners points out that his hunting tourists have strict rules to follow and that they also have an interest in game preservation. Another landowner speaks in the same way:

> There are rumours about that they wounded some animals and that they drunk too much liquor. Sometimes they might have done so, but it is not the individuals that should be blamed but those who arrange the hunting opportunities. At the same time it brings some good money to the community.
>
> (Lars, hunter and farmer in Locknevi)

Conclusion

Fiske (1991) offers a theory about human sociality in which he argues that there are four basic forms of human relationships. The first one is Communal Sharing consisting of a collective identity based on equality and inclusiveness which is typical for relatives. The second form is Authority Ranking, which are asymmetrical and hierarchical relations common among people with different status where privilege and duty is important. The third form, Equality Matching, emerges when coping with differences through reciprocal relations where it is common to share, take turn or give back 'eye for eye and tooth for tooth'. The fourth form, Market Pricing, is based on measurable values where people exchange goods after analysing the possibilities of profit and loss. Depending on the cultural context these forms of relations appear in different shapes, but always one of the four dominates. The choice of form is partly dependent on how the people involved are used to relating to each other. People often transfer relations they are familiar with to other contexts. To be able to communicate people have to agree on which kind of relationship they have in a certain situation.

Inspired by Fiske's four forms of relations, hunting tourism can be interpreted as a phenomenon of Communal Sharing and Equality Matching gradually being replaced by Market Pricing and Authority Ranking. While this change is already occurring within rural economic systems, hunting tourism appears to be speeding up the process. For example, in the distribution of meat, where it used to be the activity, the hunting, that was rewarded, it is now often only the ownership of land that qualifies for meat. When hunting becomes more expensive, ownership of land will be even more important in terms of opportunities to hunt. Hunting tourism would be impossible without a norm that gives land ownership priority over the custom that local people have the right to hunt.

Hunting tourism is an example of how economic ambitions within a community could violate cultural and social attributes. In Locknevi, both the meaning of and the social relations involved in hunting are affected in a way that contradicts the dominating norms. However, there are accepted compromises between the landowners' need to find new sources of income and the meaning of hunting for local hunters. One accepted solution is that the landowner takes part in the hunt together with the tourists. Another way is that the team invites the same paying guests every year and uses the income collectively for tenancy or equipment. These solutions strengthen communal, egalitarian relations in accordance with the norms as opposed to market pricing and hierarchical relations. In this case it is possible to maintain the organic whole of hunter-forest-game-place-history, and at the same time open up to new sources of income.

References

Abram, David (1996) *The Spell of the Sensuous*, New York: Vintage Books.

Asplund, J. (1991) *Essä om Gemeinschaft och Gesellschaft*, Göteborg: Bokförlaget Korpen.

Bauman, Z. (1990) 'Modernity and ambivalence', in M. Featherstone (ed.) *Global Culture: Nationalism, Globalization and Modernity*, London: Sage Publications, pp. 311–28.

Berman, M. (1981) *The Reenchantment of the World*, Ithaca: Cornell University Press.

Cohen, A. (1985) *The Symbolic Construction of Community*, London and New York: Routledge.

Ekman, A.-K. (1991) *Community, Carnival and Campaign: Expressions of Belonging in a Swedish Region*, Dissertation. Department of Social Anthropology, University of Stockholm, Stockholm Studies in Anthropology No. 25.

Evernden, N. (1987) *Främling i naturen*, Göteborg: Pomona.

Fiske, A.P. (1991) *Structures of Social Life. The Four Elementary Forms of Human Relations*, New York: Free Press.

Giddens, A. (1990) *The Consequences of Modernity*, Cambridge: Polity Press.

Jaktturismnäringen i Sverige (2003) Stockholm: Swedish Tourist Authority.

Johansson, E. (2000) 'Skogslöpare och vedbodsstökare', in I. Kaldare, E. Johansson, B. Fritzbøger and H. Snellman (eds) *Skogsliv: Kulturella processer i nordiska skogsbyar*, Lund: Historiska Media, pp. 44–69.

Svensson, B. (1997) 'Vardagsmiljöer och söndagskulisser, Landskapets naturliga förflutenhet och kulturella samtid', in B. Svensson and K. Saltzman (eds) *Moderna landskap. Identifikation och tradition i vardagen*, Stockholm: Natur och Kultur, pp. 21–44.

Urry, J. (2000) *Sociology Beyond Societies' Mobilities for the Twenty-first Century*, New York: Routledge.

Urry, J. (2002) *The Tourist Gaze*, 2nd edn, London: Sage Publications.

Vadnijal, D. and O'Connor, M. (1994) 'What is the value of Rangitoto Island?', *Environmental Values*, 3: 369–80.

Website addresses

<http://www.naturvardsverket.se> (accessed 28 May 2006).

<http://www2.ridsport.se> (accessed 28 May 2006).

<http://www.sgf.golf.se> (accessed 28 May 2006).

14 Arab falconry

Changes, challenges and conservation opportunities of an ancient art

Philip J. Seddon and Frederic Launay

Introduction

> Hawking is generally recognized to be the most intellectually demanding and educational form of hunting ever devised and it requires a high degree of skill and devotion from the falconer. It leads the hunter to a deep appreciation of nature, to a practical study of natural history and quite often to serious scientific research on birds of prey.
>
> (Cade 1982: 54)

Falconry, or hawking, is the hunting of birds or mammals with trained birds of prey, and is considered to be an art with ancient origins in the Middle East. Only a few species of raptorial bird are suitable for falconry as they must have a naturally aggressive hunting style. In some parts of the world, such as Central Asia and Kazahkhstan, golden eagles may be used to hunt fur-bearing animals like foxes, but traditional falconry uses the short-winged goshawks and sparrow-hawks, best suited for hunting in woods, and the long-winged falcons, principally the peregrine, the gyrfalcon, the saker and the merlin. The latter high-flying birds hunt exclusively in open areas, often attacking with a downward dive (stoop) at breathtaking speeds in excess of 290km per hour, stunning or killing their prey by hitting them with their feet or chest. However, unlike their Western counterparts, Arab falconers do not hunt their birds with downward dives but from the fist in a tail-chase pursuit. On the ground surviving prey are dispatched with an efficient bite to the neck that severs the spinal cord using a special tooth-like projection on the falcon's upper mandible. The long-winged falcons are favoured by Arab falconers in the pursuit of an art that represents links with a way of life that has changed in the upheavals of massive social and economic change.

Once restricted to a subsistence or small-scale recreational harvest of desert birds, the vast wealth that has flowed from the exploitation of massive petroleum reserves has enabled the average Gulf State citizen to gain access to all parts of the Arabian Peninsula for falconry, and allows hyper-rich Arab falconers to pursue their passion virtually without limits. As local populations of favoured quarry species have declined, falconers from the Arabian Peninsula have travelled to seek new hunting grounds in North Africa and Central Asia, to the detriment of native species in these regions. But the wealth that enables some falconers to flaunt local species protection regulations in the host countries they visit, and to mount truly massive hunting expeditions, both logistically and economically, has also meant that more environmentally and ecologically responsible Arab falconers have the means to promote and support species conservation measures.

In this chapter we explore how the economic changes wrought by petro-dollars have vastly increased the scale and impacts of recreational falconry in the Middle East. We start by examining the origins of falconry in the Middle East, and review the traditional practice of falconry in the Arabian Peninsula. With a focus on the Kingdom of Saudi Arabia we consider how the exploitation of oil has caused social and economic change, and how this in turn has changed the scale of falconry and its impacts on both falcons and their quarry, the houbara bustard. We conclude on a more up-beat note to consider the ways in which the passion of Arab falconers is being translated into effective species conservation measures.

Origins of falconry

The geographical origins of falconry are not known, but there is good reason to believe they lie in the Middle East. The earliest representation of falconry comes from the Syrian site of Tell Chuera, within the basalt rock desert that stretches south to Saudi Arabia, dating from the third millennium BC (Canby 2002). From the beginning of the second millennium BC pottery, seals, carvings and statues from Anatolia portray a rich array of falconers, falcons and the trappings of the art, such as jesses (the soft leather straps attached to the legs of the falcon), fingerless gauntlets, lures and neckbands (Canby 2002). It is assumed falconry was first introduced to Europe with the migration westward of the people from the Asian steppes in the third century AD, with the first written references to falconry in Europe dating to the fifth century AD (Prummel 1997). The bridge between the ancient art of Arab falconry and that practised in Europe, and hence the rest of the world, was the treatise on falconry *De Arte Venandi cum Avibus* (the art of hunting with birds) by the Holy Roman Emperor Frederick II of Hohenstaufen (AD 1194–1250) (Schramm 2001), a book still considered outstanding by today's practitioners.

The Arab art of falconry

Traditional Arab falconry (*qans*) necessitates the development and successful completion of three different, but complex tasks: trapping, training and hunting, the skills for each of which have been honed through countless generations of a people that were as much a part of the deserts of the Arabian Peninsula as their prey.

In the recent past the season used to start in October each year when the cooler autumn conditions see the arrival in the Arabian Peninsula of the first migrating falcons, moving south from their northern breeding grounds. Trappers congregate in camps at key points along the migration paths to lure and trap falcons using a variety of lures, including song birds, pigeons and small mammals, and traps such as the noose-covered Balchatri. Most sought after are the saker falcon (*saqr*) *Falco cherruq* and the peregrine *Falco peregrinus*, with the larger female saker (*al hurr*) and peregrine (*shahin*) being preferred.

Training of falcons begins immediately in the trapping camps where the first priority is to accustom the falcons to being held and to entice them to feed by offering them scraps of meat while being handled. The trainer (*saqqar*) will talk to the falcon, stroke their feathers, and hood (*burga*) and unhood them. The falcon quickly learns to fly to a lure (*tilwah*) made of the wings of a prey species, and this lure may be used to entice the falcon back after a flight. During the hunting proper ,which takes place during the cold winter months of November through to March, hunting parties would travel out into the steppe deserts of the Arabian Peninsula, originally on camel back, searching for recent sign, of the premier quarry the houbara bustard, though desert hares and stone curlew were also traditionally hunted.

Arab falconry without the houbara bustard is inconceivable; it's been suggested that Arab falconry would not survive in its traditional form without the houbara (Osborne 1996). The houbara is one of the most widespread of the bustards, the *Otidae*, ranging from the Canary Islands in the east, across North Africa and west as far as Mongolia and China. This last, the Asiatic, or Macqueen's bustard, ranges also from southern Arabia up to northern Kazakhstan and exists as a single large meta-population containing birds that are variously migratory (breeding in the extreme north and north-west and moving south in winter), semi-migratory, or virtually sedentary, but which are all linked by dispersal (Pitra *et al.* 2004). Traditionally Arab falconry was sustained by houbara found within the Arabian Peninsula where small resident populations were supplemented by the winter influx of migratory birds from the north. The houbara is a medium-sized bird capable of powerful flight, moving rapidly despite deceptively slow wing beats. A buff-brown and black speckled plumage provides superb camouflage in the sandy and rocky deserts.

When a houbara is flushed by a falconry party it will fly strongly, with deep wing beats flashing black and white. A falcon would be unhooded and launched, quickly sighting the fleeing prey and starting its pursuit. The thrill of pitting a saker or peregrine against a houbara is that ordinarily these slender falcons would not tackle a bird the size of the bustard but must be trained to do so. The falcon will stoop on the houbara, knocking it to the ground but seldom actually killing it with the first strike. The houbara is quickly dispatched with a bite to the neck by the falcon when on the ground. The falconer will race to the site of the kill to retrieve the falcon, and before replacing the hood will feed it some tasty morsel from a previously killed pigeon, so the falcon does not get used to eating houbara on the hunt.

Traditionally, at the end of the winter hunting season in March or April, trained falcons were released back to the wild, presumably to return to the breeding areas in the north after a foreshortened, but eventful migration. Few if any falcons were

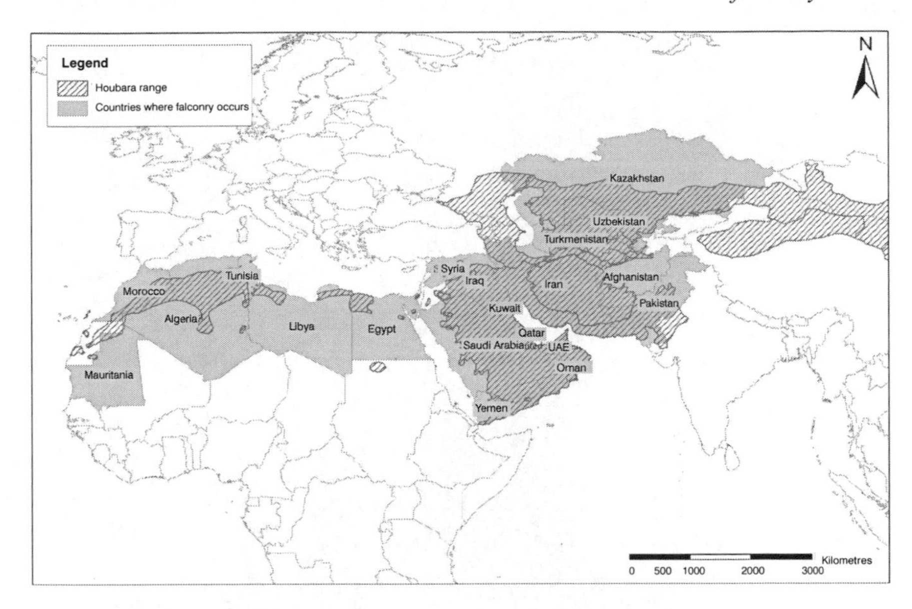

Figure 14.1 Distribution of falconry and houbara (Source: National Avian Research Centre – Environment Agency, Abu Dhabi)

kept all year round, presumably because of the inconvenience of housing and feeding them through the searing hot months of the Arabian summer. In the past exceptional birds may have been retained by more wealthy falconers, but for the average, relatively poor hunter, falconry had little impact on either prey species or the falcons themselves. The environmental impacts of falconry were to increase drastically, however, as oil-fired wealth in the region gave hunters, for the first time, the ability to remove the obstacles that had *de facto* made their art and their sport sustainable.

Social and economic change in the Middle East

> With population growth and the prosperity that has come from oil … no institutional framework has emerged that ensures that the people use their wild resources, especially the natural vegetation and wildlife, sustainably.
>
> (Child 1989)

The recent history of the Gulf States of the Arabian Peninsula, comprising the Kingdom of Saudi Arabia, the United Arab Emirates, Bahrain, Kuwait and Qatar has been shaped by three dominant factors: Islam, charismatic tribal leaders. and crude oil. Social, economic and environmental change in the region may best be illustrated by a summary of events in Saudi Arabia – the nation with the largest landmass in the Arabian Peninsula and owner of what amounts to over a third of the world's proven reserves of crude oil.

Figure 14.2 Hunter with falcon on his wrist sits in front of a recently caught houbara bustard, the main quarry of Arabian falconers (Source: National Avian Research Centre – Environment Agency, Abu Dhabi)

Oil exploration began in 1933, and it soon became evident that Saudi Arabia's Eastern Province sat over the largest pool of oil in the Middle East. By the end of 1938 something less than half a million barrels of oil had been produced; by 1944 annual production had increased to eight million barrels, reaching 60 million barrels only two years later. Today the proven reserves have been estimated at around 260 billion barrels, around a quarter of the world's oil (EIA 2005).

The impact on Saudi Arabia of the revenue generated by the sale of petroleum products has been massive and sustained (Grutz 1999). At the peak of oil prices in 1981 Saudi Arabia made US$119 billion, or USS$13.6 million an hour from oil (Yergin 1991); the main contribution to a US$150 billion GDP. More recently, 2006 forecast earnings are US$154 billion, or US$17.6 million an hour (EIA 2005).

It should be remembered that the unprecedented expansion of infrastructure and services in the last 50 or 60 years has affected a relatively small, culturally conservative population, characterised by strong family and tribal allegiances, but united beneath the umbrella of Islam, the precepts of which form the foundation of Saudi Arabian social and justice systems, and govern the daily lives of all residents. The many benefits of development, the improved transport and communications networks, increasing literacy rates and expansion of the education system, improved health care and reduced infant mortality, to name a few, have been accompanied by some almost inevitable environmental costs.

In the pre-oil days the ability of humans to have an impact on natural resources was limited by relatively low population densities, modest economic means and a widespread dependence on subsistence agriculture, including nomadism – an efficient means of exploiting fugitive and seasonally fluctuating plant resources. An increasing human population and an objective of national self-sufficiency in food production combined with increased demand for sheep and goat meat and for crop plants. With greater wealth subsistence agricultural systems could be freed from dependence on variation in annual rainfall. Overgrazing and, to a lesser extent, the loss of natural vegetation to cultivation and recreation, has threatened both wildlife and rural productivity (Child and Grainger 1990). On top of this, the prevalence of all-terrain vehicles and automatic weapons has seen unregulated hunting expand into once remote and inaccessible areas. During the last century a number of native species have become extinct, including: Asiatic cheetah *Acinonyx jabatus*; Arabian ostrich *Struthio camelus syriacus*; Arabian oryx *Oryx leucoryx* (extinct in the wild, 1970s); or suffered declines in breeding populations and range, such as Nubian ibex *Capra ibex nubiana*; Arabian leopard *Panthera pardus nimr*; houbara bustard *Chlamydotis [undulata] macqueenii*; and mountain gazelle *Gazella gazella* (Jennings 1989; Nader 1989; Harrison and Bates 1991).

Environmental impacts of Arab falconry

> The great wealth created by the gushing oil wells brought rapid change to the old ways of life in Arabia. The vast lands, once almost unvisited except by drifting Bedouin, were now all within easy reach of any falconer in search of houbara as well as other quarry such as stone curlews and desert hares. This led to overhunting and quarry numbers declined.
>
> (Upton 2001: 15)

It would be untrue and unfair to claim that the wealth created through exploitation of oil reserves has made all Arab falconry unsustainable. What it has done is to firmly place the focus of Arab falconry on recreation rather than subsistence hunting, to open up the sport to a wider group of people, to make it possible to hunt houbara outside the Arabian Peninsula, as a tourist-hunter, and to make hunting of houbara and other prey more efficient and more comfortable for the average falconer. The majority of falconry practised by Arab falconers will entail winter hunting trips of 2–14 days to desert sites within the country of residency, by groups of between 2–20 male friends, each with a single falcon. This form of domestic falconry is very widespread, but relatively low impact, although in the absence of suitable quarry species for falcons, other native species may be caught, or shot and killed, e.g. reptiles and passerine or raptorial birds. Such falconers can come from any sector of society, from relatively impoverished semi-nomadic herdsmen, through to more affluent businessmen and professionals.

Since 2001 Saudi Arabia has undertaken expansion of tourist facilities and services as part of a strategy of economic diversification. Tourism planning seeks to encourage both domestic tourists and international visitors (particularly those entering the Kingdom primarily for religious reasons) to visit sites of natural and cultural importance. The emphasis, however, is on the creation of mass tourism hubs and there are no initiatives to promote falconry as a focal tourism activity. However, in recognition of the cultural importance of falconry, the Supreme Commission for Tourism sponsors a falconry competition in the An Nuayriyah district in the Eastern Province in which falconers hunt pigeons in time trials (Waleed Al-Hemaidi, pers. comm.). There are similar initiatives in other Gulf States. As a result, however, falconry within the Arabian Peninsula remains a largely independent and unregulated pursuit, albeit one that is a primary motivator for domestic travel. It seems likely that the environmental impacts will be modest in relation to the activities of the average falconer.

However, the average falconer is not where the problem lies. The disproportionate accumulation of immense wealth by an elite few has, in some cases, enabled removal of any constraints at all in the pursuit of the art. But the oil-wealth that has vastly expanded the ability of some practitioners to have a detrimental impact on wildlife, also provides a means for great conservation gains to be made; but first the bad news.

Five changes to the practice of Arab falconry have occurred in the last approximately 50 years: (1) whereas once a falconer might fly only one or a few birds, wealthy falconers can now employ people to trap (or purchase), train and tend hundreds of falcons; (2) trained falcons for sale in the market were never cheap, but in recent decades prices for the best birds have become inflated to an absurd degree, encouraging more people to trap and trade falcons in the pursuit of a big sale; (3) high purchase prices, the employment of full-time falconers on staff, and the creation of large, climate controlled facilities to house and care for falcons means that proportionately fewer falcons are being released at the end of the hunting season; (4) there has emerged a large trade in live houbara, trapped in countries such as Pakistan and Iran and sold to Arab falconers to train their birds; and finally (5) depletion of houbara populations in the Arabian Peninsula have prompted falconers to seek new, largely untapped populations elsewhere in the range. The first three factors have resulted in concern over the impacts on wild falcon populations; the last two factors are seeing the gradual overexploitation of once remote houbara populations.

Falcons

The saker falcon is considered the traditional hunting falcon in Arab falconry and as a consequence has been most affected by increased trapping for falconry, particularly during the 1980s and 1990s. Although once widespread across the Palearctic region, with breeding grounds ranging from Austria through to Mongolia and China, saker populations have undergone dramatic declines in the last decade (Barton 2002), estimated to be between 48 nd 70 per cent between 1990 and 2003 when only 3,600 to 4,400 pairs were estimated to remain (Birdlife International 2004). The species is considered endangered with the principal threats being habitat loss and human persecution, primarily trapping for Arab falconry. Birdlife estimates for the numbers of saker falcons taken by trappers for Middle Eastern falconry annually (Birdlife International 2004) are 4,000 Saudi Arabia, 1,000 Qatar, and 500–1,000 each Bahrain, Kuwait and UAE. Assuming a mortality rate of 5 per cent during the trapping and transport process the best estimate of annual consumption of sakers is up to 8,400 birds annually (Barton 2002).

There is also reason to believe that this level of exploitation may be more damaging than it first appears. Arab falconers prefer the larger female saker so there is a sex ratio bias in birds taken, but more than that, there is preference for certain plumage types associated with certain breeding populations that may mean disproportionate impacts. Eastham *et al.* (2002) examined the conservation implications of the Arab falconer's preference for female sakers with barred dorsal plumage (*jarudi*), pale ventral plumage (*ashgar* and *abiyad*) and for large dark brown and grey barred birds (*sinjari* and *shunqar* types) that means saker populations from Mongolia, northern China and south-east Russia are under the

Figure 14.3 Off-road vehicle used for falconry in the United Arab Emirates. Source: National Avian Research Centre – Environment Agency, Abu Dhabi

greatest pressure. The pressure on saker populations has increased with trapping taking place not only during the traditional winter migration period, but also both on passage and in breeding grounds. Similar pressure is being placed on populations of peregrine and gyrfalcon from Siberia and Russia. Legislative changes, particularly the banning or regulation of commercial trade in species used in falconry following their listing in Appendix I or II of CITES (Convention on International Trade in Endangered Species), but also the increasing popularity of captive-bred falcons derived from operations funded by falconers (see below) may have reduced pressure on wild populations in recent years.

Houbara

> Where the survival of Arab hawking is likely to stand or fall is not in the availability of suitable hawks, but in the sufficiency of quarry, particularly the houbara bustard. Without houbara there can be no traditional Arab hawking.
>
> (Upton 2001: 19)

The houbara bustard is a sand-coloured steppe desert specialist that has gained a level of conservation that seems not in accord with its official conservation status. Three forms are recognised: *Chlamydotis undulata fuertaventura* restricted to the Eastern Canary Islands; *Chlamydotis undulata undulata* in North Africa, and the Asian houbara bustard *Chlamydotis macqueenii* (formerly a subspecies of *C. undulata*) extending from Egypt to China. *C. macqueenii*, but also *C. u. undula*, populations are most affected by Middle Eastern falconry. Although some estimates of the numbers of individual houbara range into the tens of thousands, precise estimates of population size are elusive. Houbara are listed as Vulnerable by Birdlife International (2004) due to global population decline of 35 per cent over the last 20 years. The principal cause of declines has been hunting by Arab falconers (Goriup 1997; Combreau *et al.* 2001, 2002; Tourenq *et al.* 2004), and associated poaching of live birds, especially from Pakistan, for training of falcons in the Arabian Peninsula.

Because of the possibility of unfavourable scrutiny from conservation agencies, the activities of the larger falconry parties originating from the Gulf States are kept largely secret and accurate estimates of the numbers of houbara taken annually throughout the range are hard to come by. The available numbers, however, paint a grim picture. Even 20 years ago Arab dignitaries hunting in Pakistan were said to take about 3,000 birds per year (Osborne 1996). In the 1980s it was estimated that 2,000–5,000 houbara were being taken by Pakistani trappers each year to supply an illegal trade to Arabia. More recent estimates by the WWF suggest at least 8,000 houbara a year are being smuggled out of Pakistan to the Gulf, where a large bird can fetch up to US$1,000 (McGirk 2003). In recent years the price of a houbara bustard on the black market in the Gulf States has exceeded the price of an average wild-caught saker falcon.

The inability to accurately assess annual falconry harvest and the imprecision of population estimates for houbara make it difficult to place estimates of off-take in any perspective. The most compelling data has come from banding and satellite tracking studies funded by the UAE government which indicated a high risk of extinction of the Asian houbara bustard within the next 50 years at current levels of hunting-related mortality; a maximum sustainable yield of around 7 per cent of the adult population was postulated and contrasted to the >20 per cent that is currently estimated be taken (Combreau *et al.* 2001).

Although there are regional and international agreements in place for the wider protection of the houbara bustard (CITES, CMS (Convention on Migratory Species)) and national hunting laws and protected areas in many of the houbara's range states, in some high profile cases extremely wealthy Arab falconers have been able to hunt with virtual impunity. The anecdotal tales of Arab falconer excesses abound – reports of hunting expeditions involving hundreds of falcons transported in specially equipped passenger jets. There are unofficial estimates that Arab falconers spend US$10–20 million per hunt (Tulepov and Asanova 2003). These astonishing estimates become more credible when you consider that a single hunt may involve in excess of 300 people, and all the requisite logistics and infrastructure to sustain the occupation of large hunting camps for weeks, or even months, at a time. Up to 14 such hunts may operate within Pakistan each winter hunting season (Tulepov and Asanova 2003). Some range states, such as Pakistan and Tunisia, and more recently Iran, Morocco, Kazakhstan and Turkmenistan, have facilitated hunting by powerful falconers from the Middle East by assigning exclusive hunting zones. Falconers employ people from local communities to act as guides and labourers within large winter hunting camps, and in return for support and protection from local authorities falconers may provide funding for the building of mosques, roads and other infrastructure and facilities, while new 4WD vehicles imported from the Middle East for the hunting season will be left behind as gifts to local leaders. While there is no doubt that this form of international recreational tourism has the potential to contribute significantly to host economies, it is unclear exactly who are the financial beneficiaries. In addition, there is some suggestion that this largesse has enabled some unscrupulous falconers to flaunt local hunting regulations, to exceed bag limits and to hunt within wildlife protected areas (Tulepov and Asanova 2003). The excesses and corruption that characterise the activities of a few falconers have influenced the general perceptions of all Arab falconry, unfairly tainting the reputation of the many responsible hunters from the Middle East. But by any measure the impacts on houbara populations of the irresponsible actions of a wealthy and powerful minority have been huge and are ongoing.

Falconry and species conservation

The wealth, regional power and international influence wielded by the elite of the Middle East have freed recreational falconry from the bounds that would normally constrain sport hunting and is resulting in the decline of wild populations of both the prey houbara and the falcons used to hunt them. But the wealth and power that have grown from oil revenues have also made possible the development of wildlife research and conservation programmes in the Middle East. Two of the most prominent falconry nations, Saudi Arabia and the United Arab Emirates, have the largest, most diverse and well-funded wildlife conservation and environmental protection programmes in the Arabian Peninsula, and in both cases these programmes have grown in the 1980s and early 1990s from projects motivated by and focused on the houbara bustard as a falconry resource. Saudi Arabia's National Commission for Wildlife Conservation and Development (NCWCD) was formed in 1986 to create and manage a network of wildlife protected areas. The first projects sought to restore populations of houbara bustard that had been depleted through over-hunting and the first protected areas were sites for protection of remnant wild populations (Seddon and van Heezik 1996) or the reintroduction of captive bred houbara (Gélinaud *et al.* 1997). Later projects focused on other native species such as the Arabian oryx, but the houbara may be considered an umbrella species for conservation efforts in the Kingdom, with over 50,000 km^2 of protected area nominally created to protect habitat for houbara, to the benefit of many other native desert dwelling species (Seddon and Abuzinada 1997). The goal of houbara conservation in Saudi Arabia is the restoration of self-sustaining populations of resident breeding birds protected within a network of protected areas, but which may one day support sustainable falconry in hunting areas outside reserves (Combreau *et al.* 1995; Seddon *et al.* 1995; van Heezik and Ostrowski 2001). With UAE, Saudi Arabia has taken the lead in regional conservation programmes for houbara.

In UAE the National Avian Research Centre (NARC), now within the Abu Dhabi-based Environmental Agency (EAD) (formerly the Environmental Research and Wildlife Development Agency (ERWDA)), was created in 1993, and again one of the first projects established was a captive breeding and research station for houbara. The focus of EAD houbara projects is less on the restoration of local houbara populations, that breed only sporadically, and more on the development of sustainable falconry that has minimal impact on wild houbara populations globally. This entails the captive breeding of houbara for use in both hunting and the training of falcons, provision of research expertise and training to other, less well-funded houbara conservation projects in other range states, the rehabilitation and release of confiscated illegally traded houbara (Bailey *et al.* 2002), houbara research and conservation in China and Central Asia, and the creation and dissemination of public awareness materials. The EAD also supports research and monitoring programmes for Eurasian populations of saker and peregrine falcons, and runs a state-of-the-art falconry hospital, the Abu Dhabi Falcon Hospital (ADFH) catering to hunting falcons from UAE, Qatar, Saudi Arabia, Oman and Bahrain (Bailey and Sullivan 2000). Similar falcon hospitals, treating in excess of 6,000 birds a year, have been established in the region using private funding, most notably in Riyadh, Saudi Arabia (Samour 1999; Naldo and Samour 2003), and Dubai, UAE (Bailey *et al.* 2003; Lloyd 2004); UAE alone now has six large hospitals entirely devoted to the treatment and care of falcons. The ADFH aims to promote awareness of health issues for captive falcons and is the most direct and successful means of making contact with falconers. Working with the EAD the

Abu Dhabi-based office of the World Wide Fund for Nature (WWF) has instituted a capacity building programme for all the UAE CITES enforcement officers, and a passport system using implantable micro-chips to regulate both the movements of birds for falconry and the trade in falcons was established by the UAE authorities (Perry 2003). The UAE projects seek to make falconers partners in conservation measures and to foster a sense of stewardship that will translate into sustainable hunting practices. The houbara bustard and saker falcon are therefore considered flagship species in UAE, around which wider conservation advances can be made (Bailey and Sullivan 2000).

Funding from governments and individuals in the Middle East has also resulted in the creation of houbara bustard captive breeding programmes in North Africa to secure self-sustaining wild houbara populations (Le Cuziat *et al.* 2005) and to support put-and-take hunting within sustainably managed protected areas that are host to a wide variety of wildlife. Middle Eastern funding has also supported the expansion of private falcon breeding projects. Due to the increasing proportion of captive-bred falcons, including hybrid birds that are now used by over 75 per cent of falconers in UAE, Kuwait, Oman and Bahrain, there is now a surplus of hunting falcons and a related marked decline in the rates at which wild falcons are trapped.

Ultimately the political support and funding for houbara-related programmes exists only because of falconry. If the houbara bustard was just one more drab desert-dwelling species it would no doubt be suffering to some extent from the ubiquitous effects of habitat loss and degradation, but would probably not be in dire straits, nor would it have been the catalyst for the development of wide reaching conservation programmes and the attendant growth of regional expertise and public awareness in environmental protection.

The future of Arab falconry

The fate of Arab falconry is entangled with the fate of the falcons used for hunting and the prey that is sought, the houbara bustard. Wild populations of both are in decline, a decline substantially due to overexploitation driven by Middle Eastern falconry. Use of captive-bred falcons for hunting may only partially alleviate pressure on wild populations since so long as even only a few wild-caught falcons can command high prices there will be stimulus for impoverished trappers to try and snare passing falcons. Similarly, use of captive-bred houbara may augment hunted populations and take pressure off illegal trapping and trade for training of hunting falcons, but while they have the means there will always be falconers that will exploit wild populations and provide a market for illegal trade. Legislation and enforcement can only go so far, and may be ineffective at altering the activities of the elite few hunters whose activities are having a disproportionately negative impact. However, there is cause for optimism as the Middle Eastern falconers themselves have proven willing to support conservation measures, to regulate their own activities, and to act to sustain in perpetuity a defining Arab tradition.

References

Al-Hemaidi, W., Assistant Deputy Secretary General, Supreme Commission for Tourism, Saudia Arabia, email correspondence, 17 July 2006.

Bailey, T. and Sullivan, T. (2000) 'The National Avian Research Center opens a new falcon hospital in Abu Dhabi', *Falco*, 15: 3.

Bailey, T., Di Somma, A. and al Muhairi, H. (2003) 'Twenty years of falcon medicine at Dubai Falcon Hospital, 1983–2003', *Falco*, 222: 15–17.

Bailey, T.A., Silvanose, C., Manvell, R., Gough, R.E., Kinne, J., Combreau, O. and Launay, F. (2002) 'Medical dilemmas associated with rehabilitating confiscated houbara bustards (*Chlamydotis undulata macqueenii*) after avian pox and paramyxovirus Type 1 infection', *Journal of Wildlife Diseases*, 38(3): 518–32.

Barton, N. (2002) 'Recent data on saker trapping pressure', *Falco*, 20: 5–8.

Birdlife International (2004) *Threatened Birds of the World 2004*, CD-ROM and <http://www.birdlife.org> (accessed 10 December 2005).

Cade, T. (1982) *Falcons of the World*, Ithaca, NY: Cornell University Press.

Canby, J.V. (2002) 'Falconry (hawking) in Hittite lands', *Journal of Near Eastern Studies*, 61(3): 161–201.

Child, G. (1989) *Socio-economic Influences on Renewable Resource Conservation, Development and Use in Saudi Arabia*, unpublished report. Riyadh: National Commission for Wildlife Conservation and Development (NCWCD).

Child, G. and Grainger, J. (1990) *A System Plan for Protected Areas for Wildlife Conservation and Sustainable Rural Development in Saudi Arabia*, Gland, Switzerland: IUCN, and Riyadh, Saudi Arabia: National Commission for Wildlife Conservation and Development (NCWCD).

Combreau, O., Launay, F. and Lawrence, M. (2001) 'An assessment of annual mortality rates in adult-sized migrant houbara bustards (*Chlamydotis [undulata] macqueenii*)', *Animal Conservation*, 4: 133–41.

Combreau, O., Qiao, J., Lawrence, M., Gao, X., Yao, J., Yang, W. and Launay, F. (2002) 'Breeding success in a Houbara Bustard *Chlamydotis [undulata] macqueenii* population on the eastern fringe of the Jun gar Basin, People's Republic of China', *Ibis*, 144 (on-line), E45–E56.1.

Combreau, O., Saint Jalme, M., Seddon, P., Rambaud, F., van Heezik, Y., Paillat, P., Gaucher, P. and Smith, T. (1995) 'A program for houbara bustard restoration in Saudi Arabia', in J.A. Bissonetta, and P.R. Krausman (eds) *Integrating People and Wildlife for a Sustainable Future*, Proceedings of the First International Wildlife Management Congress. Bethesda, MD: The Wildlife Society.

Eastham, C.P., Nicholls, M.K. and Fox, N.C. (2002) 'Morphological variation of the saker (*Falco cherrug*) and the implications for conservation', *Biodiversity and Conservation*, 11: 305–25.

EIA (2005) *Saudi Arabia Country Analysis*, August 2005. Washington, DC: Energy Information Administration, US Department of Energy.

Gélinaud, G., Combreau, O. and Seddon, P.J. (1997) 'First breeding by captive-bred houbara bustards introduced in central Saudi Arabia', *Journal of Arid Environments*, 35: 527–34.

Goriup, P. (1997) 'The world status of the Houbara Bustard *Chalmydotis undulata*', *Bird Conservation International*, 7: 373–97.

Grutz, J.W. (1999) 'Prelude to discovery', *Aramco World*, 50(1): 30–5.

Harrison, D.L and Bates, P.J.J. (1991) *The Mammals of Arabia*, 2nd edn, Sevenoaks: Harrison Zoological Museum.

Jennings, M.C. (1989) 'The birds of Saudi Arabia: past, present and future', in A.H. Abuzinada, P.H. Goriup and I.A. Nader (eds) *Wildlife Conservation and Development in Saudi Arabia*, Proceedings of the First Symposium, Riyadh 1987. NCWCD Publication No. 3, pp. 255–62.

Le Cuziat, J., Lacroix, F., Roche, P., Vidal, E., Medail, F., Orhant, N. and Beranger, P.M. (2005) 'Landscape and human influences on the distribution of the endangered North African houbara bustard (*Chlamydotis undulata undulata*) in Eastern Morocco', *Animal Conservation*, 8: 143–52.

Lloyd, C. (2004) 'A new public hospital in Dubai for falcons and exotic species', *Falco*, 24: 18.

McGirk, J. (2003) 'In an Asian desert, tribes want bird-loving hunters to fly away', *Boston Globe*, Boston, MA. 28 December: A.17.

Nader, I.A. (1989) 'Rare and endangered mammals of Saudi Arabia', in A H. Abuzinada, P.H., Goriup and I.A. Nader, (eds) *Wildlife Conservation and Development in Saudi Arabia*, Proceedings of the First Symposium, Riyadh 1987. NCWCD Publication No. 3, pp. 220–33.

Naldo, J.L. and Samour, J.H. (2003) 'Update from the Fahad bin Sultan Falcon Center, Saudi Arabia', *Falco*, 21: 13–14.

Osborne, P.E. (ed.) (1996) *The Desert Ecology of Abu Dhabi – A Review and Recent Studies*, Newbury: Pices.

Perry, L.S. (2003) 'Travelling with falcons: the new UAE falcon passport', *Falco*, 21: 12.

Pitra, C., D'Aloia, M.-A., Lieckfeldt, D. and Combreau, O. (2004) 'Genetic variation across the current range of the Asian houbara bustard (*Chlamydotis undulata macqueenii*)', *Conservation Genetics*, 5(2): 205–15.

Prummel, W. (1997) 'Evidence of hawking (falconry) from bird and mammal bones', *International Journal of Osteoarchaeology*, 7: 333–8.

Samour, J. (1999) 'The Fahad bin Sultan Falcon Centre', *Falco*, 14: 4.

Schramm, M. (2001) 'Frederick II of Hohenstaufen and Arabic science', *Science in Context*, 14(1 and 2): 289–312.

Seddon, P.J. and Abuzinada, A.H. (1997) 'Saudi Arabia expands its network of protected areas for migratory houbara bustards', *Houbara News*, 2: 4–6. IUCN, Species Survival Commission's Houbara Specialist Group newsletter.

Seddon, P.J. and van Heezik, Y. (1996) 'Inter-seasonal changes in numbers of houbara bustards in Harrat al-Harrah, Saudi Arabia: implications for managing a remnant population', *Biological Conservation*, 75: 139–46.

Seddon, P.J., Saint Jalme, M., van Heezik, Y., Paillat, P., Gaucher, P. and Combreau, O. (1995) 'Restoration of houbara bustard populations in Saudi Arabia: developments and future directions', *Oryx*, 29(2): 136–42.

Thesiger, W. (1999) *Crossing the Sands*, London: Motivate Publishing.

Tourenq, C., Combreau, O., Pole, S.B., Lawrence, M., Ageyev, V.S., Karpov, A.A. and Launay, F. (2004) 'Monitoring of Asian houbara bustard *Chlamydotis macqueenii* populations in Kazakhstan reveals dramatic decline', *Oryx*, 38(1): 62–7.

Tulepov, A. and Asanova, N. (2003) 'Wildlife hunting in Pakistan: the houbara bustard',. *TED Case Study*, *710*, Trade Environment Database. Washington, DC. Online Journal: <http://www.america.edu/TED> (accessed 10 December 2005).

Upton, R. (2001) *Arab Falconry: History of a Way of Life*, Hancock Wildlife Conservation Center, WA: Hancock House Publishers.

van Heezik, Y. and Ostrowski, S. (2001) 'Conservation breeding for reintroductions: assessing survival in a captive flock of houbara bustards', *Animal Conservation*, 4: 195–201.

Yergin, D. (1991) *The Prize: The Epic Quest for Oil, Money and Power*, London: Simon & Schuster Ltd.

Part IV

Current issues and destination development

15 Communicating for wildlife management or hunting tourism

The case of the Manitoba spring bear hunt

Michael Campbell

Introduction

Canada has long been recognized for its wealth of natural beauty and resources. This natural bounty has been the foundation of its economy and Canada's international tourism attraction, which has to a large degree included both non-consumptive and consumptive use of wildlife through such activities as hunting and fishing. Nature is a vital part of Canada's image and tourism promotions are replete with images of natural features and wildlife. These promotions also frequently include images of consumptive uses of nature and fishing in particular. While hunting images are less visible in general tourist promotions material, materials directed exclusively to hunters exist in all provinces. In Manitoba, fishing and hunting are two of the province's most significant tourism income generators surpassed only by visiting friends and relatives (VFR) tourism. However, public acceptance of consumptive uses of wildlife, and hunting in particular, for recreation has become a concern for both provincial wildlife officials and hunting tourism operators.

The increasing urbanization of Canadian society has been posited as one important factor in these changing public attitudes towards the consumptive uses of wildlife for recreation. This attitude shift is perhaps more significant to the hunting tourism industry due to the importance often attached to the perceived motives of the hunter (trophy) or the context associated with hunting as a tourism attraction. That is, the image of non-resident hunting is often that of a trophy hunter and indeed most marketing materials and Saturday morning 'big-buck' shows tend to promote this view, whether it is accurate or not.

The apparent change in the public's social mores and attitudes with respect to consumptive uses of wildlife have resulted in the opposition to and cancellation of a number of hunts across North America and Canada. Bear hunting in particular has been singled out in a number of jurisdictions and opponents have been quite successful achieving their goals of limiting and in some cases ceasing the hunt entirely (Minnis 1998). The bear hunt in Manitoba is both an important tourism product as well as an important component of the province's wildlife management strategy. Approximately 60 per cent of bear harvest licences in the province are sold to non-resident alien (tourist) hunters and much of the revenue generated from this remains in isolated and rural communities where few other economic opportunities exist.[1] The Manitoba bear hunt, and the spring bear hunt in particular, provides a unique example of how one Canadian jurisdiction responded to intense lobbying to end the hunt. This chapter explores the role of the bear hunt in tourism and wildlife management in the province, the development of the spring bear hunt conflict, and the approach the province took in resolving the issue.

Background

Hunting has been a significant component of many Canadian provinces' tourism products for a number of years. The discussion of hunting as a tourism product is an extremely complex subject incorporating such diverse disciplines as wildlife management, biology, psychology, political studies and economics. While hunting provides many direct and indirect benefits to the Province of Manitoba it also requires that provincial wildlife managers balance their goals of managing wildlife populations within social and biological carrying capacities with the public's desires and wishes and the resources available. This, in turn, requires an understanding of not only the public desires but also their attitudes and state of knowledge. While hunting in the service of wildlife management contributes directly to the provincial economy through licence sales and the provision of equipment and services to the hunter, and indirectly through reduced crop and vehicle insurance payments and helps to ensure healthy, stable wildlife populations, it is entirely possible that this rationale does not resonate with the public. This is not to suggest that wildlife management agencies should manage according to the whims of the public but that it is incumbent upon them to understand the public so that they can communicate the need to manage wildlife populations.

Over the past 30 years the number of hunters has declined across Canada and current estimates suggest 5 per cent of the population participates in hunting (DuWors *et al.* 1999). Commensurate with this decline, anti-hunting organizations have increased their efforts to both alter the public perception of hunting and affect government policy and legislation with respect to wildlife management and hunting. These developments have resulted in increasing concern on the part of fish and wildlife agents across North America, many of whom fear that the scientific management of wildlife populations is in danger due to the influence of an uninformed public. To date the bulk of fish and wildlife agencies' research has been focused upon biological and ecological concerns related to wildlife and habitat management, whereas in reality the bulk of wildlife branch activities involve the management of people. This is particularly the case where the public may be uninformed about the reason certain activities are being pursued. As such it is essential that wildlife managers understand what the public believes.

The role of Manitoba's bear hunt

The province of Manitoba in Canada occupies the geographic centre of the North American continent and encompasses five ecozones (prairie, boreal plain, boreal shield, taiga shield, Hudson's Bay plain). Black bears (*Ursus Americanus*) range across most of the province with the greatest densities of bears being found in the boreal plain region – a region characterized as the transition between prairie and the boreal forest to the north (Manitoba Conservation 2004). The boreal plains region is also a region of intensive agriculture and has been significantly modified by human activity.

Black bears are the most widely distributed of all North American bears and are not always black. They display a wide variety of color variation including black, blonde, chocolate brown, cinnamon, white and 'blue'. It is the wide range of color variation among the Manitoba bear population that attracts many non-resident hunters to the province. Indeed, despite the frequent characterization of the Manitoba hunt as a trophy hunt, very few true 'trophy' animals (based upon such criteria as Boone and Crocket) animals are taken.

Manitoba has a healthy black bear population that is estimated to be between 25,000 to 30,000 and stable to slightly increasing (Hrystienko *et al.* 2004). This population increase is occurring in the boreal plain region which is also the region of greatest bear densities. As a region with significant agricultural activity, it is also one where conflict between increasing bear populations and human activity is a significant concern. Given these conditions, the goal of Manitoba Conservation, and the Wildlife Branch of Manitoba Conservation in particular, is to manage the black bear population such that it remains below the biological and social carrying capacity. The biological carrying capacity represents the maximum bear population that the habitat can accommodate while the social carrying capacity is somewhat less concrete and will vary with the knowledge and attitudes of the population. Generally, however, the social carrying capacity is deemed to have been exceeded when the species in question becomes a nuisance. Based upon population estimates and fecundity (Hrystienko *et al.* 2004), Manitoba Conservation attempts to maintain a harvest level at or below 10 per cent. The current annual harvest of approximately 1,800 bears represents a 6 per cent harvest and is well within sustainable levels (Manitoba Conservation 2004). The majority of these harvested bears (70 per cent) are taken in the spring season over bait. In addition, over 70 per cent of the bears are taken by non-resident (alien) hunters. Manitoba provides both a spring and a fall hunting season for black bears and allows hunters to take only one animal in any given year. Failure to take a bear in the spring season allows the hunter to use the same tag for the fall hunt, thus increasing the likelihood of hunting success.

As noted above, the greatest densities of bears in Manitoba occur in the boreal plains region and as a result the majority of non-resident (alien) hunt takes place in this region of the province. Bear hunting in Manitoba generally takes place over baits where the hunter sits in a stand and monitors the bait station until a suitable target is attracted. Hunting with dogs is illegal in Manitoba.

Manitoba's geography is such that traditional methods of spot and stalk do not allow for the same level of success as the current baiting stations do. Indeed, spot and stalk is unlikely to be a successful strategy at all given the provincial geography. As such, in order to ensure the population of bears remains below the biological and social carrying capacity, and to obtain a harvest of between 6 per cent and 10 per cent per annum, far more licences than currently are issued would be needed. Most of Manitoba is relatively flat country that is heavily treed and as such provides limited sightlines for the hunter. This single fact essentially precludes spot and stalk hunting as is practised in the mountainous areas of Canada. In this regard hunting over bait also ensures the hunt is safer as the bait and hide can both be selected with knowledge of what lies behind the intended target.

Hunting over bait also allows the hunter time to select the animal he wishes to harvest and as such provides ample opportunity to determine if the bear is a female and with cubs as it is illegal to shoot female bears with cubs in the province. Furthermore, all non-resident hunters are required to engage the services of an outfitter and guide in order to hunt in the province providing additional economic inputs to the province.

Economic impact of the spring bear hunt

Hunting provides the province of Manitoba with one of its most lucrative tourism products. Currently, hunting is an important economic activity in Manitoba with local hunters accounting for CAN$24.8 million in receipts in 1996 (Environment Canada 2000). Non-resident hunters are estimated to have contributed CAN$30 million in direct revenue in 1994, the period for which the most recent figures are available (Canadian Tourism Research Institute 1995). Given the trend to increased allotments of non-resident hunting opportunities, this figure is likely to be much higher at present. When combined with fishing, this total rises to CAN$180 million and places consumptive wildlife tourism as the single greatest tourism receipt other than visiting friends and relatives. While direct figures for the economic value of bear hunting in Manitoba as a whole are unavailable, some generalizations can be made.

The province of Manitoba provides 3,200 licences annually for the bear hunt, and since 1999 non-residents have made up approximately 60 per cent of total licence holders (see Table 15.1). Given an overall hunter success rate of less than 55 per cent and the goal of maintaining harvest below 10 per cent of the population, the province allocates 3,200 licences per year and adjusts the number of non-resident licences available based on the resident demand for hunting opportunities. This has meant that over the past two decades, as resident demand has declined, there have been greater numbers of non-resident hunter opportunities available, increasing the tourism revenue. Licence sales alone for non-residents amounts to CAN$352,000 per year. Given that non-resident bear hunters must hunt with a guide with an average cost of US$1,500 for a five-day hunt the value of the hunt exceeds this by a substantial margin.[2]

Table 15.1 Manitoba bear license and harvest rates, 1993–2003

	Licenses sold		*Total # bears harvested*		*Success rate*	
Year	*Resident*	*Non-res.*	*Resident*	*Non-res.*	*Resident*	*Non-res.*
1993	1,849	953	726	770	39.3	80.8
1994	1,877	1,201	730	933	38.7	77.7
1995	2,133	1,383	820	1,087	38.4	78.6
1996	1,574	1,443	399	965	25.4	66.9
1997	1,363	1,520	442	1,077	29.1	70.9
1998	1,485	1,513	495	1,139	30.3	69.7
1999	1,243	1,718	332	1,217	26.7	70.9
2000	1,251	1,988	426	1,438	22.9	77.2
2001	1,264	1,969	384	1,366	21.9	78.1
2002	1,309	1,821	434	1,306	24.9	75.1
2003	1,306	1,920	516	1,455	26.2	73.8

Source: Manitoba Conservation

An economic analysis conducted by the Manitoba Lodges and Outfitters Association (MLOA) for the parkland region[3] in 1996 indicated that the spring bear hunt generated over CAN$2,774,640 in economic activity and CAN$719,999 in direct government tax revenues (MLOA 1999). The study also found that direct expenditures from non-resident hunters totalled CAN$3,327,071. In addition, the spring season of the hunt was identified as a more significant and profitable component of many outfitters' overall operation and, if cancelled, could render the overall operations economically untenable. The spring hunt alone resulted in 40 jobs in the parkland area and, when considered in light of the spring bear hunt's importance to overall operation of several outfitters in the region, it was estimated that cancelling this hunt would result in the loss of 100 jobs across the province (62 in the parkland region alone) (MLOA 1999).

It is worth noting that the majority of jobs supported by the spring hunt are in rural areas where few other economic activities exist and that the spring hunt provides much needed employment for rural and aboriginal populations. Perhaps more significant to the maintenance of the hunt in the face of ongoing opposition is the fact that the parkland region is also politically volatile, having two key swing ridings. Few politicians wish to alienate their constituents on this issue alone.

Development of the conflict

Much has been written about the changing relationship between humans and wild nature over the past 40 years (Dizard and Muth 2001; Organ and Fritzell 2000; Duda and Young 1996; Nash 1968), including the apparent decline in public acceptance of hunting as a legitimate activity, wildlife management strategy or pastime (Baker 1996).

Studies undertaken to gain understanding of public attitudes towards hunting (e.g. DiCamillo 1995; Fleishman-Hillard 1994; Shaw 1975), have consistently found that approximately 10 per cent opposed hunting under all circumstances, 10 per cent hunted or supported hunting in all circumstances, and 75–80 per cent neither strongly supported nor strongly opposed hunting. Subsequent studies indicated that opinions could and did change in response to how hunting was perceived or characterized, that is, the context of the activity. For example, when hunting was characterized as 'for food' opposition held around the 10 per cent level; however, when characterized as 'for trophy' or 'for sport' opposition increased to nearly 100 per cent and 80 per cent respectively (Prairie Research Associates 1999; Duda and Young 1996).

On the surface the increase in opposition appears to be related to the manner in which hunting is characterized; that is, hunting characterized as 'for sport' or 'trophy' increases opposition, whereas hunting characterized as 'for food' decreases opposition and increases support. It could also be suggested, however, that the perceived motives of hunters may be influencing public attitudes towards their behaviour. These changing views primarily reflect attitudes of the non-hunting population, though hunters also display marked (if somewhat suppressed) differences in response to activity characterization (Prairie Research Associates 1999; Duda and Young 1996). These opinions are situational and subject to change based on the perceived motives of the hunter and uses of the game. Thus, the manner in which hunting is portrayed can invoke opposition to hunting and hunters.

Similar reactions were identified when hunting is characterized as 'for recreation'. In this case at least two influences may be at work. First, as in the preceding examples, hunting for recreation might elicit a view of killing for fun, reflecting concerns about the motives of the hunter and second an interpretation of recreation that is limited to ideas of play and fun (Campbell and MacKay 2003).

One limitation of many of these earlier studies on public attitudes towards hunting was the use of non-representative samples. Many studies were conducted on populations not representative of the general population, such as college students and urban residents (Shaw 1975; Kellert 1978; Hooper 1992; Diefenbach *et al.* 1997). Another limitation of previous work relates to the manner in which the questions were asked, which sometimes caused opposition to hunters to be confused with opposition to hunting (Heeringa 1984). This is a particularly salient point given subsequent findings that opposition to hunting may be rooted in negative opinions about hunters (Rohlfing 1979). Although the debate is much more complicated, the wildlife management community continues to frame arguments in terms of pro- and anti-hunting and hunter vs. non-hunter (Minnis 1996).

This complexity is highlighted by Muth *et al.* (1998) in a study of attitudes of fish and wildlife professionals which shows support or opposition to different types of hunting to be multifaceted (i.e. bear hunt with dogs vs. grouse hunt with dogs) and may reflect greater affinity with certain hunted species, social class bias and perceptions of fair chase. Similar results were evident in attitudes towards trapping where the motivation, methods and characteristics of the trappers themselves might be more of an issue with many people than the activity itself (Muth *et al.* 1998).

In addition, some authors have recently suggested that the public is becoming increasingly sceptical of traditional arguments for allowing regulated sport hunting (e.g. wildlife management, population regulation) (Holmsman 2000), and to be acceptable hunting must be connected to the message of heritage and culture (Dizard and Muth 2001; Mahoney 2001; Organ 2001). In this context hunting is seen as an activity that links the participant to the landscape and a receding past. As such it is a re-creative process and much in contrast with concepts of recreation as 'play'. While intuitively appealing this connection has not been borne out by recent research (Campbell and MacKay 2003; MacKay and Campbell 2004).

While opposition to hunting is not a new phenomenon per se, the existence of a large and potentially influential audience receptive to the message of those opposing hunting is. The ability to engage this audience and move them to exert pressure on political leaders is increasingly viewed as the arena in which the struggle for control over wildlife issues is waged. As such the battle is one for influence over public opinion.

In almost all situations where bear hunting has been banned through ballot initiative (that is, a referendum), the rationale for the ban has been based upon two key premises. These are: (1) that hunting with dogs or over bait is unsportsmanlike or unethical and does not offer the bear a 'fair' chance; and (2) that hunting of bears in spring will result in the shooting of females with cubs and therefore the subsequent orphaning of the cubs. This is further developed to suggest that the orphaned cubs will suffer a slow and painful death through starvation or from depredation by male bears. Minnis (1998) catalogued citizen-sponsored ballot initiatives across the United States and identified 15 hunting-related initiatives (total 24 wildlife related) between 1972 and 2000. Of these hunting-related initiatives, nine dealt with bear hunting and of these seven specifically mentioned the use of bait. In total, eight of these hunting initiatives and four of the nine bear specific initiatives were successful in stopping or limiting the hunt. Ballot initiatives continue to be utilized by both pro- and anti-hunting groups throughout the United States. Despite this, the use of citizen-based ballot initiatives is not a significant factor in Canada where referendums and grass-roots politics is still an emerging trend. As such, most coordinated attempts to curb hunting to date have taken the form of lobbying and political action. Manitoba Conservation began to become concerned when action was initiated to ban the spring bear hunt in the mid-1990s and commissioned the University of Manitoba to conduct research into public attitudes towards hunting and wildlife management in the province. Concurrently in Ontario, the International Fund for Animal Welfare (IFAW) began orchestrating events to draw attention to the spring bear hunt and their view that it was unnecessary and resulted in orphaned cubs. This effort was being conducted in the midst of a provincial by-election that could weaken the standing government, providing IFAW with an excellent opportunity to meet its goals.

IFAW distributed graphic and emotionally charged videos depicting a bear hunt resulting in orphaned cubs to key ridings during the run-up to the by-election and demanded that the government cancel the spring bear hunt or face defeat at the polls. The Ontario Provincial Government, over the objections of its own Members (from rural and northern ridings) cancelled the hunt and was rewarded at the polls. Emboldened with their success in Ontario IFAW turned their attention to Manitoba and its spring hunt.

Opposition to the bear hunt in Manitoba, like Ontario, focused on the spring baited bear hunt and has followed closely the arguments put in place in other campaigns. Once again a looming election appeared to set the stage for a potentially successful effort to ban a hunting activity. The orphaning of cubs due to failure to identify females in the spring, and potential that the use of bait was not 'a fair chase' were key foci of the campaign. IFAW cooperated with the Winnipeg Humane Society (WHS) in the campaign and made use of emotional material that linked hunting to cubs starving to death from losing their mothers. In addition, protests were held in conjunction with efforts to establish a Winnie the Pooh museum in Winnipeg, in a bid to gain public support with a favourite children's character. (Note: The name Winnie the Pooh is generally accepted to be derived from Winnipeg, Manitoba's capital and the original name given to a cub orphaned when a hunter shot its mother. A Winnipeg soldier in White River Ontario while en route to join the 2nd Canadian Infantry during the First World War purchased the cub. The cub was subsequently donated to the London Zoo and it is reputed that A.A. Milne visited the zoo frequently with his son Christopher Robin and based the Winnie character on the bear named Winnipeg.)

IFAW once again distributed the videos to two potential swing ridings in Winnipeg and demanded that the parties include banning the bear hunt in their election platform. This time the focus on Winnipeg (the capital city of Manitoba: population 700,000) proved not to be as strategic as in Ontario. Neither leading political party would commit to banning the spring bear hunt despite extensive media coverage (IFAW initiated video distribution and a radio campaign while the Winnipeg Humane Society initiated a Winnie the Pooh fest and attempted to connect it with the spring bear hunt to influence the public). The key issue in this particular case was the fact that the key ridings in the election and the ones that frequently swung to determine the government were in the boreal plains, the region where the greatest density of black bears are and where the majority of non-resident hunting takes place. Neither the government nor the official opposition was willing to alienate voters in these key areas. While hardly a success for Manitoba Conservation, IFAW's failure to end the hunt did provide a reprieve and the opportunity to prepare for the future challenges that were certain to come.

In most previous cases where an interest group has moved to ban an activity such as bear hunts, the issue has been characterized as a clash of views (e.g. Locker and Decker 1995; Lush 1996) based upon basic beliefs about what is 'right' in terms of the consumptive uses of wildlife. Given that only 10 per cent of the population is likely to hold these extreme views the debate then becomes about capturing the public to one point of view or the other. In the Manitoba 1999 case the two 'sides' of the argument were never able to face off. However, the Wildlife Branch of Manitoba Conservation recognized early on that this was likely the first of many challenges to their wildlife management strategies.

In order to be better prepared for future challenges the province undertook a study of public attitudes and values with respect to hunting as wildlife management. The goal of the study was to obtain an accurate picture of the public's support or opposition to hunting along with a sense of what they believed and knew about wildlife, wildlife management and hunting.

Manitoba response

Beginning in 1999 Manitoba Conservation launched an extensive study of Manitoba residents' attitudes to wildlife and hunting as a wildlife management tool. The study was undertaken by the University of Manitoba and comprised two main methods, first an omnibus survey of Manitoba residents weighted for geographic representation was conducted in order to understand the attitudinal and normative factors influencing Manitoban's support or opposition to hunting and various wildlife management activities and, second, two series of focus groups were conducted with Manitobans expressing moderate views towards hunting to evaluate and understand how messages, media and messengers influenced the public's acceptance of a particular message regarding hunting and wildlife management.

The research design was a mail survey distributed to a regionally stratified random sample of 3,000 Manitoba households based on the provincial telephone company's database. The process included a postage-paid return envelope, incentive prize, follow-up postcard, and replacement questionnaire to non-respondents (Dillman 2000). The survey instrument design and data treatment followed the theory of reasoned action (Ajzen and Fishbein 1980). Scale items were based upon a review of the literature and the results of an elicitation survey conducted with a sub-sample of the target population to generate salient referents and beliefs about supporting hunting. This procedure resulted in 20 belief statements and 16 salient referents. From these, a series of seven-point bipolar evaluative scales were designed to assess attitudes and subjective norms. Of the 20 attitudinal items, four were related specifically to hunting as wildlife management (licence fees to support wildlife conservation; to reduce/control disease in wildlife; hunting to maintain population levels; hunting to conserve wildlife habitat) and three were specifically related to tourism and economic activity (hunting in the province as a tourism activity; hunting as a contributor to the provincial economy; hunting that requires licence fees). In addition, three attitudinal items were related to the negative management consequences of hunting (hunting that results in injury or cruelty to animals; hunting that forces wildlife to migrate; and hunting that upsets the balance of nature). Overall, results of the survey indicated that, by and large, Manitobans were supportive of hunting and believed that it contributed to positive wildlife management. That is, they believed that hunting could result in habitat conservation and improve wildlife population health, evaluating these as positive outcomes. However, they also showed strong negative evaluation of hunting that might result in negative effects on wildlife. They were, however, less likely to believe this outcome. In addition, Manitobans were slightly positive towards hunting as a tourism and economic activity (MacKay and Campbell 2004), an important consideration when one looks at the trend in licence sales for hunting bear in Manitoba.

Results of the focus groups suggested that moderate Manitobans were highly desirous of increased information about wildlife and wildlife management in the province if the information was presented in a balanced and scientific manner. They further indicated that the Wildlife Branch was obligated to counter the arguments produced by animal rights activists so long as the counter-argument was balanced and based upon science. In general, moderate Manitobans viewed most animal rights groups as extremist and were unlikely to be influenced by them. This, however, was predicated upon their ability to identify the messenger.

Based upon the study of public attitudes the Wildlife Branch undertook steps to ensure that it was prepared for future challenges to its wildlife management strategies. This was accomplished through a two-part strategy that incorporated the scientific evaluation of potential cub orphaning (Hrystienko *et al.* 2004) and the development of a communications strategy that was tuned to public attitudes and understanding of wildlife management and what and who resonated with them. That is, what did the public know about wildlife and wildlife management, what were their attitudes towards these practices, what messenger were they most likely to respond to and how should information be provided to the public with respect to these issues? Perhaps the most significant result of the analysis of public views was that the public was very receptive to information from the Wildlife Branch and felt it was the branch's responsibility to provide it. Further, they were interested in balanced, science-based information and thought the Branch was obligated to respond as such to any messages that challenged the Branche's activities and programmes (Campbell *et al.* 2001). Despite this, it was also evident from the study that the public in general was woefully uninformed about wildlife and wildlife management issues (e.g. many people believed that white-tail deer were endangered and in need of protection from hunting when in fact the population is increasing and in many areas exceeds the carrying capacity).

In response to this a three-phase communication strategy was developed for the province involving: (1) laying the foundation; (2) ongoing communications; and (3) responses to specific attacks on Branch activities. Foundation communications are identified as general information on the status of wildlife and wildlife populations in the province. Ongoing communications are designed to reinforce the foundation information and to inform the public of ongoing or current management activities. Finally, response communications are very specific communications designed to counter the messages challenging Branch activities. All of the communications developed for any of these phases follow the balanced, science-based format identified above. While the full communications strategy has yet to be completely rolled out, the basic tenets of the programme have been adhered to in all Wildlife Branch-based communications. Specific individuals were identified as spokespeople and educated in how to respond to the press and public. Messages were always presented in a balanced scientific manner and directed towards the 80 per cent of the population who were neither hunters nor animal rights activists. Where necessary, Conservation Officers (a group identified in focus groups as having high credibility) were used as messengers. To date, the approach has been very effective and when coupled with the data regarding the low level of cub orphaning (Hrystienko *et al.* 2004) has been successful in supporting the maintenance of the spring bear hunt.

When charged that the baiting of bears was not sporting the provincial spokesperson's response was that, although it was not sporting, the province was not in the business of providing sporting opportunities but sustainably managing the province's wildlife resources with limited resources. The baited spring hunt was not only effective and humane in meeting these goals it also provided valuable sustainable economic activity through tourism in areas with fewer other opportunities. In addition, the hunt as conducted was not only extremely safe but the safest option available. Communications did not discuss nuisance bears in part because nuisance bear activity is influenced by many other factors beyond spring harvests. Nevertheless, anecdotal evidence from Ontario suggests that the banning of the spring bear hunt has resulted in significant increases in nuisance bear activity in a number of Ontario municipalities and at least one municipality reported a nine-fold increase in nuisance bear activity (Hrystienko 2005).

In regards to the opposition to the spring bear hunt, the communications strategy has been extremely successful and despite near annual challenges to the spring bear hunt the Wildlife Branch has been able to maintain balance and reason in the debate over bear hunting in Manitoba. This has been accomplished in part by recognizing that the audience for their communication is not the opponents to the hunt but the 80 per cent of the population whose attitudes are more moderate and context driven. Through regular science-based and balanced discussions of wildlife issues the Branch has placed itself in the role of reasoned and balanced expert. This is in part accomplished through the rigid application of the resource management agenda. That is, the Branch is not in the business of providing or maximizing hunter opportunities but in maintaining a healthy bear population below biological and social carrying capacities. Hunting by resident and non-resident alike is simply a cost-effective, indeed profitable means of accomplishing this. As such, baited hunting is not about the sport of the hunt but about ensuring that the management of the bear population is conducted in a safe and effective manner. Maintaining this position and bolstering it with scientific information regarding actual cub orphaning has effectively neutralized the major arguments of those opposed to the hunt.

Conclusion

The bear hunt in Manitoba, and the spring hunt in particular, are very important elements of both the local tourism economy and the province's bear management strategy. As a tourism product the bear hunt alone generates in excess of CAN$5,000,000 in direct revenue and supplements the guiding and outfitting industry during a time of the year with few other hunting opportunities. As such it represents a critical component in the industry's ability to remain sustainable. As wildlife management, it is an important component of the control of the provincial bear population below the biological and social carrying capacity. Despite this there continues to be controversy over the conduct of the hunt and the need for the hunt in general. In order to continue to meet its management objectives while providing a sustainable tourism revenue stream the province will need to continue to engage those Manitobans with moderate views regarding hunting through regular and balanced communications.

Notes

1 Manitoba recognizes three classes of hunters: residents of Manitoba, non-resident Canadians and non-resident aliens. Non-resident Canadians are a very small (<5 per cent of the Canadian) total. The vast majority of non-resident hunters are classified as non-resident aliens and are drawn primarily from the United States (98 per cent). Most originate from those states with higher levels of hunting participation (e.g. Wisconsin, Michigan, North Dakota, Texas, Arkansas, Tennessee, etc.).

2 Note that the ban on the spring bear hunt in neighbouring Ontario (see below) has not appreciably increased demand for opportunities in Manitoba. This may be because the spring hunt in Ontario did not require that alien non-residents employ the use of a guide and as such represents a different market than the Manitoba hunt.

3 The parkland region of Manitoba is an administrative unit that encompasses 25,000 km^2 of which nearly 10,000 km^2 are shared by national parks, provincial parks or forest reserves. It is a largely agricultural region and has historically swung between the two primary political parties in the province.

References

Ajzen, I. and Fishbein, M. (1980) *Understanding Attitudes and Predicting Social Behaviour*, Upper Saddle River, NJ: Prentice-Hall Inc.

Baker, G. (1996) 'Hunting 1996, a year to remember', *Transactions of the North American Wildlife and Natural Resources Conference*, 60: 212–22.

Campbell, J.M. and MacKay, K.J. (2003) 'Attitudinal and normative influences on support for hunting as a wildlife management strategy', *Human Dimensions of Wildlife*, 8(3): 181–97.

Campbell, J.M., MacKay, K.J. and Ostrop, E. (2001) 'Communicating the benefits of hunting in Manitoba: results of focus group research', *HLHPRI Technical Report HLHPRI094*.

Canadian Tourism Research Institute (1995) *Manitoba Lodges and Outfitters Economic Impact Analysis*, March, 1995.

DiCamillo, J.A. (1995) 'Focus groups as a tool for fish and wildlife management: a case study', *Wildlife Society Bulletin*, 23: 616–20.

Diefenbach, D.R., Palmer, W.L. and Shope, W.K. (1997) 'Attitudes of Pennsylvania sportsmen towards managing white-tailed deer to protect the ecological integrity of forests', *Wildlife Society Bulletin*, 25: 244–51.

Dillman, D. (1978) *Mail and Telephone Surveys: The Total Design Method*, New York: Wiley.

Dillman, D.A. (2000) *Mail and Internet Surveys: The Tailored Design Method*, New York: John Wiley & Sons.

Dizard, J.E. and Muth, R.M. (2001) 'The value of hunting: connections to a receding past and why these connections matter', Presentation at 66th *N*orth American Wildlife Conference, Washington, DC. March 2001.

Duda, M.D. and Young, K. (1996) 'Public opinion on hunting, fishing and endangered species', *Responsive Management*, 1–12.

DuWors, E., Villeneuve, M., Filion, F.L., Reid, R., Bouchard, P., Legg, D., Boxall, P., Williamson, T., Bath, A. and Meis, S. (1999) *The Importance of Nature to Canadians: Survey Highlights*, Ottawa: Environment Canada.

Environment Canada (2000) *The Importance of Nature to Canadians: The Economic Significance of Nature Related Activities*, Federal Provincial Task Force on the Importance of Nature to Canadians.

Fleishman-Hillard Research Inc. (1994) *Attitudes of the Uncommitted Middle Towards Wildlife Management*, St. Louis, MO: Fleishman-Hillard Research Inc.

Heeringa, S.G. (1984) *American Public Attitudes Toward Hunting: A Review and Analysis of Published Research on American Attitudes Toward Hunting and Related Wildlife Practices*, Report prepared for the Wildlife Conservation Fund of America. Ann Arbor, MI: University of Michigan.

Holmsman, R.H. (2000) 'Goodwill hunting? Exploring the role of hunters as ecosystem stewards', *Wildlife Society Bulletin*, 28(4): 808–16.

Hooper, J.K. (1992) 'Animal welfarists and rightists: insights into expanding constituency for wildlife interpreters', *Legacy*, Nov./Dec.: 20–5.

Hrystienko, H. (2005) Personal communication.

Hrystienko, H., Pastuck, D., Rebizant, K.J., Knudsen, B. and Connor, M.L. (2004) 'Using reproductive data to model American black bear cub orphaning in Manitoba due to spring harvest of females', *Ursus*, 15(1): 23–34.

Kellert, S.R. (1978) 'Attitudes and characteristics of hunters and anti-hunters', *Transactions of the North American Wildlife and Natural Resources Conference*, 43: 412–23.

Locker, C.A. and Decker, D.J. (1995) 'Colorado black bear hunting referendum: what was behind the vote', *Wildlife Society Bulletin*, 23(3): 370–6.

Lush, B. (1996) *A Summary of Public Perceptions Regarding Hunting and Wildlife Issues*, British Columbia Wildlife Federation International Workshop, 13–16 June.

MacKay, K.J. and Campbell, J.M. (2004) 'An examination of attitudes toward hunting as a tourism product', *Tourism Management*, 25(5): 443–52.

Mahoney, S. (2001) Paper presented at 66th North American Wildlife Conference, Washington, DC, March 2001.

Manitoba Conservation (2003) *Wildlife Branch License Statistics*, Unpublished.

Manitoba Conservation (2004) *Black Bear Hunting in Manitoba: A Review*, Winnipeg, Manitoba: Wildlife and Ecosystem Conservation Branch.

Manitoba Lodges and Outfitters Association (MLOA) (1999) *Mid West Manitoba (Parkland) Region Economic Impact Study of Outfitter Operations*, Dauphin, Manitoba: MLOA.

Minnis, D.L. (1996) 'The opposition to hunting: a typology of beliefs', *Transactions of the North American Wildlife and Natural Resources Conference,* 62: 346–60.

Minnis, D.L. (1998) 'Wildlife policy-making by the electorate: an overview of citizen-sponsored ballot measures on hunting and trapping', *Wildlife Society Bulletin*, 26(1): 76–83.

Muth, R.M., Hamilton, D.A., Organ, J.F., Witter, D.J., Mather, M.E. and Daigle, J.J. (1998) 'The future of wildlife and fisheries policy management: assessing the attitudes and values of wildlife and fisheries professionals', *Transactions of the North American Wildlife and Natural Resources Conference*, 63: 604–27.

Nash, R. (1968) 'Wilderness and man in North America', in J.G. Nelson and R.C. Scace (eds) *The Canadian National Parks: Today and Tomorrow*, Studies in Land Use History and Landscape Change, Waterloo, ON: University of Waterloo, pp. 66–93.

Organ, J. (2001) Opening address 'North America's Hunting tradition', presented at 66th North American Wildlife Conference, Washington, DC, March 2001.

Organ, J. and Fritzell, E.K. (2000) 'Trends in consumptive recreation and the wildlife profession', *Wildlife Society Bulletin*, 28(4): 780–7.

Prairie Research Associates (1999) *Attitudes of Winnipeg Residents Towards Hunting*, Unpublished Report for Manitoba Conservation. Winnipeg, MB.

Rohlfing, A.H. (1979) 'Hunter conduct and public attitudes', *Transactions of the North American Wildlife and Natural Resources Conference*, 43: 405–11.

Shaw, W.W. (1975) 'Meanings of wildlife for Americans: contemporary attitudes and social trends', *North American Wildlife Conference*, 39: 15–155.

16 Catch and release tourism

Community, culture and consumptive wildlife tourism strategies in rural Idaho

Kenneth Cohen and Nick Sanyal

Introduction

This chapter explores the changing socio-cultural landscape of three small towns in rural northern Idaho, each engaged in deliberately re-envisioning their future. Faced with the closure of local timber mills, each town is identifying its unique cultural relationship to resource-based tourism. Consumptive wildlife tourism has become a major theme in re-envisioning their economic future. Transitioning from a timber extraction economy to a consumptive wildlife tourism economy takes more than vision; it takes collective action on the part of community residents to progress towards realizing that vision. Key factors in building the capacity for community-driven tourism development strategies are examined through a case study approach that illuminates the cultural barriers and opportunities that impede or promote a community in transition.

In recent decades, tourism has increasingly been touted as a primary means for replacing declining revenues from resource extraction (e.g. timber) and building a sustainable economy at the local level (Harris and Russell 2001). The communities examined within this case study, are diverse in many regards, but all have ample sport hunting and fishing opportunities that are renowned within the region and increasingly becoming known on a national level (McLaughlin *et al.* 1989). Balancing the economic, social and environmental dimensions, sustainability's triple bottom-line (Krippendorf 1982), of community economic revitalization is a challenge grounded in the local culture of the host community. Spending 18 months engaged in participatory action research in these communities revealed patterns of acceptance of and resistance to repositioning the local economy to effectively capitalize on the abundant local fishing and hunting opportunities. The experience of these communities is both illuminating and cautionary.

A saying in rural community development work is that 'all rural communities have one thing in common: they are all different'. While the three communities this case study examines are within an hour's drive of one another, they are as distinct as three individuals. Ideally, the persona of each community would be taken into consideration in tailoring a sustainable approach to community-based natural resource tourism that acknowledges local wisdom, perceives unique barriers to participation and galvanizes collective action attuned to specific local needs.

One approach for achieving these outcomes has recently emerged in the field of community-driven development – community coaching. Community coaching is an adaptive practice tailored to unique community contexts to guide systemic change via participant empowerment (Cohen 2006). This new method for co-generating community change is becoming increasingly pervasive in the rural United States. Community coaching is now practised in 200 communities within the United States (Emery *et al.* 2005). While the purposes and interventions vary, community coaching-based strategies focus on an inclusive process of community capacity building to achieve self-determination (Cohen 2006). To catalyze and support this process, community coaches guide activities and broker resources for building the capacity of community participants to achieve their desired goals. At the heart of a community coaching strategy is the belief that community development objectives, such as developing a consumptive wildlife tourism base, benefit from being grounded in and emanating from the unique socio-cultural attributes and characteristics of place that result in particular facilitators and barriers to change. Once these factors are identified through a community-based participatory approach, more effective strategies for community development can be designed.

To design effective initiatives, it is first necessary to fully understand a place, and acknowledge that the people that constitute the community inhabiting that geography are diverse and motivated by disparate reasons to call the particular locale 'home'. Thus consumptive wildlife tourism may present a general mode of economic development within the region, but in practice the proposition of building a tourism-friendly community resonates differently within different segments of the community.

One commonality among the three case study communities is a rich tradition of hunting and fishing. These pursuits, which had long been a central component of the local culture, were increasingly being advocated as a cash crop for the local economy by private and public economic development forces within each community and the region.

To better understand the response to these efforts, it is necessary to have a clear understanding of the place in which these developments are occurring. This clarity was provided by the results of the case study of the communities.

The case study

The case study conducted in these three rural communities focused on their participation in a programme funded by an agency whose mission is to reduce rural poverty in the northwest United States. The funding agency sought out communities that fit the basic profile of a rural community in decline including poverty rates of over 10 per cent and a declining population. The communities in north-central Idaho participated in this 18-month programme based on their meeting these criteria and their willingness to participate. The role of the community coach in this programme was to operationalize the programme's logic model, a set of key assumptions, strategies and actions for the cultivation of local leadership, the enhancement of social capital and the synergy of internal and external partnerships that would lead to interventions that would reduce poverty and population decline. The community coach guided the process, allowing each community to develop its own vision, strategies and actions for identifying and meeting its needs and goals.

The context

Idaho has the most abundant game populations within the lower 48 US states, with local residents hunting big game four times as often as the national average (SCORPT 2003–07). Mule deer, elk, antelope, moose, and bighorn sheep are popular species. In addition to hunting, fishing is big business in Idaho, with anglers spending approximately US$300 million in the state annually (O'Laughllin 2005). Steelhead and salmon anglers alone spent $100 million in 2001.

The communities

The three case study communities are located in an area where scenic beauty, diverse and abundant large and small game populations, and accessible year-round and seasonal fishing opportunities abound. In order to preserve the anonymity of the participants in the programme, each town has been assigned a pseudonym. Two of the communities, River Town and Hawk Town, are located on the banks of a river known for its spring and fall steelhead and salmon runs. Along with Elk Town, these communities lie in the heart of hunting and fishing territory, and each has responded to their unique setting in a distinct fashion based on their socio-cultural dimensions.

Elk Town

Elk Town is a community with a population of approximately 140. Elk Town used to be a company town, with two mills and a peak population of 1,300. With the incremental closure of the mills, the town's population dwindled. By 1984 the local school closed, signalling to all who remained that the future existence of the town itself was in peril. Presently there are two restaurants, two bars, a hotel, a general store and an airstrip. There are no sidewalks and the streets are unpaved. The town is set in a pristine location with a lake, campgrounds, scenic waterfalls and trail system that make it a scenic destination for cross-country skiers, snowmobilers and, above all, hunters.

There are more All-Terrain Vehicles on the streets than cars. The general store posts the photos of hunters with their game next to a map with pins for visitors to identify where they are from. The map is riddled with pins from throughout the United States, and the guest book routinely has signatures from international travellers.

When scheduling meetings in Elk Town, the primary obstacle to setting meeting dates was the hunting season. People frequently greet one another with 'did you fill your tag?'. Hunting is discussed not just in terms of sport, but in terms of filling the freezer. It is sport, culture and subsistence. In fact, in Elk Town it was not uncommon for someone to come to a meeting in a shirt recently stained in blood from dressing a deer. Clearly, hunting is central to life in Elk Town.

A question that caused divisions within the community was how to preserve what the locals valued most about where they lived while developing a sustainable economy. In fact, this question was central to all three communities: *how do you develop the local economy while preserving the rural character of the community?* In Elk Town that question became dichotomous. One camp advocated for the no action alternative. They perceived economic development in the terms frequently cited by the regional economic development establishment, 'grow or die'. If grow meant radical change, they'd rather die. The fact that there were only 12 children under the age of 18 left in town did not concern them. The following exemplar quotes are reflective of the prevailing sentiment within this group of residents.

> I've lived here most of my life. I remember when this town had a dance hall and a skating rink. It was really something. But now it's quiet and that's just fine with me. Why do we need new people moving in? It's just a few of the business owners that care about that, trying to make it like it's the whole town, well it's not. (Russ)
>
> I've got my spots. I don't see any reason for somebody from who knows where coming in and messin' up a good thing. Last year I didn't get enough time to even get my limit. It'd be even harder the more folks you got. (Steve)

Contrarily, another segment of the community believed strongly that the death of the town would signal the loss of a way of life built on values worth preserving and promoting. These individuals cherished their way of life in Elk Town and saw well-managed tourism as the means for ensuring that they could continue to live their chosen lifestyle while providing that same opportunity for future generations.

> I lived here many years ago. When I came back the town was dying. There are so few families left after the school closed. It's just really sad. I'm not worried about growing too much like some. There's just no land for that. What we need is a place where young families want to move. We can't let this town die. (Shirl)
>
> Those that don't want anybody moving in just don't want any change at all. Meanwhile it is changing, we're all getting older. But when they can't pay their utility bill they'll think different. I want the store to stay in business because I don't want to have to drive over an hour just to get a gallon of milk, so I've got my selfish reasons too. But if were going to keep these places going we need families moving to town. The rest of us are getting too old. (Ray)

Ultimately, through the community coaching programme the town's established and emergent leaders collectively developed and embraced a community-driven approach for generating a stronger tourism economy in Elk Town. Throughout the exploration of strengthening and promoting consumptive wildlife tourism, long-term impacts to the game population never emerged as serious considerations. Generally, the community readily identified with the hunting and fishing tourists, and welcomed their presence. While some concerns about competition for game with 'outsiders' remained an issue, many programme participants expressed confidence that their 'secret' spots would remain largely undiscovered. Rather than seeing hunters as a threat to their cultural sensibilities, participants saw hunting and fishing as key elements in traditional family life, and believed promoting that activity as a tourism base would attract the type of families that would be economically productive while preserving their conservative community values.

The programme participants in Elk Town were quick to realize that they did not have the internal community resources to achieve their objective of revitalizing the town. Consequently, the programme planning committee developed a plan to bring community groups and external partners together to develop a collaborative approach to problem solving. One of their activities included hosting a community wild-game feed dinner where they invited people who routinely come out on weekends to hunt, summer residents, and people who signed the guest book at the General Store, to join them in getting involved in planning the community's future. In addition to receiving public input, they also used the event as an opportunity to promote Elk Town as a great place for families to live.

This progressive approach was directed by local residents who hoped to extend their resources beyond the perimeter of their town. They also enhanced their partnership with the US Forest Service; expanded their community event calendar to include a fishing derby, hunting clinics, and family activities, to attract a variety of year-round visitors; received US Forest Service grant funding for a gateway informational kiosk designed to improve the safe and appropriate use of trails while encouraging visitors to explore local amenities; and generated a more optimistic atmosphere about the future of their town. While divisions remained within the community regarding the benefits and extent of consumptive wildlife tourism in Elk Town, a majority of residents embraced this mode of tourism as a vital component to their future economic and community well-being.

River Town

The issues were quite different in River Town. River Town, with a population of approximately 3,000, is located just off a highway that runs along a major river. The town is within the Indian Reservation; however, the community is 95 per cent Caucasian and has little interaction with the Tribal government. There are no indications within the downtown area that the town is on the reservation. The town is home to several federal agencies, including a national fish hatchery, and has a more stratified socio-economic make-up than the other two communities. In River Town, only one of three mills remains operative, and the closing of the mills still stings. If these towns' sense of loss were viewed in terms of Kübler-Ross's (1997) life cycle, Elk Town would be in the acceptance phase, while River Town remains in the anger phase. Significantly, it has been 'angry' for decades, given that most of the mill closures and loss of timber-related jobs occurred in the late 1980s, largely resultant from litigation stemming from the enactment of the Endangered Species Act.

Mistrust of outsiders and government agencies was a common theme in River Town, as evident in frequent remarks of participants when partnerships were explored to strengthen community initiatives designed to improve the local economy. In River Town, the prevailing sentiment was that both timber and steelhead are abundant, and the government regulations that reduced logging and now mandate the summer drawdown of the local reservoir to promote steelhead migration (resulting in loss of watercraft tourism activity on the reservoir) blatantly disregard the local community's economic well-being. In all three communities, individual property rights were of paramount concern, but in River Town the concerns were acutely directed towards government interventions at the local level.

As part of the community coaching programme, a community visioning meeting was held in each community designed to elicit what they would like their community to look like in 20 years. These meetings were heavily publicized and extremely well attended in each town. In River Town 200 residents participated in the visioning process. Through brainstorming and a nominal group technique, ideas were generated and prioritized. Finally, these ideas were crafted by a small team into a vision statement which was then read back to the audience for approval. River Town's original vision statement began with these words: 'River Town is the steelhead capital of the world. We practice catch and release tourism…' In this statement what they are referring to is catching tourists, not fish. The statement refers to the catch-and-release regulation in force in some fisheries that requires anglers to release all the fish they catch. This portion of the vision statement emanated from one gentleman in attendance who said: 'We should do catch and release tourism. You know you catch'em, turn them upside down and shake their pockets out, and then release them.' While this was said in a somewhat joking manner, and it received a lot of chuckles, the idea behind it stuck and made it into the final vision statement.

The concept of branding the town as the 'steelhead capital of the world' resonated at the community meeting. However, later when the phrase was publicized throughout the community, resistance began to emerge. The business community, represented by the local chamber of commerce, no longer supported anchoring the community's economy around consumptive wildlife tourism, citing its seasonal nature and unpredictable fish runs as too tenuous a basis for the local economy. While they envisioned consumptive wildlife tourism as a component of the economy, they saw light industry and manufacturing as a more consistent revenue generator, and one they associated with higher paying jobs. While a diversification strategy was not at all inconsistent with the programme's purposes, the unwillingness of the business community to explore the branding concept in concert with other segments of the community effectively signalled the dissipation of any chance of its success.

One of those community segments was represented by several members of the programme's steering committee that envisioned the 'steelhead capital of the world' brand as the cornerstone of a tourism economy that would include a rich mix of arts and culture revolving around the theme of the community's natural capital. They enthusiastically embraced the idea and led the effort to have a public meeting to discuss the potential of the branding idea.

This group initiated and planned a public meeting to further explore the theme and discover how the community might stake its claim to the title by generating an annual event that would signal the community's standing as the premiere steelhead fishing destination in the United States. For the meeting, the programme's steering committee invited key stakeholders, including representatives from the national fish hatchery, a state tourism consultant, local hunting and fishing guides, angling-based merchants and small business owners.

The national fish hatchery representative expressed her support for the idea and shared a model of a community in the northwest that had used a similar theme to grow an annual event that flourished. The state tourism representative shared the importance of having a committed team of leaders to catalyze and sustain the theme and event. A community member who envisioned having visual arts signal the town's steelhead theme shared her ideas for steelhead sculptures along the highway and into town, linking the traveller to key destinations within the community while identifying alternative attractions for non-anglers. At this point, the hunters and anglers in the meeting began to voice concerns about the complexity and purposes of the event. Within one week after the meeting, the guides and merchants had opted out of this initiative citing concerns about how all the 'hoopla' was going to bring in money; the chamber of commerce signalled its discontent with the singular theme; and most of the programme's steering committee members declined to commit to an event with seemingly little community support. Consequently, several months later when a community action planning process was developed through the steering committee and launched at the community level, the catch-and-release tourism statement was removed from the vision statement and the 'steelhead capital of the world' theme was struck from the lead sentence and demoted to the second paragraph along with other community attributes and resources.

It was clear to the community coach observing this fragmentation that community-driven tourism strategies based on the provision of consumptive wildlife recreation depend as much on the community's social capital as its natural capital. While River Town remains a popular destination for hunters and anglers, and external public and private marketing forces continue to promote it as a recreation destination, the opportunity for the community to capture and maximize the financial benefits of consumptive wildlife tourism remains untapped, especially given the extensive leakage of revenue diminishing River Town's financial benefit from the activity.

For a community coach, the contrasting experiences of River Town and Elk Town provide a stark comparison. The citizens of Elk Town, while having a history of community divisiveness, were able to coalesce a driving force of diverse community members to promote community self-direction. They developed an action plan built on the community's strengths and weaknesses, and sought to mitigate the weaknesses not through abdicating responsibility to an outside source, but by partnering with external resources to address the concerns from their local perspective. Elk Town faced their challenges squarely, and relied on an asset-based approach (focusing on resources rather than needs) to solve the problems once identified.

Contrarily, the citizens of River Town remained wary of external providers and did not overcome internal social divisiveness. They struck a defensive posture when their problems, such as tourism revenue leakage, were addressed by outside consultants, and blamed government policies for their economic decline. While both communities readily acknowledged the relative advantage of their abundant natural resource base, they found social capital, defined as the resource potential of social relationships (Agnitsch *et al.* 2006: 36), in each community varied greatly. Elk Town, which was the community with the older average age, less diversity, higher poverty rate, rapidly declining population, and greater social isolation, was willing to risk change to improve their lot.

Hawk Town

Hawk Town (population approximately 1,500) is unique in comparison to Elk Town and River Town. Hawk Town is in essence two communities struggling to harmoniously co-exist in the same geographic place. Hawk Town is only 25 miles away from River Town, but the distance belies the difference in these two communities. While these towns are similar across many social and economic variables, the one significant difference is that according to the 2004 census Hawk Town's population is 10 per cent Native American – and a little commonsense and a quick look around would dismiss this estimate as exceedingly low. In any case, the tribal presence in Hawk Town is pronounced beyond its numbers. Along with this increased presence is tension between tribal and non-tribal members of the community. Nowhere is this tension more evident than in issues of consumptive wildlife harvesting.

Traditionally, many of the tensions in this community go unspoken and unresolved. The community coaching programme opened up new opportunities for dialogue, surfacing some of these underlying tensions in a constructive environment where issues were clarified and the hope of resolution emerged. Nonetheless, these conversations frequently felt like they were on the brink of erupting into hostility, as the following conversation between a Tribal and non-Tribal member of the community regarding developing a new fishing site illustrates.

NON TRIBAL: 'We are having a meeting with Fish and Wildlife to talk about restoring "Little Pond" and we want to invite you to come so the Tribe is there.'

TRIBAL: 'You should have come to the Tribe before having the meeting.'

NON-TRIBAL: 'That's what we're doing. We're inviting you to the meeting. We haven't really had much of a conversation about this with Fish and Wildlife, that's what the meeting is for. We just talked to them about the meeting.'

TRIBAL: 'You're still inviting me to the meeting after talking to them and you should have come to us before the meeting.'

NON-TRIBAL: 'We haven't had the meeting. This is like semantics. We want to have a meeting with the Tribe and Fish and Wildlife so we're all getting together to talk and we want you to be there.'

TRIBAL: 'Listen, it's not about me coming to your meeting. What if I can't make it? Do you know how many meetings I have to attend? And if I'm not there then you say, "see they're not interested". We have a Natural Resource staff person. You should have contacted that person and gone to them to have your meeting. You're on the reservation, you should be coming to our meetings, not inviting us to yours. We've been talking about a project downstream from that pond and this could affect it. Did you think of that?'

NON-TRIBAL: 'Okay, we'll call the guy. But I still think we're saying the same thing. Getting everyone together and talking about it. We just want to make a nice place for everybody to fish.'

While anger was near the surface of this conversation, it never erupted and the participants continued their involvement in the programme. The Tribal member was later elected to the local Tribal Council, where he advocated for community programmes that met the needs of all of Hawk Town's citizens.

Issues of consumptive practices on the Reservation have deep historical and cultural implications for many tribal members. Such issues of fishing and hunting rights routinely lock the Tribe in state-wide legal battles. For the Tribal representatives in the community coaching programme, consumptive wildlife harvesting is not just an economic strategy, but a culturally significant practice with both historical meaning and current implications. As such, there are cultural sensitivities that extend beyond *what* a person takes, to *why* a person takes. Consumptive wildlife tourism was not rejected outright as an approach to economic diversification in Hawk Town by the Tribal participants. Rather, the issues that arose revolved around respecting the place in which it was occurring, and the rights and traditions of a people for whom wildlife consumption is a thread to a threatened identity.

In Hawk Town the community action plan priority issues included developing a resource-based tourism strategy, improving Tribal and non-Tribal relations, and developing cultural tourism to share the rich history and present practices of the Tribe. Subsequently, the programme's participants formed a non-profit umbrella organization to attract and administer grants for both Tribal and non-Tribal activities.

Discoveries

Fundamental to the practice of community coaching is building opportunities for community dialogue (Cohen 2006). That dialogue then becomes a catalyst for creating a vision. In turn, that vision becomes a launching point for community-driven planning. While each community discussed here engaged in a similar process, the results of the process varied significantly. The variance within community responses and outcomes illuminates the contextual differences in rural communities, differences that can impede tourism strategies if left unaddressed.

While tourism in general is supported in these rural communities through traditional channels of economic development (economic development commissions, chambers of commerce, state tourism initiatives) our research revealed how two primary components of sustainability, the social and environmental dimensions, may not become integrated as part of a community development strategy left to the traditional economic development framework (Cohen 2006).

The agency-funded programme assumed that all dimensions of sustainability, including that of the physical environment, would emerge as the communities engaged in community development activities, and therefore did not introduce any specific programmatic elements to guide dialogue towards issues focusing on the natural resource base. Consequently, the conversations that emerged in the context of community development seldom strayed from discussions of economic enhancement. These conversations, frequently steered by local economic development professionals, were focused on the natural resource base as an attraction, not an ecological system. With the exception of Hawk Town, where the natural environment took on a cultural dimension, issues such as impacts on species populations and variation never emerged.

It was clear that the economic future of all three communities was tied to the preservation of the natural resource base. While the environment provided for each community's future economic resource base, the word 'environmentalist' was negatively loaded. During our 18 months working in these communities, not a single community member identified themselves as an environmentalist, advocated an environmentally sensitive approach to nature-based tourism, or mentioned the preservation of the natural surrounding at any time in any way. In these case study communities, environmentalism is perceived as an outside, political movement contrary to property rights and self-determination at the local level. As such, when it was mentioned, it was used as a derogatory term.

Conclusions

This research has offered a glimpse of the highly contextual process of how consumptive wildlife tourism strategies unfolded through a community coaching based initiative, and shared the mixed results it afforded within the case study communities. Of the many lessons learned during this pilot programme, one thing is abundantly clear: successful consumptive wildlife tourism strategies do not emerge independently of the community in which they take place. There is a saying that all politics are local. It could be said that all tourism is local. The social, economic and environmental elements of sustainable tourism occur in a community context, and therefore contextual approaches to building sustainable consumptive wildlife tourism strategies would be well served by mechanisms, such as community coaching, that leverage local capacity and create a trusting environment for dialoguing and framing community issues. Clearly, we had mixed success; community coaching has its limitations, though those limitations may have been mitigated in River Town if a more nuanced approach to engaging the community had been developed prior to launching the community coaching programme (Cohen 2006).

Community coaching strategies guide the community-driven development process, building the capacity of the community participants to identify and determine a course of action based on their unique needs and interests. While community level self-determination is a goal of community coaching based initiatives, in our case it became clear that programmatically, if an element is not clearly introduced, it may never emerge. Therefore, those programmes promoting sustainable consumptive wildlife tourism strategies may need to introduce concepts, such as environmental impacts, lest they be subsumed by more traditional economic development growth strategies. The effectiveness with which these concepts are introduced can benefit from the prolonged, cultural immersion into a community associated with community coaching strategies that enable the coach to understand important socio-cultural differences and sensitivities.

To date, the community coaching programme's core groups in the communities of Elk Town and Hawk Town continue to progress towards their economic development goals, demonstrating the resourcefulness, trust and commitment that can turn a community vision into a reality.

References

Agnitsch, K., Flora, J. and Ryan, V. (2006) 'Bonding and bridging social capital: the interactive effects on community action', *Journal of Community Development*, 37: 36.

Cohen, K.A. (2006) *Community, Culture, and Change: Structure and Strategies for Community Coaching Based Initiatives*, Unpublished doctoral dissertation, University of Idaho, Moscow.

Emery, M., Hubbell, K. and Salant, P. (2005) *Coaching for Community and Organizational Change*, Symposium conducted at the Community Coaching Roundtable, Boise, ID.

Harris, C. and Russell, K.C. (2001) 'Dimensions of community autonomy in timber towns in the inland northwest', *Society and Natural Resources*, 14: 21–38.

Krippendorf, J. (1982) 'Towards new tourism policy: the importance of environmental and sociocultural factors', *Tourism Management*, 3(3): 135–48.

Kübler-Ross, E. (1997) *On Death and Dying*, New York: Simon and Schuster.

McLaughlin, W.J., Sanyal, N., Tangen-Foster, J., Tynon, J.F., Allen, S. and Harris, C.C. (1989) *1987–88 Idaho Rifle Elk Hunting Study. Volume 1: Results*, Contribution No. 499. Idaho Forest, Wildlife and Range Experiment Station.

O'Laughlin, J. (2005) *Economic Impact of Salmon and Steelhead Fishing in Idaho: Review of the Idaho River United Report* (Issue Brief No. 6), Moscow, ID: University of Idaho College of Natural Resources Policy Analysis Group.

17 Marine fishing tourism in Lofoten, Northern Norway

The management of the fish resources

Øystein Normann

Introduction

> The sea is one of the few truly wild areas left on our planet.
>
> (Orams 1999: 34)

Fishing tourism is an important segment of the water-based activities tourists may enjoy along rivers, lakes and on the coast. Economically, and by numbers, marine-based fishing tourism is the most widespread and it has been growing substantially during the last decades (Gartside 2001; Higham 2005; Hallenstvedt and Wulff 2004). The mature markets in the USA and Canada have developed over several decades, and in the USA alone recreational fishing represents 1.4 million standard work years and participation in recreation fisheries is as big as golf and American football together (Simonsen 2002; NOAA 2005). As long as marine fish are a freely available resource, recreational fishing will continue to increase with a growing population and extended leisure time (Gjøsæter and Sunnanå 2005). It will therefore be an important pull factor in many decisions to travel, whether the activity is a primary or secondary feature (Hinch and Higham 2003).

Fishing challenges both knowledge and skills and thus provides feelings of achievement and competency (Orams 1999). Kenchington (1990a) maintains that the joy of escape surpasses the reality of catching fish. The social dimension is also important as marine fishing is inclusive and normally conducted together with family and friends. Recreational angling is often used in the marketing of the tourism industry as this contributes to the positive perception of the area's environmental quality (Dean *et al.* 1990). No doubt the best marketing is done by satisfied fishers returning home with their experience and fishing stories.

Marine fishing is relatively uncomplicated, the gear is reasonably priced, the access from land or boat is normally easy and, not least, the chance of catching fish is high. Also, angling normally is not particularly physically challenging, allowing physically disabled persons to participate on equal terms as non-disabled. It also facilitates participation for a growing group of people in Western societies whose physical condition excludes them from many outdoor sports activities.

Fishing fits into the growth area of special interest tourism or niche tourism, which according to Novelli (2005: 5) refers to 'a specific product tailored to meet the needs of a particular audience/market segment'. The focus in niche tourism is shifting from consuming to participation, developing skills and collecting experiences (Tarlow 2002; Nordin 2005).

Despite its attractiveness and accessibility, marine fishing tourism (MFT) lies within a complex policy environment. The rapid growth of MFT at the Lofoten Islands in Northern Norway illustrates the complexity of the issues surrounding the activity.

Tourism in Lofoten and commercial fishing

The Lofoten Islands are an archipelago consisting of a small group of islands north of the Arctic Circle in Northern Norway (Figure 17.1). The living conditions are dominated by the warm waters of the North Atlantic Drift that sweeps up along the coast and prevents arctic conditions from penetrating further south. The total area is 1,227 square kilometres with 24,500 people living in the region. Cod fisheries have been a key factor in the settlement of the Norwegian coastal regions, not least in Lofoten. Of the 9,500 in work, 20 per cent have work related to fishing or the fishing industry and 3 per cent are engaged within tourism, which is a relatively large although highly seasonal industry (Puijk 1996). Tourism in Lofoten accelerated with the boom of the private car in the 1960s, and today all inhabited areas have tarmac road connections. The rorbus ('dwellings for those who row' (Puijk 1996: 210)) are fishermen's huts located on the waterfront and the rorbu holidays developed as a concept in the 1970s (Fossum 2000).

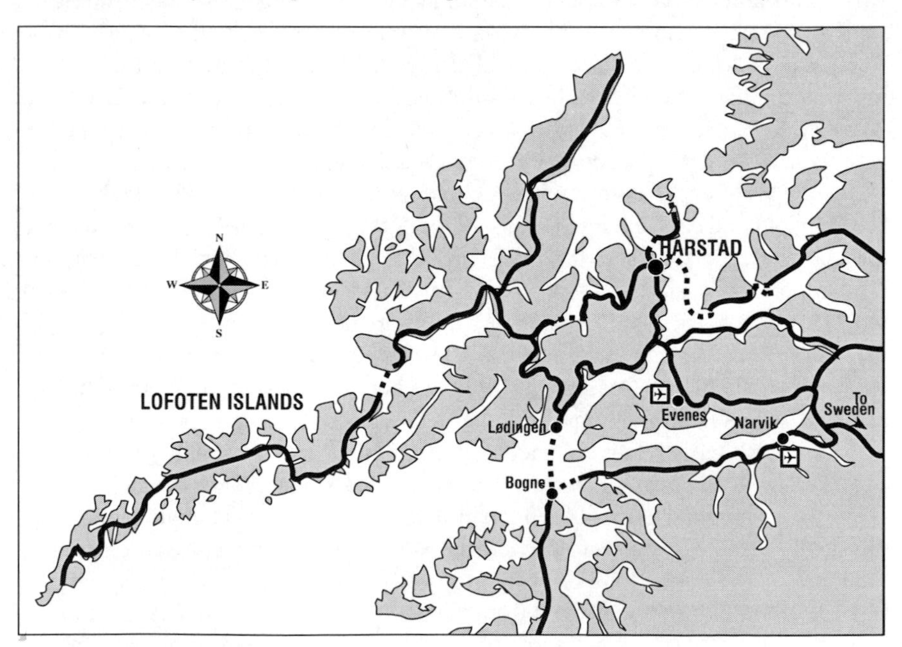

Figure 17.1 Map of the Lofoten Area (Source: Harstad University College)

The sea around Lofoten probably provides the most important coastal fishing grounds in Norway. As a spawning ground for the Arctic Norwegian cod and several other species, it has for centuries drawn thousands of fishermen from the northern part of Norway to the seasonal fisheries which take place each year between January and April. For tourism, the main resources of Lofoten are the unique scenery, the rich fish stocks, nature and cultural history.

In Lofoten the fisheries still are the strongest culture bearer as the area is so tightly connected to fishing and the fishing industry (Sagen 1999). The small-scale coastal vessels ('sjark') supply the fishing villages with the raw material for the traditional production of valuable fish products. This selective and small-scale fishing is a sustainable use that is vital both for the fisheries and the settlement in Lofoten where there has been a livelihood available for coastal dwellers since time immemorial (Nilsen 1998). The small fishing villages rely on seasonal fishing which takes place when the fish approach the coast to spawn and to find food, but also the small municipalities in the Lofoten area, as in the rest of the Western world, have seen how global capitalism can initially create work places, and then subsequently make conditions for small-scale industries impossible (Nadel-Klein 2003). However, the coastal fleet is under pressure to adapt and reorganise in response to changing conditions including reduced resources (Nilsen 1998). Consequently, fishermen are less and less a dominating user group of the Norwegian coastline (Nordstrand 2000) as is also the case in almost all former coastal communities and fishing hamlets around the North Atlantic Ocean (Nadel-Klein 2003).

The challenges facing commercial fishing also have impacts on the tourism industry as the image of Lofoten as a prospering fishing region, industry is important in the promotion of the product. The fishermen themselves are also tourist attractions, with people wanting to visit fishing villages where they can see fishermen mending their gear or landing their catch (Kurlansky 1997). The tourists' authentic experience of Lofoten is associated with elements that cannot be copied, like the rorbus, the fish racks, the fishing boats and the small fishing hamlets with the clustered houses (Puijk 1996). But these symbols are now at stake, as the coastal landscape may alter its character and associated heritage values if the traditional coastal fisheries are reduced, and partly replaced by other activities such as fish farming (NOU 28 2004).

Furthermore, in Lofoten commercial fishing and MFT are competing for the same fish resources and also are facing a third 'competitor' as conservation groups show concern for the dwindling fish stocks and in particular concern for the coastal cod. This competition makes Lofoten an interesting area for studying MFT. With 30,000 tourists fishing during a short season in a relatively concentrated area and, as the activity is increasing, this probably means that there will be a growing need for development, as well as concern for the environmental consequences. The main reasons to pay attention to this particular segment of tourism are connected to management and sustainability: neglecting the sustainability of MFT may lead to significant negative effects (Orams 1999).

Marine fishing tourism

Fishing can be considered as tourism when anglers cross borders to go fishing. Marine fishing tourists in Norway are defined as non-domestic anglers, whereas non-commercial fishing done by Norwegians in Norway is defined as recreational or leisure fishing (St.meld. 19 2004–05). These distinctions are used in this chapter.

As in the rest of the world, MFT in Norway has grown rapidly during the last decade to become an important part of the Norwegian tourism industry and to local economies. From 35,000 in 1995, more than 250,000 tourists were predicted to come to Norway to go fishing in 2005, seven out of 10 of these being German (Hallenstvedt and Wulff 2002).

The simplest description of recreational fishing is to define it as 'catching fish for fun' taking little notice of its many specialised forms, which depend on where and how it is conducted, the fish species present, time of year and the intention of the angler. Kappel (2003) applies a definition which is based on the equipment used, stating that recreational angling is done with rod or line, excluding the use of nets and long lines. MFT may accordingly be defined as salt-water rod or line fishing conducted by tourists.

The physical dimension of fishing, its competition with natural and environmental forces (fish, water, wind), rules adhered to regulations (size of fish, timing, quotas, etc.) and internal and external rewards, qualifies MFT also for the term sport tourism (Higham 2005). Gammon and Robinson's (1997) (in Hinch and Higham 2003) division between sport tourism and tourism sports may also be applied to 'fishing tourism' and 'tourism fishing'. Fishing tourism signifies

Figure 17.2 Participants at the European Boat Angling Championship in Harstad, 2006 (Photo: Ø. Normann)

the importance of the activity. The sport is the primary motivation of travel and fishing is vital to the choice of time and destination. Tourism fishing is the fishing activity conducted by the majority of tourists who enjoy the activity as one of several other incidental or sporadic travel activities during their vacation stay. The two groups may be further subdivided as illustrated in Table 17.1.

Fishing tourism – tourism fishing

Anglers may be identified and grouped according to various parameters. Demography, geography, interests and approach are often applied to segment fishing tourists (Borch *et al.* 2000) (see Table 17.1). The tourists practising fishing during their vacation may be segmented also according to their commitment to the activity.

Based on Norwegian conditions, Hallenstvedt and Wulff (2002: 31–3) estimate the dedicated group to constitute 20 per cent of the tourism fishers, while the family fishers and the fishing tourists constitute 40 per cent each. Although fishing activities are demanded by only 15 per cent of the around 200,000 tourists visiting Lofoten (Viken *et al.* 2004) it still means that 30,000 tourists go fishing, of which 6,000 belong to the dedicated group of fishing tourists. The trophy hunters are the desired group among the sport fishing tourists due to their spending and they may also buy tackle not available at home (Dean *et al.* 1990). The most specialised forms of MFT give the highest revenue, but they depend on high quality and limited resources and therefore require good planning to be sustainable. The 'black sheep' are the fishers with a foraging focus whose aim is to return home with as much fish fillet as possible (Borch *et al.* 2000).

The relatively high proportion of family fishers and fishing tourists could be explained from the development in the travel market towards authentic experiences, which include also heritage, local culture and closeness to nature (Poon 2003).

Table 17.1 Marine fishing anglers grouped according to activity, motives and approach.

Activity	*Typology and motives*	*Approach*	
Sport fishing Dedicated professionals	**Sport-fishers** Trophy hunters **Food-fishers** Foraging focused	Product-related approach	Shaped in accordance with specific tourists' needs and wants.
Fishing sport Amateurs	**Family fishers** Social motives **Fishing tourists** Opportunity motivated	Geographical and demographic approach	A maritime setting in a traditional fishing community which gives relevance to the tourists' activity. Location, population, culture and setting are important

After Gammon and Robinson 1997; Borch et al. 2000; Novelli 2005: 9; St.meld. 19 (2004–05).

Most coastal sites in Norway, and the Lofoten area in particular can offer this. The possibility to go fishing also contributes to the positive perception of the area's environmental quality. Recreational angling is accordingly used in the marketing of the destination (Dean *et al.* 1990).

Demography – spending – demands

The large and influential group of 'younger-older-people' want to be active (Poon 2003), and it is no surprise that most fishing tourists belong here (Hinch and Higham 2003). Their wish to get a break leads to active relaxation and fishing is attracting many. These 'Baby Boomers' (aged 50 to 65) are also the largest share of family vacationers, where entertainment and escape are important motivation factors (Nordin 2005). Hallenstvedt and Wulff (2002) state that six out of 10 fishing tourists in Norway are between 35 and 55 years of age and that they mostly

Figure 17.3 After the fish is weighed and recorded it is left behind. Sport-fishers don't care much about foraging (Photo: Ø. Normann)

travel in groups. This is also a rather large market segment which is easy to target (Simonsen 2002). Of the tourists in Lofoten 53 per cent are between 45 and 65 years old (Viken *et al.* 2004).

Most marine activities are conducted by upper socio-economic groups (Orams 1999), and when it comes to fishing this is often reflected in the cost of equipment and clothing. A high motivation reduces the focus on price, which makes this a group willing to pay more than the average tourist by about 17 per cent (CGEandY 2003). Economically, fishing tourism has many advantages as it engages several sectors of the tourism industry and local workforce, which in turn direct income to the local communities with positive multiplier effects. Especially when conducted from a charter boat, recreational MFT has a rather high diurnal spending (Simonsen 2002). Sport fishermen in the Nordic countries also demonstrate a high consumer surplus (i.e. willingness to pay more than the market price) (Toivonen *et al.* 2004).

The fish species

Inshore and coastal species have long been the core of recreational and commercial fisheries and MFT adds to the existing harvesting pressure, complicating the issue of resource allocation (Gartside 2001). However, much remains to be discovered about these species as marine science in Norway has been focused on the near coast to an only small degree (Maurstad and Sundet 1998).

Due to the amount of fish in the sea and the number of species available, European tourists regard Norway as an ideal place to conduct maritime fishing. In fact Norway can boast the European weight record for 11 species (Hallenstvedt and Wulff 2001) and among these the cod is by far the most popular.

Of the two cod species of Northern Norway, Norwegian Arctic cod ('skrei') and coastal cod, it is the coastal cod that is caught by the majority of maritime fishing tourists who go to Norway. The coastal cod is rather stationary and recent research has revealed that it is constituted of several local stocks (Rinde *et al.* 1998) of which three-quarters are located north of 67° (Maurstad and Sundet 1998). A gradual reduction of the coastal cod stock during the last decade due to a failing recruitment (NOU 28 2004) is of great concern to all stakeholders who depend on the resource, including the tourism industry. It should, however, be noted that the cod as a species is not threatened with extinction, it is not even endangered. What marine biologists are saying is that several of the stocks are too small to support a high sustainable fisheries yield.

As an important part of the large marine tourism industry, MFT also claims its share of the coastal fish stocks (Orams 1999). Cooke and Cowx (2004) highlight the lack of global statistics on recreational fishing harvest, but based on Canadian data they estimate recreational fish harvest to be around 12 per cent of the global take. In the USA it is estimated that it constitutes 9 per cent of the total harvest of finfish (Pritchard 2004) and even more on the most popular species which are targeted by the recreational fishery. Their conclusion along with others is that both commercial and recreational fishing takes a substantial part of the fish harvested and should accordingly be regulated (Coleman *et al.* 2004). Similarly, Gjøsæter and Sunnanå (2005) argue that MFT is an industry and consequently should be compared to and evaluated in accordance with other industries that exploit the coastal resources. If treated in line with commercial fishing, fish caught by tourists could be recorded in total harvest quotas (Gartside 2001).

The knowledge of the volume of fish harvested by fishing tourists in Norway is poor (Borch 2005b), the estimates varying from 8,000 tons (CGEandY 2003) to 15,000 tons (Hallenstvedt and Wulff 2001). A recent estimate made by Essens Management (2005) indicates that professional fishers and domestic recreational fishers harvested respectively 75 per cent and 20 per cent of the total annual catch of 5,800 tons of coastal cod. The share landed by the tourist fishers is between 2 per cent and 5 per cent (595 tons and 900 tons) (Essens 2005: 2).

In addition to the coastal cod, several other species can be caught during the summer season in the coastal waters of Northern Norway and Lofoten. These species, and local knowledge about their occurrence and how, where and when to catch them, together constitute a valuable resource for the development of a sustainable MFT.

Marine fishing tourism enterprises

In Norway there are more than 1,000 enterprises offering accommodation, boat hire and fishing equipment. Of these, 70 per cent were established between 1990 and 2000 and half of the employed are women (Hallenstvedt and Wulff 2002). In addition, 10,000 boats between 15 and 70 feet are available for hire (Borch 2005a). The fishing equipment includes rods, reels, lures, life jackets and, increasingly, radios, depth recorders, echo sounders and positioning devices (GPS) (Dean *et al.* 1990). Enterprises connected to MFT in Lofoten are normally small one-person or family-run operations and there is often a correlation between the special interest of enthusiasts who develop their hobby into a tourism business venture (Novelli 2005). However, it still seems like the enterprises are continuously lagging behind expectations and demands from the sport fishing tourist who requires well-equipped boats and preferably a guide. The conditions, as offered today, are more attractive to the foraging fishers who want as much fish as possible for private consumption. Facilities for gutting with water, freezing facilities and even vacuum packing are offered accordingly. The core product is the fish fillet (Nordstrand 2000).

There are no large-scale advantages in the fishing industry (Nilsen 2005) and probably not in the MFT industry either, due to the nature of the resources it too is based on. A concentration of fishing tourism enterprises in a few limited areas may in the long run place too much strain on local fish stocks (CGEandY 2003) and frequent harvesting of limited stocks may eventually result in depletion to the point where recovery is impossible (Warnken and Byrnes 2004).

Many of the enterprises developed for marine fishing tourists in Norway are located near the airports and the ports of call of the ferries from the continent. This localisation does not necessarily relate to the occurrence of healthy fish stocks, and the result may be overfishing on locally weak fish stocks, with displeased customers (Hallenstvedt 2005). Sustainable fisheries must leave declining stocks to exploit the healthy (Eikeland 1998) and in commercial fishing the catch must cover the costs or else the fisher will move to new fishing grounds if the catch declines. Marine fishing tourists have few opportunities to move far, and will consequently further exploit depleted stocks if the enterprise is located at an unfavourable site (Kenchington 1990a).

Despite these reservations, a recent evaluation of MFT in Norway indicates that the value added to the community per kilo fish caught might be as much as 10 times that of the traditional fishery (CGEandY 2003). On average, fishing tourists spend between NOK 240 to 400 (€30 to €50) per kilo of fish caught (St.meld. 19 2004–05). MFT is, in other words, a profitable way to make the most out of a vulnerable resource, thus fishing tourism might be considered as an alternative trace of the 'value chain'. The economy it stimulates spreads much further and creates benefits through the multiplier effect as it also generates use of other facilities in the region, more jobs, household income and taxes (Gaffney 1990). This may eventually lead to a demand to allocate the resources to the industry that contributes the highest return to the community (Mikkelsen 2003).

Conflicts and convergent interests

Divergent interests

The interactions between MFT and other uses of coastal and marine environments can cause complex and passionate disputes over resource management (Kenchington 1990a). Like a cuckoo in the nest, the MFT enterprises emerge as a new competitor to commercial fishing, prying on the vulnerable commodity which until recently solely has belonged to commercial fishing, distributed by complicated rules. The four areas of conflict between the fishing industry and the tourism industry are the distribution of the fish resource, the competition for space (fishing grounds and space in harbours), different environmental views and cultural antagonism between the industries (Puijk 1996). The debate is hampered by the scarcity of facts about the amount of fish that is actually caught by marine fishing tourists (Jacobsen 2005), and consequently to what extent MFT is a threat to the coastal cod and the other fish species.

Kenchington (1990b) identifies three approaches to multiple use management which illustrates the field of concerns that has to be addressed (see Figure 17.4). First, it is the traditional harvesting use; second, the classical conservation interests; and finally, a recent utilitarian approach connected to recreation, tourism and education. He further argues that decisions on biological resources should be regarded as an issue of cultural and political choice where the key factors in a decision should be sustainability and 'benefits to human society as a whole in addition to the benefits to individuals and groups within that society' (Kenchington 1990b: 50).

In Norway the conflicts between the tourism industry and commercial fishing seem to decline towards the north, due to fewer people and more space available

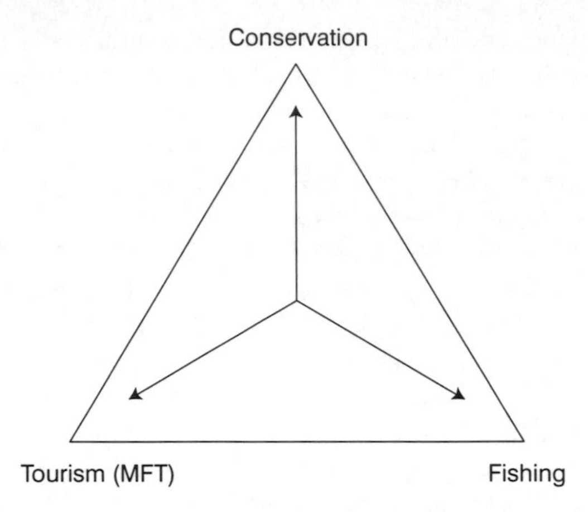

Figure 17.4 Approaches to multiple use management (Source: Kenchington 1990b)

(Borch *et al*. 2000). In Lofoten people in the fishing villages are familiar with the periodical influx of non-locals, and foreigners are just a modern form of strangers. The main local economic sectors, fishing and tourism, are to a large degree interrelated with each other. First, fishing is a central element in tourism and, second, tourists are housed in fishermen's cabins (rorbus). In fact, most people in the fishing trade look at the tourism industry as a natural part of the coastal industries in Lofoten (Sagen 1999).

Boyd and Hall (2005) argue that tourism should collaborate with existing resource-based industries. In Lofoten, it is evident that the tourism industry strongly depends upon and is heavily building its product on the fishing traditions of the region (Fossum 2000). The importance of the fishermen's competency as boatmen and carriers of tradition can hardly be overestimated. It would therefore be of no gain for the tourism industry if the commercial fishing is phased out and reduced to a staged performance. This will reduce the authenticity and consequently reduce the attraction value of an MFT destination.

However, Gjøsæter and Sunnanå (2005) have identified some important interactions between fishermen and tourists that need to be considered. The resource conflict is connected to several conditions. First, it is a problem that tourists catch undersized fish. A solution to this is to increase information and teach how to handle fish in case of releasing. Second, maps and GPS make it possible to avoid areas with small fish and call on fishing grounds with bigger fish and sustainable stocks. Third, attention should be drawn to species which can endure exploitation, like haddock, saithe, mackerel, redfish and tusk, all of which are both game fish and excellent for food. Next, information about fishing tackle, spoon bait and hooks used for the different species could be given in addition to what times fishing ought to be conducted. Finally, focus should be placed on the joy of cooking and eating your own fish while on holiday (Gjøsæter and Sunnanå 2005).

Convergent interests

According to Kurlansky (1997) the fishing industry may be facing the hunter's dilemma; to have work fish must be caught, but catching enough for a living will destroy the fish stock. A solution to this could be a combination of fishing and tourism. Flexibility by diversifying on fish species or participating in other trades, like tourism, is a necessity to survive in times of change and resource depletion (Nilsen 1998). In a situation where the fish stock is weak and declining and the quotas might be reduced, an ideal solution for many fishermen with small vessels would be to sell fish as an experience, which would also give a better price on each kilo of the fish (Nordstrand 2000). The advantage of the smaller vessels is to be found in their flexibility, i.e. their ability to readjust to new conditions in a short time (Nilsen 2005). Growth of the charterboat fishing sector in peripheral areas can contribute to economic multiplier effect benefits, as well as attract new tourists (Gartside 2001).

Nordstrand (2000) regards MFT as a success in the tourism industry, but notes that the right to fish and access to areas rich in fish are the basic foundation for the development of MFT. To achieve legitimacy depends on the local community getting a fair share of the resultant economic development (Mikkelsen 2003). The challenge is to gain control of and profit from the growth of the charterboat fishing activity and organise so that the benefits of tourism spending accrue to the local economy.

However, there is a problem that commercial fishers may lose their legal right to fish if they do not catch their quotas (Eikeland 1998). This can be solved if professional fishermen with an ordinary fish-quota could take tourists out to sea to help catch it (Mikkelsen 2003). Local fishermen's knowledge is important to maintain sustainable development as they can provide information about the health of the local cod stocks, i.e. whether it is stable or in decline (Maurstad and Sundet 1998). As important as the amount of fish caught, is the concern about where and when not to fish, to avoid disrupting basic ecological functions (Wilson *et al.* 1994).

The worry about the future of fish stocks is a concern that will be a constant for all forms of fishing, whether commercial, recreational or conducted by tourists. Although small, the impact of MFT represents a new and additional pressure on local stocks as harvesting what used to be seasonal is becoming continuous on small local stocks (Hallenstvedt 2005). This might therefore be the straw that breaks the camel's back.

Conclusion

The competition for marine fishing tourists is global. The search for a quality fishing experience leads to a demand for virgin and exotic waters (Dean *et al.* 1990) and the location of the destination becomes secondary to the fishing possibilities and quality it beholds (ETC Research Group 2005). Fishing tourists might as well go to New Zealand as Ireland or Norway to fulfil their passion. The

Lofoten area has a unique opportunity to pursue the steady growth of interest for MFT and combine this with international fishing adventure travel while managing valuable resources for sustainable yield. What is essential is to identify trade-offs between economic benefits and conservation and MFT could contribute to this (Hall 2005).

Although there are some skerries in the sea it also seems sensible that fishermen adapt to the new possibilities and take their share of the yield derived from MFT. Their experience, knowledge and professionalism will also improve the product sought by tourists. This could even lead to better monitoring, protecting fragile stocks, while producing satisfied customers and stimulating revisits. If the tourism and fishing industries join forces and manage to work together, this could also develop and strengthen traditional coastal culture, vital both to the population and the tourism industry (Nordstrand 2000). As noted by Gaffney (1990), any destination serious about pursuing the opportunities of fishing tourism should consider investing at least as much time, money and effort in the development of such an activity as they presently invest in the commercial fishing industry.

> If the cod us should fail, what have we then,
> What should we from here to Bergen send?
> The cargo vessels would sail empty.
> (Petter Dass, 1647–1707, cleric and poet)

References

Borch, T. (2005a) 'The economic value of marine fishing tourism: the case of Norway', paper presented at 4th World Recreational Fishing Conference, Trondheim (Norway) 12–16 June.

Borch, T. (2005b) 'Managing marine fishing tourism', paper presented at the 11th International Symposium on Society and Resource Management (ISSRM 2005), Östersund (Sweden) 16–19 June.

Borch, T., Ellingsen, M.-B. and Midtgard, M.R. (2000) Fisketurisme i Nord-Norge – bedriftsutvikling og kompetanseheving. NORUT Samfunnsforskning.

Boyd, S. and Hall, C.M. (2005) 'Nature-based tourism in peripheral areas: making peripheral destinations competitive', in C.M. Hall and S. Boyd (eds) *Nature-based Tourism in Peripheral Areas: Development or Disaster?* Clevedon: Channel View Publications.

Caddy, J.F. and Agnew, D.J. (2003) 'Recovery plans for depleted fish stocks: an overview of global experience', invited plenary lecture at ICES Annual Science Conference 24–27 September 2003 in Tallinn, Estonia. <http://www.ices.dk/products/CMdocs/2003/INVITED/INV2PAP.PDF> (accessed 4 November 2005).

Cap Gemini Ernst and Young (CGEandY) (2003) Vurdering av turistfiske som inntektskilde. Norwegian College of Fishery Science, Tromsø University. <http://www.rbl.no/images/fiskrapport.pdf> (accessed 23 November 2005).

Coleman, F.C., Figueira, W.F., Ueland, J.S. and Crowder, L.B. (2004) 'The impact of United States recreational fisheries on marine fish populations', *Science*, 305(5692): 1958–9.

Cooke, S.J. and Cowx, I.G. (2004) 'The role of recreational fishing in global fish crisis', *Bioscience*, 54(9): 857–9.

Dean, J., Cyr, E., Dean, S., Jehangeer, I. and Nallee, M. (1990) 'The Marlin fishery and its role in the economy of Mauritius', in M.L Miller and J. Auyong (eds) *Proceedings of the 1990 Congress on Coastal and Marine Tourism*, Vol. 2, Corvallis, OR: National Coastal Resources Research Institute.

Eikeland, Sveinung (1998) 'Flexibility in the fishing commons', in S. Jentoft (ed.) *Commons in a Cold Climate. Coastal Fisheries and Reindeer Pastoralism in North Norway: The Co-management Approach*, MAB Series, Vol. 22, Paris: UNESCO.

Essens Management (2005) *Har turistfisket innvirkning på bestanden av kysttorsk?* Trondheim, Norway: Notat.

ETC Research Group (2005) *European Tourism Insights 2004, Including Outlook for 2005,* European Travel Commissionn (ETC), Brussels, January 2005. <www.etc-corporate.org>.

Fossum, H. (2000) 'Utror og in-sted. Reiseliv og reiselivsorganisering i Lofoten 1960–1995', Hovedfagsoppgave, second degree level in History. Tromsø University.

Gaffney, R. (1990) 'The case for the development of a recreational bill fishery', in M.L. Miller, and J. Auyong (eds) *Proceedings of the 1990 Congress on Coastal and Marine Tourism*, Vol. 2, Corvallis, OR: National Coastal Resources Research Institute.

Gammon, S. and Robinson, T. (1997) 'Sport and tourism: a conceptual framework', *Journal of Sport Tourism*, 4(3): 8–24.

Gartside, D. (2001) *Fishing Tourism. Chatering Boat Fishing, Wildlife Tourism Research Report Series: No. 12*, Australia.

Gjøsæter, Jakob and Sunnanå, Knut (2005) *Turistfiske utenfor kvoter og reguleringer: Kyst og Havbruk 2005*, Bergen: Havforskningsinstituttet. <http://www.imr.no/__data/page/5679/3.2_Turistfiske_utenfor_kvoter_og_reguleringer.pdf> (accessed 23 November 2005).

Hall, C.M. (2005) *Tourism: Rethinking the Social Science of Mobility*, Harlow: Pearson Education Ltd.

Hallenstvedt, A. (2005) 'Recreational fishing/tourism fishing', contribution/presentation at seminar on coastal cod. Tromsø 1–2 September 2005, in A.K. Veim, *Rapport fra 'Kysttorskgruppen'*. <http://www.fiskeridir.no/fiskeridir/ressursforvaltning/rapporter/rapport_fra_kysttorskgruppen_2005> (accessed 6 October 2005).

Hallenstvedt, A. and Wulff, I. (2001) *Fisk som agn: utenlandsk turistfiske i Norge*, Norwegian College of Fishery Science, University of Tromsø.

Hallenstvedt, A. and Wulff, I. (2002) *Turistfiske som inntektskilde*, Rapport utarbeidet for Norges Turistråd, Norwegian College of Fishery Science, University of Tromsø.

Hallenstvedt, A. and Wulff, I. (2004) *Fritidsfiske i sjøen 2003*, Norwegian College of Fishery Science, University of Tromsø. <http://www.fiskerifond.no/files/projects/attach/fritidsfiskesjoen2003.pdf> (accessed 18 October 2005).

Higham, J. (2005) *Sport Tourism Destinations: Issues, Opportunities and Analysis*, Oxford: Elsevier/Butterworth-Heinemann.

Hinch, T. and Higham, J. (2003) *Sport Tourism Development*, Clevedon: Channel View.

Jacobsen, J.K.S. (2005) *Utenlandske bilturisters fiske i saltvann i Norge 2004*, TØI rapport 788/2005. Oslo. <http://www.toi.no/attach/a1104396r599590/788_2005.pdf> (accessed 10 December 2005).

Kappel, J. (2003) 'The socio-economic value of recreational angling in Europe', PPT presentation. EAA presentation. <http://www.eaa-europe.org/2003/PP%20presentations/EAA_EFTTA_25_March_2004_FINAL.pps> (accessed 30 November 2005).

Kenchington, R.A. (1990a) 'Tourism in coastal and maritime environments: a recreational perspective', in M.L. Miller and J. Auyong (eds) *Proceedings of the 1990 Congress on Coastal and Marine Tourism*, Vol. 1, Corvallis, OR: National Coastal Resources Research Institute.

Kenchington, R.A. (1990b) *Managing Marine Environments*, New York: Taylor & Francis.

Kurlansky, M. (1997) *Cod*, London: Jonathan Cape.

Maurstad, A. and Sundet, J.H. (1998) 'The invisible cod: fishermen's and scientists' knowledge', in S. Jentoft (ed.) *Commons in a Cold Climate. Coastal Fisheries and Reindeer Pastoralism in North Norway: The Co-management Approach*, MAB Series Vol. 22, Paris: UNESCO.

Mikkelsen, E.I. (2003) 'Tradable rights schemes for allocation of resources between user groups? Fisheries, aquaculture and tourism in Norwegian coastal zone', paper for Rights and Duties in the Coastal Zone, Stockholm 12–14 June. <http://www.nfh.uit.no/dok/mikkelsen.pdf> (accessed 17 November 2005).

Nadel-Klein, J. (2003) *Fishing for Heritage: Modernity and Loss Along the Scottish Coast*, Oxford: Berg.

National Oceanic and Atmospheric Administration (NOAA) (2005) *A Vision for Marine Recreational Fisheries*, NOAA Recreational Fisheries Strategic Plan FY2005-FY2010.

Nilsen, R. (1998) 'The coastal survivors: industrialization, local adaptions and resource management in the North Norwegian fisheries', in S. Jentoft (ed.) *Commons in a Cold Climate. Coastal Fisheries and Reindeer Pastoralism in North Norway: The Co-management Approach*, MAB Series Vol. 22, Paris: UNESCO.

Nilsen, R. (2005) 'Kystfiskets skjulte fortrinn (The hidden advantages of coastal fishing)', *Dagbladet*, 15 February 2005.

Nordin, S. (2005) *Tourism of Tomorrow: Travel Trends and Forces of Change*, U 2005: 27, Ostersund, Sweden: ETOUR. <http://www.etour.se/download/18.58f7691048241fd54800062/Tourism+of+tomorrow.pdf> (accessed 23 October 2005).

Nordstrand, K.B. (2000) 'Fisketurisme eller turistfiske', in M. Husmo and J.P. Johnsen (eds) *Fra bygd og fjord til kafébord?*, Trondheim, Norway: Tapir.

Norges Offentlige Utredninger (NOU) (2004) *Lov om bevaring av natur, landskap og biologisk mangfold*. NOU 2004: 28. Oslo: Norwegian Ministry of the Environment.

Novelli, M. (ed.) (2005) *Niche Tourism: Contemporary Issues, Trends and Cases*, Oxford: Elsevier.

Orams, Mark (1999) *Marine Tourism*, London and New York: Routledge.

Poon, A. (2003) *A New Tourism Scenario: Key Fututre Trends*, *The Berlin Report*, Tourism Intelligence International. <http://www.fedhasa.co.za/wc/BERLIN%20REPORT.pdf> (accessed 12 October 2005).

Pritchard, E.S. (ed.) (2004) *Fisheries of the United States 2003*, National Marine Fisheries Service, Office of Science and Technology, Fisheries Statistics Division. <http://www.st.nmfs.gov/st1/fus/fus03/2003_fus.pdf> (accessed 30 September 2005).

Puijk, R. (1996) 'Dealing with fish and tourists', in J. Boisssevain (ed.) *Coping with Tourists*, Oxford: Berghahn Books, pp. 204–26.

Rice, J.C., Shelton, P.A., Rivard, D., Chouinard, G.A. and Fréchet, A. (2003) 'Recovering Canadian Atlantic cod stocks: the shape of things to come?', paper at ICES Annual Science Conference 24–27 September 2003, Tallinn, Estonia.

Rinde, Eli, Bjørge, Arne, Eggereide, Arne and Tufteland, Geir (1998) *Kystøkologi*, Oslo: Universitetsforlaget.

Sagen, A.O. (1999) 'Fiskeri- og reiselivsnæring – hånd i hånd mot felles mål?', Hovedfagsoppgave, second degree level in Geography. University of Bergen.

Simonsen, S. (2002) *Danmarks Turistråd: Aktiv Ferie Alliance*, Desk Research 2002: 1. Attended at <http://hotel.eucnord.net/jotp/turisme/resources/rapport%20om%20Aktiv%20Ferie.pdf> (accessed 15 October 2005).

St.meld. 19 (2004–05) *Marin næringsutvikling. Den blå åker*, White paper to the parliament on marine industrial development. The blue farmland. Norwegian Ministry of Fisheries and Coastal Affairs.

Tarlow, P. (2002) 'Tourism in the twenty-first century', *The Futuris*, 36(5): 48–52.

Toivonen. A.-L., Roth, E., Naverud, S., Gudbergsson, G., Appelblad, H., Bengtsson, B. and Tuunainen, P. (2004) 'The economic value of recreational fisheries in Nordic countries', *Fisheries Management and Ecology*, 11(6):1–10.

Viken, A., Akselsen S., Evjemo, B. and Hansen, A.A. (2004) *The Tourism Survey in Lofoten 2004*, FoU R 27/2004 Telenor. <http://www.telenor.com/rd/pub/rep04/R_27_2004.pdf> (accessed 24 April 2005).

Warnken, J. and Byrnes, T. (2004) 'Impacts of tourboats in marine environments', in R. Buckley (ed.) *Environmental Impacts of Ecotourism*, Wallingford: CABI Publishing.

Wilson, J.A., Acheson, J.M., Metcalfe, M. and Klaban, P. (1994) 'Chaos, complexity and community management of fisheries', *Marine Policy*, 18(4): 291–305.

World Wildlife Fund (WWF) (2003) *Årets julegave – Oljefritt Lofoten*. <http://www.wwf.no/pdf/WWF%20Bakgrunnsdokument%20Lofoten.pdf> (accessed 30 September 2005).

World Wildlife Fund (WWF) (2005) *Naturindeks for Norge – langsiktige trender i norsk natur* (Norwegian nature index – long-term trends in Norwegian nature). (Engl summary). <http://www.wwf.no/pdf/Naturindeks-fagrapport.pdf> (accessed 8 November 2005.

18 Footprints in the sand

Encounter norms for backcountry river trout anglers in New Zealand

Carl Walrond

Introduction

Trout angling is generally thought of as a solitary experience and research has shown that non-catch related aspects of the experience (peace and solitude, the natural environment) are important, especially in backcountry settings. Concerns have been expressed about the ability of backcountry river fisheries to cope with increased levels of use. High-spending short-stay anglers form a small but significant niche tourism sector in a number of international settings.

This chapter summarises research carried out on backcountry river trout anglers in New Zealand's South Island and relates the findings to the current trout-tourism market and policy environment. The research focused on quantifying social carrying capacities by adopting a normative approach to determine anglers' expectations, tolerances and preferences of different encounter levels.

Comparing the results with earlier research in New Zealand and North America it is clear that backcountry river trout anglers in New Zealand comprise a special segment of visitors who are among the least tolerant of crowding of all backcountry recreational groups. The implications of this are discussed in terms of visitor experience and the image of New Zealand as an angling destination.

Background to the crowding problem

New Zealand has an abundance of wild trout fishing rivers where the angler has been able to stalk large brown and rainbow trout in solitude amidst a comparatively pristine backdrop of native bush and mountains. From the 1980s onwards more and more anglers sought this experience as road and aerial access improved and more publications depicted the rewards of fishing these waters (Hayes *et al.* 1997). As population growth and development has placed pressures on fisheries in North America and elsewhere, New Zealand with its small population and relative lack of pollution has come to be seen as an un-crowded angling paradise.

The statutory body responsible for managing New Zealand's sport fisheries is Fish and Game New Zealand (FandGNZ). The increased use of some rivers over recent years has resulted in growing anecdotal accounts of crowding (Deans pers. comm.; Watson pers. comm.). Complaints came from both resident and tourist anglers and conflicts were often between groups of foot access and helicopter access anglers. Complaints on crowding came mainly from the easily accessible backcountry rivers around the tourist centre of Queenstown in the Otago region and Nelson Lakes National Park in the Nelson region. More often anglers were faced with finding a car already at the access point on their chosen backcountry river – one that they may have travelled many kilometres to get to. Increasing use of four wheel drive vehicles and improved road access to remote areas has made it easier to fish some rivers (Deans pers. comm.). When access to upstream reaches is via one road end and when an angler walks kilometres upstream in the course of a day's fishing sometimes spotting fewer than a dozen fish, the knowledge that another angler has got there first can spoil the fishing trip. To some anglers even the sight of footprints in the sand is enough to ruin their day (Hayes *et al.* 1997).

The angling experience on a backcountry river almost always involves fly fishing. One or two anglers move upstream sight fly fishing (fishing to visible fish) or conducting a combination of sight and blind (fishing likely looking water) fly-fishing. The potential for conflict has long been understood: 'Anglers also look ahead to the next good fishing lie; someone disturbing that spot may also encroach on their activity' (Shelby and Heberlein 1986: 141). Angler and author Ernest Hemingway described the issue in his 1925 short story *The Big Two-Hearted River*: 'Nick did not like to fish with other men on the river. Unless they were of your party, they spoiled it' (Hemingway 1987). As the angler or anglers move upstream, fish are either caught or spooked (scared by angling activity) thus affecting the water behind them for several hours, or possibly even days according to some anglers' observations (Jellyman and Graynoth 1994; Hayes *et al.* 1997).

A controversial issue has been the increase in guided fishing in New Zealand's rivers over the last two decades. Guides are tourism operators who take their clients fishing. Their experience and knowledge greatly increase the catch rates of tourist anglers. Guides have received a fair share of criticism from some resident anglers especially in relation to helicopter access (Watson, pers. comm.). To some extent it can be argued that guides and their clients (mainly tourist anglers) are easy targets for what is a much larger issue – that of increased use of backcountry rivers by both resident and tourist anglers.

While there is a belief held by managers that use levels on backcountry rivers have increased, the problem facing managers is that they do not have accurate estimations of use levels on backcountry rivers. However, one survey carried out by the National Institute for Water and Atmospheric Research (NIWA) estimated use levels on these rivers to be in the range of 0–500 anglers per season (NIWA 1997). Another measure that may indicate increased levels of use are licence sales (this assumes that anglers are not fishing more than they used to). Reviewing national licence sales over the 1980s and 1990s reveals no real growth (Britton pers. comm.). So if use levels were increasing it would seem that anglers were shifting their use to the backcountry. Although speculative, another contributing factor may be the decline in lowland river fisheries throughout the country which have suffered heavily due to the increasing intensity of agriculture. One way to show the growth of use by tourist anglers is the growth in guided trout fishing that has occurred since the 1980s in New Zealand. And while it is difficult to quantify, in some regions, and on some rivers, use by tourist anglers seems to be quite significant.

At the heart of all these concerns is the impact that increasing tourism is having on the outdoor recreational experience of New Zealand residents. Loss of access to fishing waters and increased use on backcountry fisheries together serve to erode the angling opportunities and quality of the fishing experience that New Zealand resident anglers have traditionally taken for granted.

Unplanned expansion of the guided fishing industry is also not in the guides' best interests either. The general feeling among angling guides is that they do not need any more clients during the summer period and would like to be able to spread use to the shoulder periods. Tourist and resident anglers alike are attracted to New Zealand backcountry rivers by the pristine nature of these rivers, by the feeling of solitude, and by seeing and catching large wild trout in clear waters (Walrond 1995, 1997). Some anglers consider this experience the pinnacle of fly-fishing. There is a risk that as angling use of backcountry rivers increases, the very features upon which both the guided fishing industry and resident anglers' enjoyment is based, will be undermined.

Tourism promotion and trout fishing

Tourism New Zealand, previously known as the New Zealand Tourism Board (NZTB), is the country's national tourism organisation. It has actively promoted New Zealand as a trout fishing destination and many tourist anglers visit each year (Franklin 2002). Prior to the current '100 per cent Pure New Zealand' branding of Tourism New Zealand, the former 'New Zealand Way' branding used backcountry fishing photographs in its mix of images. For example, *Summer Way,* the magazine of the New Zealand Way Ltd, in its December 1996 issue had an A3 size cover page of a lone angler on a backcountry river under the caption 'The Angler's Eldorado' (NZTB 1996a). Iconic photographic imagery of backcountry settings and profiles of tourist operators make up the bulk of the five-page spread. These images are the most appealing, featuring gin clear water, snow capped mountains and large trout. The tourist anglers' expectations of the New Zealand angling experience are created through this marketing imagery and prose, and this is what tourists come to expect when they arrive. The ability to match image with reality will be the fundamental factor in continuing to attract anglers the long distances that they must travel to fish in New Zealand.

The statutory managers of the trout fishery, FandGNZ, have been concerned over the uncontrolled expansion of tourism in New Zealand and in the 1990s criticised the NZTB for planning a major international promotion of fishing in New Zealand, including backcountry rivers. There are also political tensions within the FandGNZ movement regarding whom the resource is primarily managed for. Currently the resource is managed for the licence-holder irrespective of whether they are a resident or tourist. In some countries residents have greater rights regarding access to hunting or fishing opportunities over tourists (Ass and Skurdal

Figure 18.1 Backcountry fly-fishing on the Tauherenikau River, New Zealand (Photo: C. Walrond)

1996). Larger populations and smaller resources in these countries mean that some form of rationing becomes necessary. So far New Zealand's small population and large resource has served to soak up user pressure.

Recent figures indicate that of international visitors to New Zealand, approximately 61,000 fished in a river or lake (Ministry of Tourism 2006). Each year, FandGNZ issue about 120,000 fishing licences – for both domestic and international visitors (FandGNZ 2006). Much of this use occurs on the central North Island lake fisheries which sustain high levels of use and harvest. The backcountry angling experience is on the opposite end of the spectrum with the majority of rivers having use rates of <500 angler days per year with over 90 per cent of fish released after capture (NIWA 1997). The NZTB estimates that 5,000–6,000 trout anglers yearly are drawn to New Zealand specifically for angling, spending some $25 million p.a. (NZTB 1996b). More recent data have shown that trout anglers are the highest spending group of tourists (NZTB unpublished data 1998). Clearly, trout anglers are a very small percentage (approximately 2.6 per cent) of total visitor arrivals (2.35 million for the year ended December 2004) (Statistics New Zealand 2005). But given the push for smaller numbers of tourists providing greater revenue, this is a sector that provides considerable revenue for its size.

What are norms?

Norms are standards shared by members of a social group that are used to evaluate whether activities or environments are good or bad (Vaske *et al.* 1986). The crucial point with norms is that they involve value judgements (Shelby and Heberlein 1986). For some recreation experiences, such as a game of tennis, the norms are explicit (that is, either 2 or 4 people play) but with some activities the 'right' number of players are not formalised by rules but rather by etiquette and beliefs (Vaske *et al.* 1986). For example, an angler in the New Zealand backcountry can go for days without encountering anyone else. There are no formal rules regarding the appropriate number of encounters in wilderness settings but the implicit norms regarding wilderness experiences suggests that the number of encounters should be low. These implicit norms can be determined by research and are referred to as encounter norms (for a coverage of recreational encounter norm research see Vaske *et al.* 1986; Donnelly *et al.* 2000; Vaske and Donnelly 2002).

Encounter norms

A number of key points can be drawn from the literature on norms. First, a range of specific types of norms can be identified. For example, recreational users have norms for acceptable distances from individuals, encounters with others, at campsites, on rivers, on tracks and waiting times at rapids (Shelby and Heberlein 1986).

Second, individuals are able and willing to identify norms when asked. Third, encounter tolerance curves and similar approaches provide a basis for quantifying where, and to what extent, different groups of users share the same normative standards. Fourth, although there is considerable variance for different activities, use levels and settings, there are consistencies in norms in some experiences and settings. For example, norms for encounters in remote backcountry settings tend to be low.

Norms can be categorised into three types: no tolerance, single tolerance and multiple tolerance. No tolerance norms are characterised by a mode at zero and a high degree on consensus. A single tolerance norm can be characterised by a mode at some level greater than zero, with a sharp decline above this level.

As each norm type differs, so do the implications for management. No tolerance norms send a clear message to set at zero the acceptable level of impact (Whittaker and Shelby 1988). For single tolerance norms, where users show a level of consensus above zero, some measure of agreement is necessary. The median response is most often used.

With multiple tolerance norms, where two groups of users have different standards, developing a measure becomes even more complex and measures of central tendency are clearly misleading. As different user groups envision different experiences, developing a standard may mean choosing between the two, or zoning the resource to provide a range of experiences (Vaske *et al.* 1986).

Angler social carrying capacity

Existing research on social carrying capacity presents some insight into angler tolerance for encounters and angler perceptions of crowding. Research has shown on lower density backcountry rivers encounters are relatively important (Shelby 1981; Whittaker 1990; Shelby and Vaske 1991). Manning (1979) studied four rivers in Vermont. Anglers reported that after two encounters with other users their satisfaction decreased significantly. Anglers in these cases could be classified as purists; they had the strictest notions about how things 'should' be (Shelby and Heberlein 1986). Much of the North American research has looked at conflict between the different user groups (e.g. canoeists v. anglers). In New Zealand there appear to be fewer conflicts between different user groups due to the size of the resource, and low population (Martinson and Shelby 1992).

Greenstone River and Caples River surveys

The first, and only, comprehensive research on trout angler encounter norms in New Zealand was carried out on the Greenstone, a backcountry river near Queenstown in Otago, over the 1994/95 angling season.

The example in Figure 18.2 is drawn from data collected in the Greenstone River survey in which 146 interviewed anglers were willing to give a maximum tolerable encounter level. The encounter tolerance curve is drawn by subtracting the per cent of intolerant anglers from the per cent of tolerant anglers. Results indicated that anglers are sensitive to levels of encounter and a norm of three

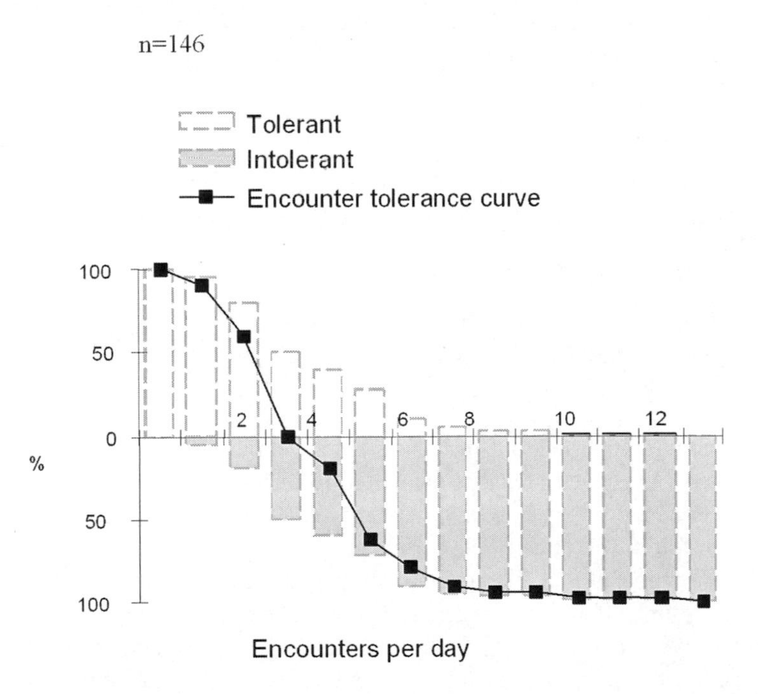

Figure 18.2 Angler encounter tolerance curve: the Greenstone River (Source: Walrond 1995)

encounters per day was identified. Once angling experiences result in three or more encounters per day the majority of anglers have exceeded their norm. At the level of use that season (estimated at around 500 angler days) only one quarter of all anglers encountered more than three anglers in their day's angling (Walrond 1995). Further research carried out on the Caples River, a tributary of the Greenstone, over the next season gave similar results with a norm of four encounters per day identified (Walrond 1997).

Anglers were also asked to rate a list of motivations. Non-catch related motivations were more highly rated than catch related motivations on both the Greenstone and Caples. Specifically, high rankings for motivations of 'Peace and solitude', 'The natural environment' and 'Absence of human structures' all indicate that anglers were after a specific type of experience (Walrond 1995, 1997).

The Otago and Nelson-Marlborough postal study areas

Over 1996/97 and 1997/98 anglers were surveyed in two FandGNZ regions: Nelson-Marlborough and Otago. Over the two fishing seasons a total of 817 Otago and Nelson-Marlborough anglers were surveyed, 635 were identified as resident and 180 as tourist. Of the total, 87 anglers were guided. Three distinct subgroups were surveyed: residents; non-resident unguided and guided anglers – the latter groups being tourists. Anglers also filled in diaries collecting data on their use of rivers and actual encounter levels.

Attitude towards encounter

Anglers were asked how they felt about encountering other anglers on backcountry rivers. They were asked to choose from the following three options. Encounter is:

- negative
- positive
- it can be both

They were then requested to give reasons for the choice they made. Close to half of all anglers (49 per cent) felt that encountering an angler or anglers could be both a positive or negative experience, 36 per cent felt it was a negative experience, and 12 per cent felt it was a positive experience.

Of those who felt that encounter was a positive experience, the main reason cited was 'share information/experience' (66 per cent). The only other reasons to score above 5 per cent were 'indicates fish' (7 per cent), and 'resource used/ everyone has the same rights' (6 per cent). Interestingly, some of the responses such as 'competition for water', 'disturbed water', were negative but were still listed by anglers who felt encounter was a positive experience. This suggests that these anglers still may see some negative aspects to encounter.

The two main reasons cited by those who felt that encounter was a negative experience were 'disturbed water' (52 per cent) and 'solitude/peace and quiet' (34 per cent). The only other significant reason was 'competition for water' (6 per cent). 'Competition for water' refers to anglers actually encountering one another on the river or in a hut, and competing for the water that is available to fish.

Anglers were asked how they would feel about various levels of encounter on a five point pleasantness scale. Results for all anglers on no road access backcountry rivers can be seen in Figure 18.3.

Interestingly, the research revealed statistically significant differences between angler subgroups based on residency, and if guided or unguided. Guided anglers – the group that has the highest proportion of international visitors – are the least tolerant of encounters. While it is important to recognise that there are significant differences between these subgroups the universal theme is that expected, tolerated and preferred encounters per day for all subgroups of backcountry river anglers in New Zealand are low (Walrond 2002). There is a remarkable degree of consensus, which is not evident for many other backcountry recreational users (Patterson and Hammitt 1991; Williams *et al.* 1992; Hammitt and Rutlin1995).

Motivations

Low encounter tolerances are closely related to angler motivations. Previous research has shown that on backcountry rivers motivations of 'peace and solitude', 'experiencing the natural environment' and 'escape' come to the fore (Teirney and Richardson 1992; Walrond 1995, 1997). Similar results were found in the

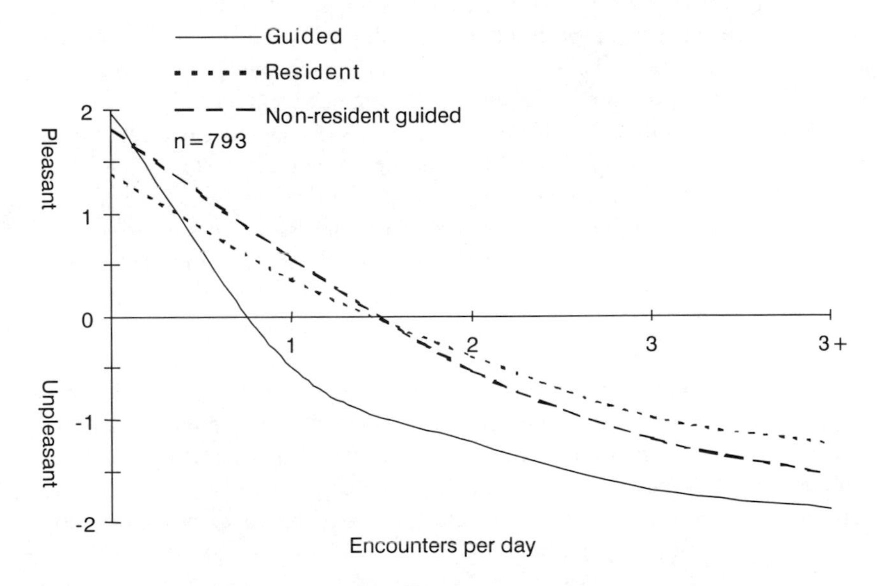

Figure 18.3 No road access backcountry rivers – feelings toward encounter (guided, resident, tourist unguided) (Source: Walrond 2002)

postal survey. Catch-related motivations ('catching large fish', 'catching many fish') were not ranked as highly as non-catch related motivations ('peace and solitude', 'experiencing the natural environment' and 'spotting trout'). Catch-related motivations were more important for the guided angler subgroup. It is emphasised that these are relative rankings and catch was still deemed to be important (Walrond 2002).

Previous research on backcountry rivers found catch to be the most important factor in determining how much value anglers placed on individual fisheries (Teirney and Richardson 1992). The relative ranking of catch below other factors such as peace and solitude and the natural environment is in line with North American trout angler studies (Fedler 1979; Chipman and Helfrich 1988). With catch and spotting trout being important motivations it is easy to understand how behavioural changes in trout could influence the quality of the backcountry angling experience.

Consequences of encounter

Encounters between trout anglers on New Zealand backcountry rivers can greatly reduce their ability to achieve their desired goals. For many anglers it may not be an encounter that is negative, but what it represents. For example, on a small backcountry river an encounter may mean no fishing as there are no readily accessible substitutes, and because angling activity disturbs the fish for later arriving anglers.

Compare this to an encounter on a hiking track that usually lasts a few seconds and does not affect the hikers' ability to walk down the track. With hiking there is a very fast distance-decay effect. If hikers are walking towards you the encounter is over within a few seconds. In the New Zealand backcountry problems with crowding have been more prevalent where hikers are concentrated for some time at huts (Kearsley and O'Neill 1994). For anglers this distance-decay effect might last for days and an encounter may not even occur. For anglers the knowledge that someone else has fished the same water prior to them can be enough to detract from the experience. This appears to be due to behavioural changes of trout.

The research results have been used to formulate a conceptual model. Figure 18.4 is a representation of the backcountry angling experience relating specifically to angler encounter. Factors that were found to have influenced perceptions of encounter in the research are marked with an asterisk (*). Those without an asterisk were not addressed in the current research but anecdotal evidence suggests they may be important influences on perception of encounter. The model illustrates the three dimensions of this experience (angler, trout and environment). Variables that influence angler perception of encounter can come from any one of these dimensions. All motivations listed (on the left) were found to be statistically significant in this research (Walrond 2002).

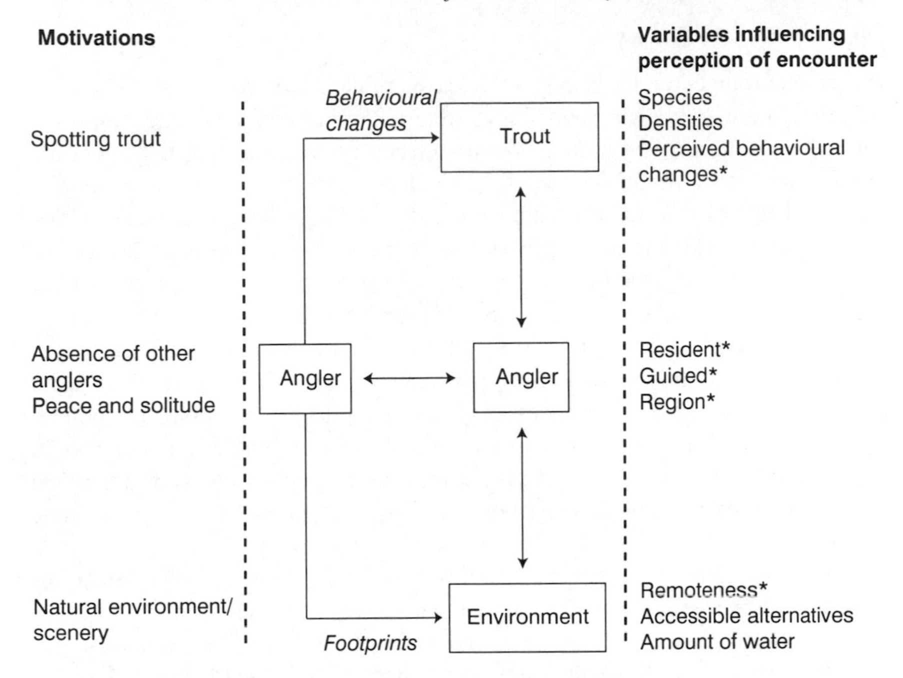

Figure 18.4 A model of angler social carrying capacity for New Zealand backcountry river trout fisheries (Source: Walrond 2002)

Angler–angler interactions

The prime backcountry angling motivations (on the left) apply to all three dimensions of the model (angler, trout, environment). The central portion of the model illustrates the interactions between the three dimensions. Direct angler–angler experiences are defined as encounter and what this represents in terms of disturbed water and competition for fishable water. Encounters are not always negative; some anglers find aspects of encounters such as swapping information a positive experience.

Angler–environment interactions

The research results mirrored the conventional wisdom that in more remote areas encounter norms are lower (Heberlein and Vaske 1977; Manning 1979; Shelby and Heberlein 1986; Vaske *et al.* 1986). Perceptions of use can also be evident in the environment in the form of cars parked along the river (along road access to backcountry rivers), footprints and helicopter noise (in more remote backcountry rivers).

Angler–trout interactions

Angler–environment interactions can leave evidence of recent use (such as footprints). Recent use has implications in terms of behavioural changes in trout. This is represented in the model as angler–trout interactions. Another important point is that encounter in this context also represents interactions with another species (trout) as well as those between anglers. The behavioural changes that occur in trout following angling pressure in the backcountry can influence the anglers' success and thus their experience and perceptions of use levels of the fishery. Researchers at the Cawthron Institute in New Zealand have found that on remote rivers these behavioural changes are typified by trout hiding under rocks for up to two days following angling pressure (Hayes and Young 1999). If trout are not visible then this also impacts on one of the prime motivations of 'spotting trout'. Further, if trout are not actively feeding they are unlikely to be caught. Angler–trout interactions can have a lag time of up to two days and thus affect the quality of the recreational experience for following anglers (Hayes and Young 1999).

The results of the angler surveys support the view that angler–trout interactions are one of the prime reasons for low angler expectations, tolerances and preferences for encounter. This differs from other backcountry recreational experiences such as hiking where one of the main influences of perceptions towards encounter stem from a normative belief that the appropriate numbers of people in wilderness environments should be low (Shelby and Heberlein 1986; Higham 1996). It is possibly more the effects of encounter and what these represent to the angler in terms of the impact on their ability to spot and catch fish, rather than the actual encounter, that leads to the high degree of norm consensus for backcountry river trout anglers.

With so many different variables influencing the nature of the recreational experience the model demonstrates the complexity of the backcountry angling experience and the challenges that this offers for fisheries and tourism management agencies. The model also offers future researchers a framework to further test some of the variables that are thought to influence angler perception of encounter.

Recent attempts at management

Following the angler surveys carried out over the 1990s Fish and Game Otago initiated a pilot programme over the 2004/05 fishing season aimed at maintaining the quality of the experience on the Greenstone River in Otago – a river which receives a great deal of tourist fishing use. All anglers holding a whole season licence wishing to fish the river had to apply for a special licence at no extra fee which required that they fill out a survey giving their opinions of the pilot programme. In addition, the upper reaches of the Greenstone River were subject to an open ballot system for the summer of 2005. Each permit allowed two anglers a three- or four-day block to fish the upper river. Anglers surveyed were roughly divided in half as to whether they supported or opposed the new system and many could not make up their minds. The approach of the fisheries management agency to ration use shows a more hands-on management in the case of the Greenstone River. Other backcountry rivers, however, are essentially open to all those anglers who hold fishing licences and any division of water must be done informally through anglers meeting on rivers (Hayes 2005).

Over the 2005/06 fishing season fisheries managers again managed the upper Greenstone River using an allocation system. Allocation of the right to fish one of three stretches of water was operated on a first-come first-served basis. Fishing the Caples, lower Greenstone and upper Oreti Rivers also required anglers to apply for a backcountry licence which was not restricted but required anglers to provide information on their use and perceptions. Similar plans to directly allocate use on the upper Greenstone via a first-come first-served internet booking system is being implemented over the 2006/07 season.

Conclusion

These angler surveys make a contribution to the literature on encounter norms. Anglers in backcountry environments have very clear ideas about the number of other anglers they prefer and tolerate encountering. It is concluded that social carrying capacities are very low (<3 encounters per day) for backcountry river trout anglers. There was a high degree of consensus among the anglers as to expected, preferable and tolerable levels of encounter. This degree of consensus was considerably higher than social carrying capacity research carried out in North America for recreational users such as hikers and boaters in backcountry environments. This suggests that there is something unique about the New Zealand backcountry fishing experience. Low angler carrying capacities are likely to be related to perceived and actual behavioural changes in trout caused by angling pressure.

In many countries access to high-quality fishing rivers is tightly controlled and fishers must pay to fish. Trout fishing almost anywhere in New Zealand has traditionally been open access as long as the angler held a fishing licence. As use of this resource has grown by tourist anglers since the 1970s, overall use levels also appear to have grown. Managing the quality of the experience while still attracting high-spending tourist anglers to New Zealand remains a considerable challenge for fisheries managers. Can New Zealand's rivers still manage to provide tourists and residents with quality fishing experiences if use levels increase?

References

Ass, O. and Skurdal, J. (1996) 'Fishing by residents and tourists in a rural district in Norway: subsistence and sport-conflict or coexistence?', *Nordic Journal of Freshwater Research*, 72: 45–51.

Britton, M., personal communication, June 1998. Fish and Game New Zealand. Assistant director. National Office. Wellington.

Chipman, B.D. and Helfrich, L.A. (1988) 'Recreational specialisation of Virginia river anglers', *North American Journal of Fisheries Management*, 8: 390–8.

Deans, N., personal communication, May 1999. Manager. Fish and Game New Zealand – Nelson-Marlborough region.

Donnelly, M., Vaske, J., Whittaker, D. and Shelby, B. (2000) 'Toward an understanding of norm prevalence: a comparative analysis of 20 years of research', *Environmental Management*, 25(4): 403–14.

Fedler, A.J. (1979) 'Effects of manipulating trout stocking patterns on recreational and tourist fishing in Maryland', Unpublished report. Annapolis, Maryland: Maryland Department of Natural Resources, University of Maryland College Park.

Fish and Game New Zealand (FandGNZ) (2006) *Facts and Figures*. <http://www.fishandgame.org.nz/> (accessed 23 November 2006).

Franklin, A. (2002) *Trout Tourism: Investigating the growth potential for international and national trout tourists to Tasmania through a comparative study of New Zealand and Tasmanian trout fisheries*, Wildlife Tourism Research Report Series No. 24. Status Assessment of Wildlife Tourism in Australia Series. N.p.: Co-operative Research Centre for Sustainable Tourism.

Hammitt, W.E. and Rutlin, W.M. (1995) 'Use encounter standards and curves for achieved privacy in wilderness', *Leisure Sciences*, 17(4): 245–62.

Hayes, J. (2005) 'The untracked domain', *Fish and Game New Zealand*, Special Issue 21: 10–16.

Hayes, J. and Young, R. (1999) 'Angling pressure and trout catchability', *Fish and Game New Zealand*, 26: 18–27.

Hayes, J., Deans, N. and Walrond, C. (1997) 'Angling pressure on backcountry rivers', *Fish and Game New Zealand*, 16: 32–9.

Hemingway, E. (1987) *The Short Stories of Ernest Hemingway*, New York: Scribner's Sons.

Heberlein, T.A. and Vaske, J.J. (1977) *Crowding and Visitor Conflict on the Bois Brule River* (Report WRC 77-104), Water Resources Centre, University of Wisconsin.

Higham, J.E.S. (1996) *Wilderness Perceptions of International Visitors to New Zealand. The perceptual approach to management of international tourists visiting wilderness areas within New Zealand's conservation estate*, Unpublished Ph.D. thesis. Centre for Tourism. University of Otago. Dunedin, NZ.

Jellyman, D. (1991) 'Angler opinions on the behaviour of headwater trout', *Freshwater Catch*, 45: 9–11.

Jellyman, D. and Graynoth, E. (1994) *Headwater Trout Fisheries in New Zealand*, New Zealand Freshwater Research Report No. 12. Christchurch: NIWA Freshwater.

Kearsley, G.W. and O'Neill, D. (1994) 'Crowding, satisfaction and displacement: the consequences of growing tourist use of southern New Zealand's conservation estate', *Proceedings Tourism Down-under: A Tourism Research Conference*, 6–9 December, 1994. Palmerston North: Department of Management Systems, Massey University, pp. 177–84.

Manning, R.E. (1979) 'Behavioural characteristics of fishermen and other recreationists on four Vermont rivers', *Transactions of the American Fisheries Society*, 108: 536–41.

Martinson, K.S. and Shelby, B. (1992) 'Encounters and proximity norms for salmon anglers in California and New Zealand', *North American Journal of Fisheries Management*, 12: 539–67.

Ministry of Tourism (2006) *International Visitor Survey Data Set*. <http://www.tourismresearch.govt.nz/> (accessed 23 November 2006).

National Institute for Water and Atmospheric Research 9NIWA) (1997) *National Anglers Survey Database*, Unpublished.

New Zealand Tourism Board (NZTB) (1996a) *Summer Way*, Wellington: New Zealand Tourism Board.

New Zealand Tourism Board (NZTB) (1996b) Unpublished tourist-fishing statistics.

New Zealand Tourism Board (NZTB) (1998) Unpublished tourist-fishing statistics.

Patterson, M.E. and Hammitt, W.E. (1991) 'Backcountry encounter norms, actual reported encounters, and their relationship to wilderness solitude', *Journal of Leisure Research*, 22: 259–75.

Shelby, B. (1981) 'Encounter norms in backcountry settings: studies of three rivers', *Journal of Leisure Research*, 13: 129–38.

Shelby, B. and Heberlein, T. (1986) *Carrying Capacity in Recreational Settings*, Corvallis, OR: Oregon State University.

Shelby, B. and Vaske, J.J. (1991) 'Using normative data to develop evaluative standards for resource management: a comment on three recent papers', *Journal of Leisure Research*, 23(2): 173–87.

Statistics New Zealand (2005) *External Migration (December 2004 year)*,Wellington: Statistics New Zealand.

Teirney, L. and Richardson, J. (1992), 'Attributes that characterise angling rivers of importance in New Zealand, based on angler use and perceptions', *North American Journal of Fisheries Management*, 12: 693–702.

Vaske, J. and Donnelly, M. (2002) 'Generalizing the encounter–norm–crowding relationship', *Leisure Sciences*, 24(3–4): 255–69.

Vaske J., Shelby, B., Graefe, A.R. and Heberlein, T.A. (1986) 'Backcountry encounter norms: theory, method and empirical evidence', *Journal of Leisure Research*, 18: 137–53.

Walrond, C. (1995) *Wilderness Fisheries Management: A Case Study of the Social Carrying Capacity for the Greenstone River*, Unpublished thesis. Dunedin: University of Otago.

Walrond, C. (1997) *The Caples River Survey*, Unpublished social survey. Nelson: Cawthron Institute.

Walrond, C. (2000) 'Encounter norms for backcountry trout anglers in New Zealand', *International Journal of Wilderness*, 6(2): 29–33.

Walrond, C. (2002) *Encounter Levels: A Study of Backcountry River Trout Anglers in Nelson-Marlborough and Otago*, Unpublished PhD thesis. Centre for Tourism, University of Otago, Dunedin, New Zealand.

Watson, N. (1999) Personal communication. May 1999. Manager. Fish and Game New Zealand – Otago region. Dunedin.

Whittaker, D. (1990) *Selecting Indicators: Which Impacts Really Matter?*, Paper presented at workshop on defining wilderness quality: the role of standards in wilderness management, 10–11 April, Fort Collins, Colorado.

Whittaker, D. and Shelby, B.B. (1988) 'Types of norms for recreational impacts: extending the social norms concept', *Journal of Leisure Research*, 20: 261–73.

Williams, D.R., Roggenbuck, J.W., Patterson, M.E. and Watson, A.E. (1992) 'The variability of user-based social impact standards for wilderness management', *Forest Science*, 38(4): 738–56.

19 Australia as a safari hunting destination for exotic animals

Stephen J. Craig-Smith and Gordon McL. Dryden

Introduction

This chapter does not purport to cover all potential aspects of hunting as it relates to Australia: it does not address the issues of hunting as an essential food source for indigenous Australians and it does not address issues surrounding the hunting of Australian native species either for consumption or population control. What this chapter does focus on are aspects relating to the hunting of exotic animals in Australia which have recently received considerable government interest. Although accurate figures do not exist, there is both a domestic and an international market for Australian hunting. Some indication of the relative numbers and importance of domestic and international involvement is given later within the chapter. Much of this chapter is based on a major survey undertaken by the authors for the Rural Industries Research and Development Corporation based in Canberra (Dryden and Craig-Smith 2004).

The driving force behind much of this government interest is two-fold. On the one hand there is the realisation that hunting tourism, if properly controlled and regulated, has the potential to develop into a small but profitable niche market for Australian tourism and focus on rural and regional Australia where tourism income is most needed. On the other hand it is recognised that the control of exotic animals in Australia is both essential for environmental conservation and costly to execute. Whilst no one is suggesting that all exotic animal control in Australia can be managed by handing over responsibility to a potential tourist group there is scope for a win–win situation to arise by widening Australia's appeal to a small but expanding group of special interest high spending tourists and providing help in exotic animal population control.

Safari hunting is a form of recreation whereby animals are hunted and some form of trophy is taken. The nature of the trophy varies from preservation of the entire carcass after treatment by a taxidermist to the collection of animal parts such as horns to mere photographic pictures of the shot animal. Not all safari hunting, therefore, is focused on a kill, some hunters are happy to shoot tranquillising darts, measure and photograph the temporarily stunned animal and return it to the wild after that. If the animal is permanently removed from the environment this is consumptive hunting; if it is released back into the wild unharmed this is non-consumptive hunting. Most of the issues in this chapter, however, focus on a kill which brings the activity into the realm of consumptive hunting.

Before looking at the issues in detail it is useful to be clear on a few basic terms. An exotic animal is one which has been deliberately or inadvertently introduced. In Australia most of these have been introduced by Europeans either as a food source (e.g. the pig), a beast of burden (e.g. the camel) or for the specific use of hunting (e.g. the fox). In this analysis a wild animal is one which has existed in the wild for at least two generations (i.e. it may be descended from a domesticated animal which was released, or escaped from confinement, but it is not the actual animal which has escaped or been released) and as such has never been subjected to the management normally given to domestic animals. A feral animal on the other hand is one which has escaped or been released from confinement. The term game is applied to any animal which is hunted.

As a general rule the larger and more impressive the animal the greater the safari hunter's interest. For many years the 'Big Five' of Southern Africa have enjoyed a certain cachet as have the 'South Pacific 15'. Serious hunters are prepared to travel the world in search of their desired animal or animals and there is interest in shooting one each of Southern Africa's big five or one each of the South Pacific's 15. Although safari hunting is frequently associated with the African rather than the Australian continent, there are in Australia a number of large exotic animals such as the horse, the camel, the deer and the buffalo, in addition to smaller game such as foxes, rabbits and hares. All these animals have been introduced into Australia by European immigrants, all of these are causing environmental damage to Australia's natural environment, many are in keen competition with native species, most are not classified on any list of endangered or threatened animals and many are legally classified as vertebrate pests. Recent interest in Australia becoming a safari hunting destination revolves around the consumptive hunting of these types of animal (Department of the Environment, Sport and Territories 1998).

Potential for commercial hunting in Australia

Commercial hunting in Australia includes both private recreational hunting and safari hunting. Private hunting can become commercial when the hunter pays for access to the hunting area and/or to the animal being hunted. In cases where the recreational hunter does not know of suitable places to hunt s/he may use the services of an outfitter to locate the hunting venue and/or access to the desired quarry. Safari hunting is a more organised form of hunting whereby the hunter is assisted by a guide, with or without the assistance of an outfitter and there is a more complete tourist experience than just the hunt itself. Safari hunts, therefore, typically include the provision of serviced accommodation and may be associated with more conventional tourism such as visiting national parks and other rural tourist attractions.

There are two basic types of hunting: free range and estate. Free range hunting occurs where animals are allowed to roam through their natural range and are therefore not confined to specific areas. The hunter may select for or against certain animal characteristics and so implement a form of *de facto* culling. Free range (or fair chase) hunting demands that the hunter expends considerable effort to find and stalk the quarry and where the animal has a reasonable chance of not being found or of not being killed. A fundamental principle of free range hunting is that the hunt area be large – often several thousand hectares. One example of this form of hunting in Australia is found in Arnhem Land in the Northern Territory (Australian Buffalo Hunters undated).

Estate or Game Park hunting (not large in Australia because of laws specifically banning this type of hunting in some jurisdictions) involves breeding and management programmes which vary in intensity. In some cases these approximate to fauna parks with many species of animal. Breeding programmes are even used to replace animals taken or to preserve numbers if there is a likelihood of the species becoming endangered. This is not such an issue, however, where feral animals are concerned. In other cases, the estate is less intensively managed with hunters having access to those animals which are naturally present on the property and where there is little management of them. In Australia there are few establishments which approximate to estate or game park hunting because in some states and territories such establishments are actually banned.

Australia clearly has much to offer the potential safari hunter. There are a number of large animal species of interest to hunters, the exotic species are in direct competition with native animals and have in many cases to be controlled for conservation reasons, hunting has the potential to introduce much needed income into remote and regional areas and there is already a hunting industry established albeit relatively small on the world stage.

There are, however, a number of factors surrounding consumptive hunting in Australia which are highly contentious; the moral view held by many urban dwellers against killing animals for sport, the understandable reticence of hunting groups to make public their stand and reason for hunting, and the Australian international tourism image widely marketed abroad of an animal and environment friendly destination, make any increase in safari hunting tourism a vexed issue. Clearly, the focus on exotic species defuses much public condemnation which would arise if such suggestions were made concerning all Australian animals. The study referred to here was commissioned to investigate some of these issues and suggest how best to go forward if safari hunting is to be exploited to its maximum potential.

Hunting and the law in Australia

Any analysis of exotic hunting in Australia must work within Australia's legal framework. Legally, the hunting of exotic animals is permitted in all states and territories and set out below are some of the major legal considerations. Hunting in Australia is regulated by the relevant state and territory legislation although under the Commonwealth Constitution, where there is an inconsistency between federal and state law, the federal law prevails. As recently as 1997 a Senator presented a motion calling for the prohibition of all recreational hunting in Australia but the motion was not adopted (Trone 2004).

In almost all states and territories legislation relating to hunting involves acts concerning nature conservation and acts concerning animal cruelty. In the Australian Capital Territory the *Nature Conservation Act* 1980 prohibits the killing of native animals but does not appear to prohibit the killing of exotic ones and the *Animal Welfare Act* 1992 prohibits any act of cruelty against an animal and prohibits game parks where animals are confined so that they can be hunted for sport or recreation.

New South Wales is the only state where a specific statute regulates game hunting. The *Game and Feral Animal Control Act* 2002 provides for hunting of game animals on both public and private land and pest animals on private land. There are two classes of game animal: (1) deer; and (2) pigs, dogs, cats, goats, rabbits, hares and foxes. A licensing system is administered by a Game Council which issues licences regulating who can hunt and where the hunting may take place. The provisions of the *Prevention of Cruelty to Animals Act* 1979 controls how animals can be hunted and it prohibits game parks where animals are confined for sport or recreational hunting.

In the Northern Territory the *Territory Parks and Wildlife Conservation Act* 1976 protects wildlife but it provides for the control of non-indigenous feral animals and the destruction of feral animals in a park, a reserve, area of essential habitat or a sanctuary. All hunting has to abide by the provisions of the *Animal Welfare Act* 1999.

Hunting in Queensland is regulated by the *Nature Conservation Act* 1992 and the *Animal Care and Protection Act* 2001. The 2001 Act precludes animals being released from captivity for hunting with no acclimatisation period. The acclimatisation period is necessary to allow the animal to adapt to its new environment and thus give it a chance of escaping. Feral animals, however, are exempted from the above acts. This means there is no criminal liability for actions which would otherwise constitute an animal cruelty offence. The *Lands Protection (Pest and Stock Management) Act* 2002 provides for the management of pest species. Under this Act landowners must take reasonable steps to keep their lands free of pests.

South Australia has the *National Parks and Wildlife Act* 1972 which controls what can and cannot be hunted. A landowner does not need a permit to destroy unprotected animals which are damaging crops, stock or other property. The state's *National Parks and Wildlife (Hunting) Regulations* 1996 restrict what a hunter can hunt and how the hunt is conducted. The Act concerning animal cruelty in South Australia is the *Prevention of Cruelty to Animals Act* 1985.

In Tasmania wildlife is protected by the *Nature Conservation Act* 2002 but many feral animals are not included within it. Game reserves are allowed but must be managed to 'provide for the taking, on an ecologically sustainable basis … game species for commercial or private purposes' and 'to encourage appropriate tourism, recreational use and enjoyment, particularly sustainable recreational hunting' (Trone 2004). In Tasmania the Act concerning animal cruelty is the

Animal Welfare Act 1993. Under this Act it is illegal to take part in a match in which an animal is released from captivity for the purpose of being killed.

In Victoria wildlife is protected by the *Wildlife Act* 1975 which protects many animals but not pest animals. Feral animals of any type may be declared pest animals. If wildlife is adversely affecting agricultural crops special permission may be granted to kill that wildlife but a wildlife licence is required. A game licence is required to hunt game and no killing is allowed in the closed season. The *Prevention of Cruelty to Animals Act* 1986 does not apply to hunting carried out in accordance with a code of practice.

Western Australia protects its wildlife with the *Wildlife Conservation Act* 1950 and animal cruelty by the *Animal Welfare Act* 2002. What can be hunted and by what means are covered by this legislation.

As a general rule all states and territories have legislation which protects animals, and legislation prohibiting cruelty to animals. Under this legislation what species can and cannot be hunted is stipulated and by what means animals can be hunted and killed is strictly controlled. Feral animals are not as strictly regulated as native species. Hunting is generally allowed but estate hunting is banned in many states.

Australian firearms regulations tend to be stricter in Australia than they are in other South Pacific nations and different state and territory legislation can at times be a little tedious especially for international visitors. The regulations are not sufficiently different, however, to dictate the location of hunting businesses. Head offices of hunting businesses tend to be located in the state or territory of residence or the owner. These organisations tend to operate in all parts of Australia and some operate overseas in New Zealand, Canada, the USA, South Africa and New Caledonia.

Exotic animals suitable for hunting

Populations of exotic wild animals in Australia have developed from intentional introduction of these animals since the start of European colonisation. In terms of effects on agricultural enterprises and native fauna and flora or in terms of capacity to support safari hunting the following are most significant: fox (*Vulpes vulpes*), rabbit (*Oryctolagus cuniculus*), European brown hare (*Lepus capensis*), donkey (*Equus asinus*), horse (*Equus caballus*), swamp buffalo (*Bubalus bubalis*), cattle (*Bos* pp.), pig (*Sus scrofa*), goat (*Capra hircus*), deer (*Cervus* spp., *Axis axis, Dama dama*), camel (*Camelus dromedaries*), 'domestic cat' (*Felis catus*) and 'domestic dog' (*Canis familiaris*). The difficulties encountered with the increase in these populations in the wild have been well documented by Joyce (1985) and Balogh (2000).

Wild pigs are found in all Australian states but the main concentrations are found in Queensland, New South Wales and the Northern Territory (Ramsay 1994). O'Brien (1987) considers pigs to be a special problem because they are an agricultural pest, they are endemic and exotic and they pose a disease hazard. They are a concern regarding foot and mouth disease, tuberculosis (McInerney

Figure 19.1 Trophy fallow deer, Queensland (Photo: M. Daddow)

et al. 1995), leptospirosis (Mason *et al.* 1998) and Q-fever (Wong *et al.* 2001). Population densities of 1.6 pigs per square kilometre have been reported in Kosciuszko National Park (Saunders 1993), and two pigs per square kilometre on agricultural land in eastern Australia (Saunders *et al.* 1990). Considerable damage has been attributed to the wild pig population including complete removal of vegetation, disturbed and trampled soil and the dispersal of woody seeds. Control measures to date have included hunting (recreational hunting with firearms and dogs, harvesting for game meat and helicopter shooting), poisoning, trapping, fumigation, and habitat refuge destruction.

Buffalo were introduced into the Northern Territory in 1843 and by 1985 had expanded to over 341,000 head or 1.5 buffalo per square kilometre. In Northern Australia they are found in all major habitats. Bowman and Panton (1991) have also estimated that there are 365,000 head of wild cattle as well. The presence of buffalo and wild cattle increases the difficulty of containing potential outbreaks of disease as does the wild pig population. Buffalo lead to habitat destruction and overgrazing. The control of buffalo in Northern Australia will probably not result in the extinction of the herd unless there is a large expenditure of effort and money (Bowman and Panton 1991).

It has been estimated that there are over 350,000 wild horses in Australia based on surveys carried out in the 1970s and 1980s (Dobbie *et al.* 1993). Most of these are in Queensland (about 100,000) and the Northern Territory (about 200,000). In the Top End are an estimated 29,000 donkeys. Choquenot (1990) and Garrott (1991) suggest that the potential population growth of wild horses and donkeys are 15 to 20 per cent and 23 to 28 per cent per year respectively. Horses and donkeys compete with other grazing animals, spread weeds and damage fencing. Currently they are both controlled by shooting and trapping at water holes.

There are about 4 to 5 million goats in Australia (P.J. Murray, pers. comm.) occurring in all states except the Northern Territory (Parkes *et al.* 1996). Potentially, wild goat populations may increase by 42 per cent per year (Mahood, cited in McCloy and Rowe 2000). Wild goats cause overgrazing, encourage the growth of undesirable plant species and damage soil. Control methods to date include aerial shooting, commercial mustering, and trapping at watering holes. Helicopter shooting may cost up to Aus$61 per goat killed (Bayne *et al.* 2000).

Some 60,000 wild deer live in Australia. Most of these are descended from herds introduced in the 1870s. Those which formed substantial populations are hog (*Axis porcinus*) (hog do not constitute a large population but are important because they form one of the very few huntable populations in the southern hemisphere), red (*Cervus elaphus*), sambar (*Cervus unicolour*) and fallow (*Dama dama*) in Victoria (Bentley 1998); chital (*Axis axis*) red, fallow and rusa deer in Queensland (Roff 1960); fallow deer in Tasmania (Murphy 1995); rusa (*Cervus timorensis*), fallow and sambar in New South Wales (Bentley 1998) and red, and fallow deer in South Australia (Bentley 1998). Deer are thought to have adverse effects on plant diversity and abundance, compete for food with cattle and may be a possible transmission for the cattle tick. Most population control of deer involves hunting.

Australia has the world's only large population of feral camels (Dorges *et al.* 1992). Camels were first imported in 1840, and many of the extant wild herds have developed from releases in the 1920s from camel stud farms (Williams 1999). The current population of wild camels is estimated at around 200,000 grown from around 20,000 camels in the mid-1960s (Northern Territory Conservation Commission, cited by Ellard and Seidel 2000). Half of these are in the pastoral and desert areas of Western Australia, 20 to 25 per cent in the northern area of South Australia and western Queensland, and 25 to 30 per cent are in the southern regions of the Northern Territory. Wild camel populations may grow at 7 to 12 per cent per year in good conditions (Dorges and Heucke 1989, cited in McCloy and Rowe 2000). Although camels rank very low as a vertebrate pest in pastoral regions of Australia they are declared a pest in Western Australia because of the damage they cause to fencing and water troughs. Camels are also a potential carrier of various diseases and parasites. In Western Australia wild camel populations are generally controlled by shooting and a proportion of the wild herd is taken by game meat hunters (about 100,000 carcasses were sold for domestic consumption in Australia in 1997) and recreational hunters.

The current organisation of hunting in Australia

In terms of membership the most important hunting associations in Australia at the present time are the Australian Deer Association (ADA), the Sporting Shooters Association of Australia (SSAA) and the Safari Club International Downunder Chapter (SCI). There is also an Australian branch of the US based Buckmasters Club. There are ADA state organisations in all states and territories except Western Australia, and there are subsidiary branches in most states. The SSAA is represented in all states and territories. In the absence of an accepted peak body for recreational hunters in any formal sense these specific bodies have adopted that role and lobby Australian governments and overseas organisations. In addition to these large organisations there are some further 55 unaffiliated recreational hunting clubs, over 80 per cent of which are focused on New South Wales.

Commercial safari companies

There is both a domestic and an international market for hunting in Australia. Exact figures are hard to find, but a survey of commercial safari companies by Dryden and Craig-Smith (2004) revealed that the number of domestic hunters outnumbered the number of international clients in most cases.

There are at least 70 companies and/or individuals who advertise commercial hunting throughout Australia. The majority of these are based in Queensland (41 per cent), Victoria (22 per cent) and New South Wales (16 per cent). Only 10 per cent are located in the Northern Territory, but these are important because the safari hunting provided by these eight companies is probably the closest that Australia can provide to the big game hunting experience available in Southern Africa. Safari company headquarters location did not appear to be influenced by state or territory hunting regulations but rather by the location of residence of the person or persons who established the company in the first place.

From the national survey conducted by the authors in 2003 it would appear that most companies host between 11 and 30 domestic clients and fewer than five international clients (mainly from the USA, Germany and New Zealand) each year. Trophy charges per animal vary from under Aus$100 to over Aus$650. More than 75 per cent of international clients spend time on general tourism activities other than hunting whilst in Australia which would suggest that their financial contribution outweighed their lower number when compared with domestic hunters. Safari hunter clients tend to hunt for periods of one to two weeks, but rarely longer than that. Some 20 per cent of clients visit Australia only to hunt but most safari companies estimate that more than 75 per cent of clients also spend time on general tourism activities.

International clients tend to prefer to hunt deer, pigs, goats, buffalo and cattle (including banteng cattle). European visitors tend to prefer pigs and buffalo whilst North American hunters prefer buffalo, pigs and deer. Sometimes companies have to decline international hunters' requests for a particular species most often because the law does not allow hunting of that particular species. Some safari companies and recreational hunting clubs favour retaining the populations of many exotic species but with strict controls on population size. Some intervention (such as controlling herd sex structure) may be desired to improve trophy quality.

Few hunting companies expressed any difficulty with taxidermy, accommodation and the other services of professional hunters but there is less satisfaction about the law governing hunting and the export of trophies. Several companies expressed concern about Australian firearms law which they feel does not compare well with similar laws in New Zealand and New Caledonia (which are Australia's main competitors for international clients).

Almost all companies surveyed felt strongly that government tourism agencies should pay greater attention to the promotion of safari hunting; at present only the Northern Territory Tourist Commission promotes safari hunting on any scale. Many respondents asked that hunting receive its pro rata share of promotional funding with other outdoor or sporting tourism.

Recreation hunting clubs

There are more than 50 recreational hunting clubs in Australia, the largest being the Sporting Shooters Association, Field and Game Australia, and the Australian Deer Association. Over 80 per cent of the smaller clubs are located in New South Wales and 17 per cent in Victoria. Whilst some hunting clubs are relatively small others are of considerable size. The Sporting Shooters' Association, established in 1948 for instance, has over 120,000 members and sees as its main function the promotion of shooting sports and the protection of firearm owners' rights. The Australian Deer Hunters Association has over 4,000 members with branches in every state outside Western Australia. Whilst some hunting clubs comprise groups of hunters who club together to purchase the hunting rights to certain blocks of land, this practice is nowhere near as common as it is in parts of Europe. As a general rule local club control over hunting rights is strongest in Tasmania and gets progressively less strong the further north one goes in Australia.

Recreational hunters take deer (100 per cent of clubs), pigs, goats and hares/rabbits (each 86 per cent of clubs) and members of smaller clubs hunt buffalo and camels (21 per cent of clubs), horses/donkeys (14 per cent) and cattle (7 per cent). Other species include foxes and feral cats. Somewhat surprisingly, over a third of clubs surveyed had no policy on the eradication of wild exotic animals, some clubs wanting to retain all exotic species, some wanting to eradicate all exotic species and others wanting to retain only deer, buffalo, pigs, hares and blackbuck but to eradicate foxes, cats and rabbits. Many clubs (86 per cent) supported the control of wild exotic animal populations, mainly to control population sizes and to maintain or enhance trophy quality. The recreational clubs suggest that more commercial hunting would increase competition and so restrict access in the future to good hunting land and access to huntable animals. On the other hand, they also felt that more commercial hunting might encourage changes to the law which might improve hunting access generally.

Independent hunting is less commonly practised in Australia by international hunters, especially in the Northern Territory, because of the difficulties in gaining access to land belonging to indigenous peoples. In the Northern Territory the Murwangi Community Aboriginal Corporation (MCAC) for instance, has set up a tourist venture whereby buffalo, feral bull and pig hunting is made available to potential hunters. Safari hunters pay a daily fee to MCAC for hunting and pay a trophy fee per animal, with the exception of feral pigs. The motivation for this indigenous group to open up their land for strictly limited hunting was to secure a premium return from a small but well regulated and paying client group. In this locality the hunters are mainly male business persons comprising approximately 40 per cent domestic hunters and 60 per cent international hunters. In areas such as this it is not possible for hunters to wander on the land independently to shoot.

At present there is no Australian association for professional hunting guides (professional hunters) although Australians may apply to join the International Professional Hunters Association. Similarly, there is no peak body for the Australian hunting industry. This is something that needs to be urgently addressed and is one of the objectives of current government funded research.

Tourism and safari hunting in Australia

An important element of any analysis of the viability of safari hunting of Australian exotic animals must take into consideration the vital role tourism can play in supporting its economic logistics. From a tourism perspective, safari hunting of any kind presents considerable opportunities.

Consumptive hunting and tourism have considerable potential with specific groups. It is highly unlikely that most tourists will become hunters but it is easy to turn hunters into tourists and it is from this angle that tourism and consumptive hunting should be viewed. In many ways a considerable number of hunters can already be viewed as tourists albeit that they are not called that very often. Anyone who embarks on a journey away from home base for a period of 24 hours or more is actually a tourist and many hunting trips involve durations greater than one day. The tourism industry has been slow to capitalise on this potentially lucrative market possibly for fear of putting off those tourists who may be likely to take offence at consumptive hunting.

There are many potential advantages in attracting hunting tourists. Hunters are prepared to spend considerable sums of money in pursuit of their interests and this is a potentially very lucrative market segment. Hunters are often looking for specific animal types or hunting environments and are prepared to travel considerable distances to satisfy their needs. The Australian hunting environment cannot be replicated in, say, North America or South Africa and therefore there is a potential world market to be tapped. Hunting can be used as a very cost-effective method of exotic animal population control whereby the tourist covers much of the control costs in exchange for the hunting experience.

Within Australia hunting is geographically widespread with some hunting activity in all states and territories. In spite of different state and territory legislation, hunting tends to follow geographical areas related to the particular animal being hunted rather than be conducted on strictly state or territory lines.

To date the industry has been strongly focused on promoting hunting within Australia but, as with all forms of tourism, there are advantages of leverage by expanding the national tourism product to become part of a larger scale attraction. There are considerable opportunities to be realised by developing further the hunting of the Pacific 15 making Australia just one part of a wider hunting region including New Zealand and New Caledonia.

For many tourism organisations to become involved with consumptive tourism, however, certain safeguards must be strictly enforced. Just as a tourist killed by a crocodile can be very costly to the industry so could a hunting accident or publicity around anything even remotely conceived as bordering on animal cruelty. Any tourism organisation or company involved in consumptive hunting will have to ensure no adverse publicity eventuates and it may be necessary to segregate both the management operation of consumptive hunting tourism from other forms of tourism. Segregation of tourist types is not unusual because mass tourists and ecotourists are generally incompatible. Segregation could be achieved in one of two ways. Either there could be a temporal segregation whereby hunting tourists may use an area at particular times of day, week or season and other types of tourist could use the same area at other times. Alternatively, specific areas could be set aside specifically for hunting tourists and other areas for other types of tourist. A combination of the two strategies could be adopted where appropriate.

For hunting tourism to succeed both in the non-consumptive and consumptive forms, close co-operation will be necessary between the two types of operation, namely the tourism industry and the hunting industry. Examples of such co-operation can be gained from an examination of the wine industry and the tourism industry. The tourism industry should be able to provide the marketing and distribution networks and the hunting industry should be able to provide possible markets and knowledge on animal conservation and population control. If this is a little optimistic then perhaps the input of wildlife people and/or zoologists could help.

For tourism to be able to support exotic animal hunting, closer working relationships need to be developed between the two industries and appropriate areas need to be identified where exotic animal hunting can be carried out to the advantage of the hunters, to the advantage of exotic animal population control, to the advantage of conservation management and to the advantage of tourism industry operators.

Exotic safari hunting in Australia can be recognised as a legitimate industry but it is still in its early days of development. The next step is to establish a peak body to represent hunting organisations, companies and related interests and for this new body to work in close co-operation with the tourism industry and government tourism marketing organisations. The recent study on which much of this chapter has been based is working to these ends.

Acknowledgements

Both authors of this chapter would like to acknowledge the assistance rendered towards this study by N.A. Finch, J. Trone and C. Arcodia.

References

Australian Buffalo Hunters (undated). <http://www.gateway.net.au/~buffalo/main.htm> (accessed 10 May 2003).

Balogh, S. (2000) *Practical Solutions to Pest Management Problems*, Proceedings of the NSW Pest Animal Control Conference. Orange, New South Wales, Australia 25–27 October 2000. Orange, NSW: NSW Agriculture.

Bayne, P., Harden, R., Pines, K. and Taylor, U. (2000) 'Controling feral goats by shooting from a helicopter with and without the assistance of ground-based spotters', *Wildlife Research*, 27: 517—32.

Bentley, A. (1998) *An Introduction to the Deer of Australia*, Melbourne: Australian Deer Research Foundation.

Bowman, D.M.J.S. and Panton, W.J. (1991) 'Sign and habitat impact of banteng (*Bos javanicus*) and pig (*Sus scrofa*), Cobourg peninsula, northern Australia', *Australian Journal of Ecology*, 16: 15–17.

Choquenot, D. (1990) 'Rate of increase for populations of feral donkeys in northern Australia', *Journal of Mammalogy*, 71: 151–5.

Department of Environment, Sport and Territories (1998) Submission in SRRARTRC report on the *Commercial Utilisation of Australian Native Wildlife*, Canberra: Senate Planning Unit, Parliament House.

Dobbie, W.R., Berman, D.McK. and Braysher, M.L. (1993) *Managing Vertebrate Pests: Feral Horses*, Canberra: Australian Government Publishing Service.

Dorges, B. and Heucke, J. (1989) *The Impact of Feral Camels in Central Australia*, Report to the Conservation Commission of the Northern Territory.

Dorges, B., Heucke, J. and Klingel, H. (1992) 'Behaviour and social organisation of feral camels in central Australia', in W.R. Allen *et al.* (eds) *Proceedings of the First International Camel Conference*, Newmarket: R&W Publications pp. 317–18.

Dryden, G. and Craig-Smith, S. (2004) *Safari Hunting of Australian Exotic Wild Game*, Report Publication No. 04/108. Canberra: Rural Industries Research and Development Corporation.

Ellard, K. and Seidel, P. (2000) *Development of a Sustainable Camel Industry*, Publ. No. 99/118. Canberra: RIRDC.

Garrott, R.A. (1991) 'Feral horse fertility: potential and limitations', *Wildlife Society Bulletin*, 19: 52–8.

Joyce, M.B. (1985) 'The grazing industry's point of view on competition between grazing animals, wild and feral animals', in B.L. Moore and P.J. Chenoweth (eds) *Grazing Animal Welfare Symposium*, Brisbane: Queensland Division, Australian Veterinary Association, pp. 102–6.

Mahood, I.T. (1985) *Some Aspects of Ecology and the Control of Feral Goats* (Capra hircus L.) *in Western NSW*, MSC Thesis. Macquarie University, Sydney.

McCloy, L. and Rowe, P. (2000) *Assessing the Potential for a Commercial Camel Industry in Western Australia*, Publ. No. 00/123, Canberra: Rural Industries Research and Development Corporation.

McInerney, J., Small, K.J. and Caley, P. (1995) 'Prevalence of *Mycobacterium bovis* infection in feral pigs in the Northern Territory', *Australian Veterinary Journal*, 72: 448–51.

Mason, R.J., Fleming, P.J.S., Smythe, L.D., Dohnt, M.F., Norris, M.A. and Symonds, M.L. (1998) '*Leptospira interrogans* antibodies in feral pigs from New South Wales', *Journal of Wildlife Diseases*, 34: 738–43.

Murphy, B.P. (1995) 'Management of wild fallow deer in Tasmania: a sustainable approach', in G.C. Grigg, P.T. Hale and D. Lunney (eds) *Conservation Through Sustainable Use of Wildlife*, Brisbane: Centre for Conservation Biology, University of Queensland, pp. 307–11.

Murray, Peter J., School of Animal Studies, The University of Queensland.

O'Brien, P.H. (1987) 'Socio-economic and biological impact of the feral pig in New South Wales: an overview and alternative management plan', *Australian Rangeland Journal*, 9: 96–101.

Parkes, J., Henzell, R. and Pickles, G. (1996) *Managing Vertebrate Pests: Feral Goats*, Canberra: Australian Government Publishing Service.

Ramsay, B.J. (1994) *Commercial Use of Wild Animals in Australia*, Canberra: Australian Government Publishing Service.

Roff, C. (1960) 'Deer in Queensland', *Queensland Journal of Agricultural Science*, 17: 43–58.

Saunders, G. (1993) 'The demography of feral pigs (*Sus scrofa*) in Kosciuszko National Park, New South Wales', *Wildlife Research*, 20: 559–69.

Saunders, G., Kay, B. and Parker, B. (1990) 'Evaluation of a warfarin poisoning programme for feral pigs (*Sus scrofa*)', *Australian Wildlife Research*, 17: 525–33.

Trone, J. (2004) Section 5 *The Regulatory Environment* in *Safari Hunting of Australian Exotic Wild Game*, Report Publication No. 04/108. Canberra: Rural Industries Research and Development Corporation.

Williams, O.J. (1999) *Capture and Handling of Camels Destined for the Abattoir*, Alice Springs, NT: Central Australian Camel Industry Association, Rural Industries Research and Development Corporation.

Wong, R.C.W., Wilson, R., Silcock, R., Kratzing, L.M. and Looke, D. (2001) 'Unusual combination of positive IgG autoantibodies in acute Q fever infection', *Internal Medicine Journal*, 31: 432–5.

20 Conclusion

Consumptive wildlife tourism – sustainable niche or endangered species?

Brent Lovelock

The significance of the consumptive wildlife tourism industry was brought home to me one day when I interrupted my son playing a hunting game on our home computer. The game involves selecting a firearm (with the option of customising it), mode of transport, hunting guides and other accoutrements, and most importantly – choosing your hunting destination. This particular game offered various hunting destinations from North America, Australasia, Asia and Africa. Each hunting destination offers its unique challenge to the player in terms of game species, terrain, season and regulations. While 'just' a virtual game, obviously the game maker draws on an understanding of hunting and, indeed, the social, political and biological constraints acting upon the consumptive wildlife tourist. But more importantly, the game reflects the global nature of CWT. As the contributors to this book have demonstrated, CWT deals with complex and challenging experiences, and despite its 'niche' status, is a significant sector with strong global connections and potential to contribute to sustainable outcomes in a large range of destinations.

Furthermore, CWT is probably the only common tourism activity for which participants will pay upwards of US$10,000 for a single experience, which may be relatively short-lived – e.g. obtaining a single trophy fish in a single day from an exclusive fishing 'beat'. This sum equates to what people will pay for a first-class cabin on an extended ocean cruise – which would be of considerably longer duration. Of course there are other activities for which people would pay this type of money (e.g. a brief sub-space flight; a luxury overnight package in the world's best hotel; or a return trip on the Orient Express). Arguably, however, from the evidence presented in this book at least, no other single experience has the potential to be as ecologically sound, or to benefit such a range of stakeholders as does a CWT experience.

Conversely, perhaps no other tourist activity is as controversial as CWT, apart from child sex tourism. The controversial nature of the activity has not been an historical constant, indeed, over time there has been a transformation in the way that the consumptive wildlife tourist is perceived. The heroic figure in Hannam's chapter on Victorian tiger-hunting in India – all-masculine, all-British, all-powerful, protector of villagers, and upholder of order, is now seen as a despicable tiger-murderer. The very nature of CWT, that it does indeed involve killing or harvesting in some form the target species, makes this sector contentious in a world where animal rights issues are increasingly gaining air-play in the world media and popular political support.

The animal rights issue did emerge as a significant theme in this book – primarily as an obstacle to the development of CWT for some of the destinations addressed in this book. Akama observed the issue of animal rights and preservationist paradigms from a north–south perspective. Such paradigms often have colonial origins, and have been perpetuated by post-colonial relationships. Yet many of the conservation and tourism policies derived from these paradigms, as Akama concluded, fail to meet either ecological or social needs. In Kenya, ironically, a non-consumptive wildlife tourism approach has led to an accelerated destruction of wildlife habitats along with increasing land-use conflicts. Similarly, Mbaiwa, who examined safari hunting in a number of southern African settings, opined that anti-hunting sentiments will harm the small remote communities that have developed some dependencies upon CWT – from trophy fees, guiding employment, bush meat or protection of crops and lives (over 200 people are killed by wild animals each year in Tanzania alone (Dickinson 2004)).

That we live in interesting times with respect to our overall relationship with animals is demonstrated by Franklin. In his treatise on the 'animal question' Franklin argued that in a time when tourism proponents and participants are placing increased credence upon the embodied experience (e.g. Pons 2003; Crouch 2000) – 'doing' rather than just 'seeing' – consumptive forms of wildlife tourism may contribute far more effectively to an ongoing relationship with the animals and their environments than will non-consumptive means.

In some respects this call for more engagement with the objects of the tourism gaze is a reflection of the wider debate for a greater connection in many aspects of our lives – with the food we eat, in our relationships and in our careers. That we can observe celebrity chef Jamie Oliver out on an Umbrian hillside shooting wild boar for his next recipe, or other celebrity 'survivors' foraging on a desert island is illustrative of an imperative that runs parallel to the animal rights movement. That is, an imperative to re-establish an embodied connection with the 'natural' in a world where industrialisation, urbanisation and industrialised food production have led to disenchantment and alienation. That such an imperative could point us towards eating slow, eating local and, indeed, eating Bambi, offers opportunities for the perpetuation of CWT in the face of animal rights' intransigence on this activity. It remains to be seen if we will see any further convergence of these two parallel yet contradictory paths.

While many forms of tourism are contested and ultimately political, what is clear is that CWT proponents will need to be proactive in terms of building a constituency and maintaining support for their form of tourism. Campbell highlighted in his chapter the political nature of CWT, describing the battle for influence over public opinion. Recently, the Dutch *Party for Animals* became the first animal rights party to have elected members in a European Union nation

(BBC 2006). Others are following, for example the newly formed British political party Animals Count, is currently contesting elections in Wales. Both parties are calling for a total ban on hunting and sport fishing.

A second major theme to emerge in this book concerns the relationships between CWT and other users of the fish and game resource. CWT is not only contested in terms of animal rights, but also, in some cases the rights of other, non-touristic users of the resource. These include local hunters and fishers who may be impacted by outsiders using their grounds and waters. Chapters by Figgins, writing on red-deer hunting in New Zealand, a land where hunting is a traditional free-resource, Walrond writing of backcountry trout angling and Cohen and Sanyal's Idaho case studies all highlight the impact on local hunting or fishing as potential hurdles to the development of a CWT industry. Gunnarsdotter's chapter demonstrated how conflict emerges when locals act as agents of change, for example, local land-owners providing opportunities for foreign hunters. On a more positive note, Sillanpää's historical analysis of the Scandinavian Sporting Tour provided a good example of how an influx of consumptive tourists may ultimately lead locals to recognise the true value of their own fish and game resource and to develop appropriate institutional responses and protections.

Other resource users that have a stake may include suppliers and developers of hydro-electricity on waterways, forest-owners whose trees are damaged by game, competing industrial users of the resource and even insurance companies that pay out millions in claims for game-related automobile accidents each year. Mattsson *et al.* in their chapter suggested that the application of welfare economics may help resource-managers to aggregate the welfare of hunters, fishers, tourism industry members as well as other parties who may be carrying costs associated with wildlife and fish populations. It will be interesting to observe if such a welfare economics approach can also address the needs of a further stakeholder, conservationists, who in some locales have argued for a reduction or eradication of game species on ecological grounds (Lovelock 2006).

With regard to competition between tourist and industrial users of the resource, there is no doubt that this is becoming an increasingly debated issue. Normann in his coverage of marine fishing in Arctic Norway, however, observed that although tourist and industrial fishers may be in competition, there are certain complementarities – for example, marine fishing tourism can benefit from the skills and authenticity arising from the experience of commercial fishers. This was aptly demonstrated to me whilst on a recent fishing-charter trip to Stewart Island in the very south of New Zealand. The skipper of our vessel utilised an extensive inter-generational knowledge acquired through his family's 120 years of boat-building and commercial fishing in the locale, to secure us a safe and productive fishing anchorage out of the teeth of a howling Nor-wester: a perfect combination of skill, safety and authenticity.

The role of CWT in wildlife conservation was noted as being highly significant by a number of contributors. CWT was demonstrated to have roles in sustaining ecosystem robustness, addressing 'pest' species, and ensuring sustainable populations. Current forms of 'conservation-hunting' have often emerged from a historic legacy of harm that CWT has wrought on a number of game species, through unmanaged over-harvesting. It is thus observed that only by having appropriate institutional arrangements in place and a will to enforce, will a truly sustainable relationship develop between CWT and the game resource result. Increased wealth and mobility have posed threats to traditionally sustainable hunting and fishing practices. Seddon and Launay in their study of Arab falconry noted the huge impacts on both predator and prey species of this practice in post-oil-discovery Saudi Arabia. An important point to draw from this study is that the increased mobility afforded by modern wealth and technology allows the transportation of unsustainable hunting and fishing practices to foreign lands where a lack of ecological knowledge may dangerously be accompanied by a lack of ecological concern or patriotism. The irony in Seddon and Launay's study is that the very practitioners of the unsustainable practices that have led to the vulnerability of game/prey may now, through their wealth and influence, be the potential protectors of these species.

It is also clear that local communities will support and become involved in CWT if they can see positive economic and socio-cultural benefits in doing so. This was certainly demonstrated in the case of Nunavit in Foote and Wenzel's Canadian example. The conservation hunting practised in the Canadian arctic is lucrative, compared with non-consumptive bear viewing, and contributes to local culture through providing a form of employment that helps maintain traditional practices. The hunters who pay for this experience benefit in terms of the intense cultural exchange that occurs throughout the hunting experience. This aspect of cultural exchange is worthy of further exploration, as the culturally-specific human–animal relationships evident in indigenous communities may form a unique and important component of a CWT experience. For example, in Craig-Smith and Dryden's Australian case, it is conceivable that indigenous Australians could contribute immensely to a CWT product – in terms of sharing their culture's traditional environmental knowledge of fish and game species, the surrounding mythologies and aspects of their contemporary way of life – not to mention their tracking and hunting skills. In a similar manner the Maori of New Zealand could provide a unique hosting opportunity for a CWT experience there. The engagement of local providers may also go some way to help address the high levels of economic leakage reported in the industry in some settings (Mbaiwa). Whatever the involvement of indigenous communities, Cohen and Sanyal's Idaho study pointed to the need to engage in an early and meaningful manner over the development of CWT, as this often impacts upon indigenous communities' resource base and therefore invariably involves interaction with these communities in practice.

A number of chapters alluded to destination competitiveness for CWT, and an identifiable message to gain from these works is that competitiveness is ultimately linked to the institutional arrangements within each destination. It is not just the presence or numbers of game, size of the fish or length of the antlers that will contribute to success as a CWT destination – there are a number of other supply-side factors that are important. Figgins' chapter, comparing red deer hunting in

Scotland and New Zealand, argued strongly that it is the socio-economically created attributes – regulatory bodies and their policies, fencing, breeding and technology – that will ultimately be more significant in determining success. This is reinforced in chapters by Craig-Smith and Dryden who noted the importance of building an appropriate legislative framework for CWT, and Barnes and Novelli who demonstrated the importance of property rights and policy in creating economic benefits of CWT; a simple matter of legislating for guided-hunting only (for tourists) or applying a differential fee for a fishing licence may make all the difference between a form of CWT that is sustainable and one that is destined to fail.

Ultimately, proponents of CWT must keep in mind that this is low-volume high-yield tourism, and as Preston-Whyte describes, an often intensely personal and complex relationship between the fisher or hunter and their environment. As Walrond observes, participants demand a certain standard – and it should be noted that this is not always measured in terms of trophy size; a wilderness fishing experience, for example, may be destroyed by the vision of a single preceding footprint in the sand. Such demands point to the need for the CWT sector to work together, to identify best practice and to set standards (without homogenising the industry) and to share and gain from relationships with the wider tourism industry – something that in some destinations at least (Lovelock and Milham 2006) the sector needs to learn.

The final question to emerge from this collection is how CWT may fit into the destination 'mix' of activities and attractions. Some contributors, despite being supportive of CWT, Akama, for example, argued that destinations that sell themselves on wilderness or purity will find difficulties being associated with CWT in the marketplace. Craig-Smith and Dryden also argued in their Australian feasibility study, that industry trepidation over promoting CWT was triggered by a concern that it would impact upon non-consumptive wildlife tourism. They promoted a segregation of CWT from non-consumptive forms of wildlife tourism on the ground. Other contributors reinforced the perception that participants in both forms do hold different values and motivations. Dawson and Lovelock illustrated that non-consumptive marine tourists score more highly on environmental values scales, and suggested that different approaches (e.g. environmental interpretation) may be required for each group. Destinations may need to resort to non-consumptive practices temporally or spatially when consumptive practices are deemed unsustainable or incompatible. Such a complementarity of consumptive and non-consumptive uses is seen by some as a critical aspect of competitiveness for wildlife tourism (e.g. Tremblay 2001).

Anthropologists tell us that the Neanderthals died off because they failed to adapt their hunting practices to the new environmental realities they were facing (Mayell 2004): for CWT destinations to survive, they too need to adapt their hunting (of both wildlife *and* tourists!) and related practices to face the realities of a changing environment in the broadest sense of the word. Destinations and CWT sector members need to manage their institutional environments (i.e. political, legislative, marketing and community) along with the biological environment in order to achieve the sustainable forms of CWT that Foote and Wenzel describe in terms of a 'triple-bottom line'.

Further research will be important in contributing to a better understanding of CWT sector dynamics, and ultimately to a more robust industry. Fruitful areas of research will include obtaining a firmer grasp on the size and value of CWT markets (e.g. the emerging Asian demand; and CWT for women) – and a clearer picture of the economic and socio-cultural exchanges that occur between consumptive tourists and local communities. And considering the biological dependencies of CWT, it is suggested that more research is needed into the impacts of global environmental change upon target populations and their ecologies. Furthermore, considering the vulnerability of the sector to biosecurity breaches (e.g. avian influenza and game bird shooting; or the invasive Didymo algae in New Zealand's trout fisheries) some analysis of the actual and potential impacts of such breaches would be timely. On a related note, a study of the touristic consumption of genetically engineered game as a sub-sector of this industry may be useful, and challenge what we take to be 'natural' – or industrial.

More information is needed on the motivations of CWT participants in specifically touristic contexts (to complement the wealth of research on general recreational hunting and fishing), as well as data that will inform management on the extent to which consumptive and non-consumptive forms of tourism may co-exist in the same destinations and settings. The priority will be to ascertain the level of acceptance for CWT among other visitors, potential visitors and, indeed, the wider constituency. Additional research should explore the value of present and future linkages between the CWT sector and other sectors with a view to enhancing destination competitiveness while maximising overall welfare.

References

BBC (2006) New animal rights party launched. BBC News, Sunday 3 December 2006 <http://news.bbc.co.uk/1/hi/uk/6203204.stm> (accessed 23 February 2007).

Crouch, D. (2000) 'Places around us: embodied lay geographies in leisure and tourism', *Leisure Studies*, 19(2): 63–76.

Dickinson, D. (2004) *Toothache 'made lion eat humans'*. BBC News, Tuesday 19 October 2004. <http://news.bbc.co.uk/2/hi/africa/3756180.stm> (accessed 24 December 2006).

Lovelock, B.A. (2006) '"If that's a moose, I'd hate to see a rat!": visitors' perspectives on naturalness and the consequences for ecological integrity in peripheral natural areas of New Zealand', in D.K. Muller and B. Jannson (eds) *Tourism in Peripheries: Perspectives from the North and South*, Wallingford: CABI, pp. 124–40.

Lovelock, B.A. and Milham, J. (2006) *Summary of Results: Hunting Tourism Industry Survey 2005*, Unpublished report to the New Zealand Professional Hunting Guides Association.

Mayell, H. (2004) 'Climate change killed Neandertals, study says', National Geographic News, 9 February 2004 <http://news.nationalgeographic.com/news/2004/02/0209_040209_neandertals.html> (accessed 23 February 2007).

Pons, P.O. (2003) 'Being-on-holiday – tourist dwelling, bodies and place', *Tourist Studies*, 3(1): 47–66.

Tremblay, P. (2001) 'Wildlife tourism consumption: consumptive or non-consumptive?', *International Journal of Tourism Research*, 3: 81–6.

Index